DIGITAL FILTERS: ANALYSIS AND DESIGN

McGraw-Hill Series in Electrical Engineering

Stephen W. Director, Carnegie-Mellon University, *Consulting Editor*

NETWORKS AND SYSTEMS
COMMUNICATIONS AND INFORMATION THEORY
CONTROL THEORY
ELECTRONICS AND ELECTRONIC CIRCUITS
POWER AND ENERGY
ELECTROMAGNETICS
COMPUTER ENGINEERING AND SWITCHING THEORY
INTRODUCTORY AND SURVEY
RADIO, TELEVISION, RADAR, AND ANTENNAS

Previous Consulting Editors

Ronald M. Bracewell, Colin Cherry, James F. Gibbons, Willis W. Harman, Hubert Heffner, Edward W. Herold, John G. Linvill, Simon Ramo, Ronald A. Rohrer, Anthony E. Siegman, Charles Susskind, Frederick E. Terman, John G. Truxal, Ernst Wever, and John R. Whinnery

Communications and Information Theory

Stephen W. Director, Carnegie-Mellon University, *Consulting Editor*

DIGITAL FILTERS:
ANALYSIS AND DESIGN

Andreas Antoniou

Professor and Chairman, Department of Electrical Engineering
Concordia University

McGraw-Hill Book Company

New York   St. Louis   San Francisco   Auckland   Bogotá   Düsseldorf
Johannesburg   London   Madrid   Mexico   Montreal   New Delhi
Panama   Paris   São Paulo   Singapore   Sydney   Tokyo   Toronto

DIGITAL FILTERS: ANALYSIS AND DESIGN

1 2 3 4 5 6 7 8 9 0 FGRFGR 7 8 3 2 1 0 9

Library of Congress Cataloging in Publication Data

Antoniou, Andreas, date
    Digital filters.

    (McGraw-Hill electrical engineering series:
Communications and information theory section)
    Includes bibliographical references and index.
    1. Digital filters (Mathematics)   I. Title.
II. Series.
TK7872.F5A64   621.38′043   78-13998
ISBN 0-07-002117-1

This book was set in Monophoto. The editor was Frank J. Cerra;
the production supervisor was Donna Piligra; the cover was designed by Rafael Hernandez.
Fairfield Graphics was printer and binder.

TO MY MOTHER AND FATHER

CONTENTS

Appendixes

Index

PREFACE

This book attempts to put together theories, techniques, and procedures which can be used to analyze, design, and implement digital filters. The digital filter is viewed as a system which can be implemented by means of software or hardware. In other words, the book is concerned both with the construction of algorithms (or computer programs) that can be used to filter recorded signals and also with the design of dedicated digital hardware that can be used to perform real-time filtering tasks like the many required in communications systems.

The prerequisite knowledge is a typical undergraduate mathematics background of calculus, complex variables, and simple differential equations. Section 5.5 entails a basic understanding of elliptic functions. Since this topic is normally excluded from undergraduate curricula, Appendix A provides a brief but adequate treatment of these functions. Chapter 11 requires a basic understanding of random variables and processes, which are reviewed in Chap. 10.

Chapter 1 introduces the digital filter as a discrete-time system which can be linear or nonlinear, causal or noncausal, and so on. Time-domain analysis is then introduced at an elementary level. The analysis is accomplished by solving the difference equation of the filter by induction. Although this technique is unlikely to be employed in practice, it is of pedagogical value as it provides the newcomer with a solid grasp of the physical nature of a digital filter. The chapter concludes with an alternative and more advanced time-domain analysis based on a state-space characterization.

The basic mathematical tool for the analysis of digital filters, the z transform, is described in Chap. 2. Its application in the time-domain and frequency-domain analysis of linear, time-invariant filters is considered in Chap. 3.

The realization of digital filters, namely the process of translating the transfer function into a digital-filter network, is discussed in Chap. 4. Later in this chapter a very useful topological theorem, known as Tellegen's theorem, is described. It is then used to develop the concepts of reciprocity, interreciprocity, and transposition. The chapter concludes with a convenient sensitivity analysis based on Tellegen's theorem.

Transfer-function approximations for digital filters of the recursive type are almost invariably obtained indirectly by using analog-filter approximations (e.g., Tschebyscheff, elliptic, etc.). Such approximations are considered in Chap. 5. The elliptic approximation is treated in detail because this is the most efficient of the available approximations. Although the derivation of this approximation turns out to be quite involved, a simple and easy-to-apply formulation is possible, as demonstrated at the end of Sec. 5.5.

Up to Chapter 5, the digital filter is treated as a distinct entity with its own methods of analysis. In Chap. 6, a theoretical link is established between discrete-time and continuous-time signals, which leads to a direct interrelation between the z transform and the Fourier transform. By this means, the enormous wealth of analog techniques can be applied in the analysis and design of digital filters; in addition, digital filters can be applied in the processing of analog signals.

Chapter 7 deals with the approximation problem for recursive filters. Methods are described by which a given continuous-time transfer function can be transformed into a corresponding discrete-time transfer function, e.g., impulse-invariant-response method and bilinear transformation method. A detailed procedure which can be used to design Butterworth, Tschebyscheff, and elliptic filters satisfying prescribed specifications is found in Chap. 8. This can be readily programmed.

Nonrecursive filters and their approximations are considered in Chap. 9. The use of window functions is described in detail. Emphasis is placed on Kaiser's window function, which has proved very versatile. In Chap. 10 the concept of a random process is introduced as a means of representing random signals. Such signals arise in digital filters because of the inevitable signal quantization. The effects of finite word length in digital filters along with appropriate, up-to-date methods of analysis are discussed in Chap. 11. The topics considered include coefficient quantization, product quantization, scaling, and limit-cycle bounds.

A relatively recent innovation in the domain of digital filtering has been the introduction of wave digital filters. Filters of this type can have certain attractive features, e.g., low sensitivity, and are considered in Chap. 12. The chapter includes step-by-step procedures by which filters satisfying prescribed specifications can be designed. The chapter concludes with a list of guidelines that can be used in the choice of a digital-filter network (or structure).

Chapter 13 presents the discrete Fourier transform and the associated fast Fourier-transform method as mathematical tools in the software implementation of digital filters. Much effort is spent in clearly conveying the exact interrelations between the discrete Fourier transform and (1) the z transform, (2) the continuous Fourier transform, and (3) the Fourier series because if these interrelations are not thoroughly understood, the user of the fast Fourier-transform method is likely to finish with masses of meaningless numbers.

Chapter 14, which concludes the book, deals with the hardware implementation of digital filters. It starts with a brief review of boolean algebra and then proceeds to a detailed description of the various types of combinational and sequential circuits that may be used as components in the implementation. Later the main integrated-circuit families are discussed and their features compared. Subsequently, in Sec. 14.7, three specific approaches to the implementation are described and their merits and demerits discussed. The chapter concludes with a brief outline of past, present, and future trends in the domain of digital filtering.

Most of the techniques are illustrated by examples and selected sets of problems are included throughout the book. Also 17 useful computer programs are included in Appendix B, which can be used by the filter designer in real-life designs and by the student in developing projects and in solving some of the challenging problems included.

The book can serve as a text for a sequence of two one-semester courses on digital filters for senior undergraduate or first-year graduate students. It could also serve for a one-semester course for groups of students with an adequate background of discrete-time system theory, analog filters, the Fourier transform, and the theory of random variables and processes. Such a course could comprise the following:

1. Review of Chap. 6
2. Brief discussion of analog-filter approximations (Chap. 5) with the emphasis placed on application rather than derivation
3. Chapters 7 to 9
4. Chapter 11, possibly in conjunction with Sec. 4.7
5. A choice from Chaps. 12 to 14, depending on the course orientation desired

The book should also be of interest to the filter designer, in particular Appendix B, and also to the scientist who has his own discrete-time signal to process.

The book is supported by a comprehensive Solutions Manual and also by a Tape Cartridge, which contains the programs given in Appendix B. The latter will be made available to institutions adopting the book upon the payment of a nominal handling fee.

I wish to thank V. Bhargava, A. G. Constantinides, R. Crochiere, L. B. Jackson, O. Monkewich, A. Papoulis, W. Saraga, and L. Weinberg for reading parts of the manuscript and for suggesting numerous improvements; S. W. Director, R. F. J. Filipowsky, and S. K. Mitra for their detailed reviews and suggestions; J. C. Callaghan and M. N. S. Swamy for encouraging and supporting this work; June Anderson for typing the manuscript; Concordia University for supporting this work; and the National Research Council of Canada for supporting the research that led to many of the new results presented. Last but not least, I wish to thank my wife Rosemary and also Anthony, David, Constantine, and Helen for their sacrifices and understanding.

Andreas Antoniou

INTRODUCTION

Signals arise in almost every field of science and engineering, e.g., in acoustics, biomedical engineering, communications, control systems, radar, physics, seismology, and telemetry. Two general classes of signals can be identified, namely continuous-time and discrete-time signals.

A *continuous-time signal* is one that is defined at each and every instant of time. Typical examples are a voltage waveform and the velocity of a space vehicle as a function of time. A *discrete-time signal*, on the other hand, is one that is defined at discrete instants of time, perhaps every millisecond, second, or day. Examples of this type of signal are the closing price of a particular commodity on the Stock Exchange and the daily precipitation as functions of time.

A discrete-time signal, like a continuous-time signal, can be represented by a unique function of frequency referred to as the *frequency spectrum* of the signal. This is a description of the frequency content of the signal.

Filtering is a process by which the frequency spectrum of a signal can be modified, reshaped, or manipulated according to some desired specification. It may entail amplifying or attenuating a range of frequency components, rejecting or isolating one specific frequency component, etc. The uses of filtering are manifold, e.g., to eliminate signal contamination such as noise, to remove signal distortion brought about by an imperfect transmission channel or by inaccuracies in measurement, to separate two or more distinct signals which were purposely mixed in order to maximize channel utilization, to resolve signals into their frequency components, to demodulate signals, to convert discrete-time signals into continuous-time signals, and to bandlimit signals.

The digital filter is a digital system that can be used to filter discrete-time signals. It can be implemented by means of software (computer programs) or by

1

means of dedicated hardware, and in either case it can be used to filter real-time signals or non-real-time (recorded) signals.

Software digital filters made their appearance along with the first digital computer in the late forties, although the name digital filter did not emerge until the midsixties. Early in the history of the digital computer many of the classical numerical analysis formulas of Newton, Stirling, Everett, and others were used to carry out interpolation, differentiation, and integration of functions (signals) represented by means of sequences of numbers (discrete-time signals). Since interpolation, differentiation, or integration of a signal represents a manipulation of the frequency spectrum of the signal, the subroutines or programs constructed to carry out these operations were essentially digital filters. In subsequent years, many complex and highly sophisticated algorithms and programs were developed to perform a variety of filtering tasks in numerous applications, e.g., data smoothing and prediction, pattern recognition, electrocardiogram processing, and spectrum analysis. In fact, as time goes on, interest in the software digital filter is becoming progressively more intense while its applications are increasing at an exponential rate.

A bandlimited continuous-time signal can be transformed into a discrete-time signal by means of sampling. Conversely, the discrete-time signal so generated can be used to regenerate the original continuous-time signal by means of interpolation, by virtue of Shannon's sampling theorem. As a consequence, hardware digital filters can be used to perform real-time filtering tasks which in the not too distant past were performed almost exclusively by analog filters. The advantages to be gained are the traditional advantages associated with digital systems in general:

1. Component tolerances are uncritical.
2. Component drift and spurious environmental signals have no influence on the system performance.
3. Accuracy is high.
4. Physical size is small.
5. Reliability is high.

A very important additional advantage of digital filters is the ease with which filter parameters can be changed in order to change the filter characteristics. This feature allows one to design programmable filters which can be used to perform a multiplicity of filtering tasks. Also one can design new types of filters such as adaptive filters. The main disadvantage of hardware digital filters at present is their relatively high cost. However, with the tremendous advancements in the domain of large-scale integration, the cost of hardware digital filters is likely to drop drastically in the not too distant future. When this happens, hardware digital filters will replace analog filters in many more applications.

ELEMENTARY ANALYSIS

1.1 INTRODUCTION

A digital filter, like an analog filter, can be represented by a network which comprises a collection of interconnected elements. Analysis of a digital filter is the process of determining the response of the filter network to a given excitation. Design of a digital filter, on the other hand, is the process of synthetizing and implementing a filter network so that a set of prescribed excitations results in a set of desired responses.

This chapter is an introduction to the analysis of digital filters. First, the fundamental concepts of time invariance, causality, etc., as applied to digital filters are discussed. Then an elementary time-domain analysis is described. The chapter concludes with a more advanced time-domain analysis based on a state-space approach.

1.2 TYPES OF DISCRETE-TIME SIGNALS

A continuous-time signal can be represented by a function $x(t)$ whose domain is a range of numbers (t_1, t_2), where $-\infty \leq t_1$ and $t_2 \leq \infty$. Similarly, a discrete-time signal can be represented by a function $x(nT)$, where T is a constant and n is an integer in the range (n_1, n_2) such that $-\infty \leq n_1$ and $n_2 \leq \infty$. Alternatively, a discrete-time signal can be represented by $x(n)$ or x_n. In this book we shall be using the first two notations, namely $x(nT)$ and $x(n)$.

As for continuous-time signals, two types of discrete-time signals can be identified, namely nonquantized and quantized signals. A nonquantized signal is

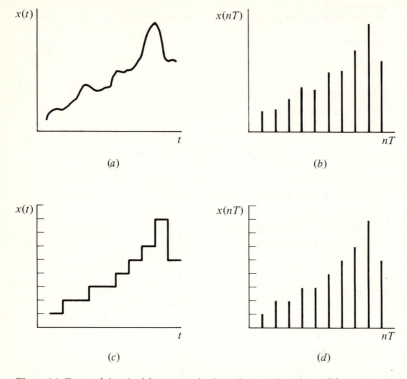

Figure 1.1 Types of signals: (a) nonquantized continuous-time signal, (b) nonquantized discrete-time signal, (c) quantized continuous-time signal, (d) quantized discrete-time signal.

one that can assume any value within a specific range, whereas a quantized signal is one that can assume only a finite number of discrete values. The ambient temperature as a function of time is a nonquantized signal. The ambient temperature, however, as measured by a digital thermometer is a quantized signal. The various types of signals are illustrated in Fig. 1.1.

1.3 THE DIGITAL FILTER AS A SYSTEM

A digital filter can be represented by the block diagram of Fig. 1.2. Input $x(nT)$ and output $y(nT)$ are the excitation and response of the filter, respectively. The response is related to the excitation by some rule of correspondence. We can indicate this fact notationally as

$$y(nT) = \mathscr{R}x(nT)$$

where $\mathscr{R}$ is an operator.

$x(nT)$ ⊶ DF ⊸ $y(nT)$

Figure 1.2 Digital filter.

Like other signal-processing systems, digital filters can be classified as time-invariant or time-dependent, causal or noncausal, linear or nonlinear [1].†

Time Invariance

A digital filter is said to be time-invariant if its internal parameters do not change with time. This means that a specific excitation will always produce the same response independently of the time of application.

Formally, an initially relaxed filter in which $x(nT) = y(nT) = 0$ for all n less than zero is said to be time-invariant if and only if

$$\mathscr{R}x(nT - kT) = y(nT - kT)$$

for all possible excitations. The behavior of a time-invariant filter is illustrated in Fig. 1.3a and b.

Example 1.1 (a) A digital filter is characterized by the equation

$$y(nT) = \mathscr{R}x(nT) = 2nTx(nT)$$

Check the filter for time invariance. (b) Repeat part (a) if

$$y(nT) = \mathscr{R}x(nT) = 12x(nT - T) + 11x(nT - 2T)$$

SOLUTION (a) The response to a delayed excitation is

$$\mathscr{R}x(nT - kT) = 2nT[x(nT - kT)]$$

The delayed response is

$$y(nT - kT) = 2(nT - kT)[x(nT - kT)]$$

Clearly

$$\mathscr{R}x(nT - kT) \neq y(nT - kT)$$

and, therefore, the filter is time-dependent.

(b) In this case

$$\mathscr{R}x(nT - kT) = 12x[(n - k)T - T] + 11x[(n - k)T - 2T] = y(nT - kT)$$

i.e., the filter is time-invariant.

† Numbered references will be found at the end of each chapter.

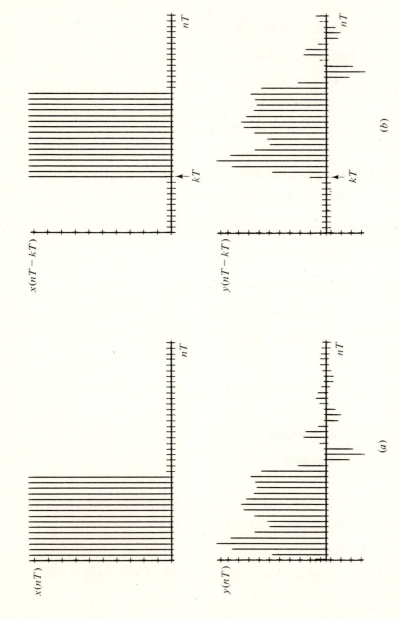

Figure 1.3 Time invariance: response to (*a*) an excitation $x(nT)$ and (*b*) to a delayed excitation $x(nT - kT)$.

Causality

A causal digital filter is one whose response at a specific instant is independent of subsequent values of the excitation. More precisely, a digital filter is causal if and only if

$$\mathscr{R}x_1(nT) = \mathscr{R}x_2(nT) \qquad \text{for } n \le k$$

for all possible pairs of excitations $x_1(nT)$ and $x_2(nT)$ satisfying the constraint

$$\begin{aligned} x_1(nT) &= x_2(nT) &&\text{for } n \le k \\ x_1(nT) &\neq x_2(nT) &&\text{for } n > k \end{aligned} \qquad (1.1)$$

This criterion is illustrated in Fig. 1.4a and b for a causal filter.

Example 1.2 (a) A digital filter is represented by

$$y(nT) = \mathscr{R}x(nT) = 3x(nT - 2T) + 3x(nT + 2T)$$

Check the filter for causality. (b) Repeat part (a) if

$$y(nT) = \mathscr{R}x(nT) = 3x(nT - T) - 3x(nT - 2T)$$

SOLUTION (a) Assume that $x_1(nT)$ and $x_2(nT)$ satisfy Eq. (1.1). For $n = k$

$$\mathscr{R}x_1(nT) = 3x_1(kT - 2T) + 3x_1(kT + 2T)$$
$$\mathscr{R}x_2(nT) = 3x_2(kT - 2T) + 3x_2(kT + 2T)$$

but since

$$3x_1(kT + 2T) \neq 3x_2(kT + 2T)$$

it follows that

$$\mathscr{R}x_1(nT) \neq \mathscr{R}x_2(nT) \qquad \text{for } n = k$$

i.e., the filter is noncausal.
   (b) For this case

$$\mathscr{R}x_1(nT) = 3x_1(nT - T) - 3x_1(nT - 2T) \qquad \text{and}$$
$$\mathscr{R}x_2(nT) = 3x_2(nT - T) - 3x_2(nT - 2T)$$

If $n \le k$, then $n - 1, n - 2 < k$ and so

$$x_1(nT - T) = x_2(nT - T) \qquad x_1(nT - 2T) = x_2(nT - 2T)$$

for $n \le k$ or

$$\mathscr{R}x_1(nT) = \mathscr{R}x_2(nT) \qquad \text{for } n \le k$$

i.e., the filter is causal.

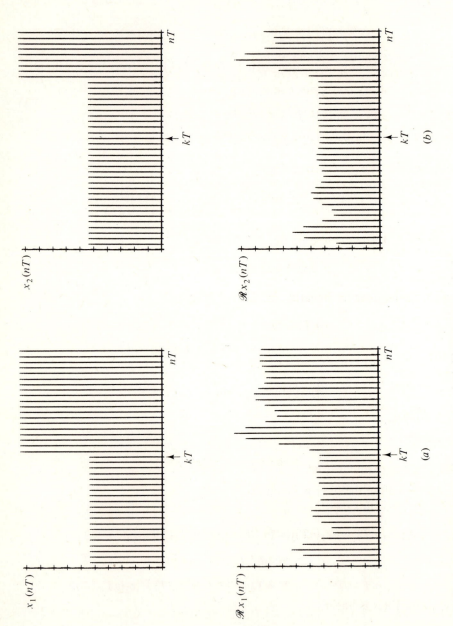

Figure 1.4 Causality criterion: (*a*) response to $x_1(nT)$, (*b*) response to $x_2(nT)$.

Linearity

A digital filter is linear if and only if it satisfies the conditions

$$\mathscr{R}\alpha x(nT) = \alpha\mathscr{R}x(nT)$$

$$\mathscr{R}[x_1(nT) + x_2(nT)] = \mathscr{R}x_1(nT) + \mathscr{R}x_2(nT)$$

for all possible values of α and all possible excitations $x_1(nT)$ and $x_2(nT)$.

The response of a linear filter to an excitation $\alpha x_1(nT) + \beta x_2(nT)$, where α and β are arbitrary constants, can be expressed as

$$y(nT) = \mathscr{R}[\alpha x_1(nT) + \beta x_2(nT)] = \mathscr{R}\alpha x_1(nT) + \mathscr{R}\beta x_2(nT)$$

$$= \alpha\mathscr{R}x_1(nT) + \beta\mathscr{R}x_2(nT)$$

Therefore, the above two conditions can be combined into one as

$$\mathscr{R}[\alpha x_1(nT) + \beta x_2(nT)] = \alpha\mathscr{R}x_1(nT) + \beta\mathscr{R}x_2(nT)$$

Example 1.3 (*a*) The response of a digital filter is of the form

$$y(nT) = \mathscr{R}x(nT) = 7x^2(nT - T)$$

Check the filter for linearity. (*b*) Repeat part (*a*) if

$$y(nT) = \mathscr{R}x(nT) = (nT)^2 x(nT + 2T)$$

SOLUTION (*a*) For a constant α other than unity

$$\mathscr{R}\alpha x(nT) = 7\alpha^2 x^2(nT - T)$$

whereas $\qquad \alpha\mathscr{R}x(nT) = 7\alpha x^2(nT - T)$

Clearly

$$\mathscr{R}\alpha x(nT) \neq \alpha\mathscr{R}x(nT)$$

and therefore the filter is nonlinear.

(*b*) For this case

$$\mathscr{R}[\alpha x_1(nT) + \beta x_2(nT)] = (nT)^2[\alpha x_1(nT + 2T) + \beta x_2(nT + 2T)]$$

$$= \alpha(nT)^2 x_1(nT + 2T) + \beta(nT)^2 x_2(nT + 2T)$$

$$= \alpha\mathscr{R}x_1(nT) + \beta\mathscr{R}x_2(nT)$$

i.e., the filter is linear.

In this book we shall be concerned almost exclusively with linear, causal, and time-invariant filters.

1.4 CHARACTERIZATION OF DIGITAL FILTERS

Analog filters are characterized in terms of differential equations. Digital filters, on the other hand, are characterized in terms of difference equations. Two types of digital filters can be identified, nonrecursive and recursive filters.

Nonrecursive Filters

The response of a nonrecursive filter at instant nT is of the form

$$y(nT) = f\{\dots, x(nT - T), x(nT), x(nT + T), \dots\}$$

If we assume linearity and time invariance, $y(nT)$ can be expressed as

$$y(nT) = \sum_{i=-\infty}^{\infty} a_i x(nT - iT) \tag{1.2}$$

where a_i represent constants. Now by assuming causality and then using the causality criterion defined earlier, we can show that

$$a_{-1} = a_{-2} = \cdots = 0$$

and so
$$y(nT) = \sum_{i=0}^{\infty} a_i x(nT - iT)$$

If, in addition, $x(nT) = 0$ for $n < 0$ and $a_i = 0$ for $i > N$,

$$y(nT) = \sum_{i=0}^{n} a_i x(nT - iT) + \sum_{i=n+1}^{\infty} a_i x(nT - iT)$$

$$= \sum_{i=0}^{N} a_i x(nT - iT) + \sum_{i=N+1}^{n} a_i x(nT - iT)$$

$$= \sum_{i=0}^{N} a_i x(nT - iT) \tag{1.3}$$

Therefore, a linear, time-invariant, causal, nonrecursive filter can be represented by an Nth-order linear difference equation. N is the *order* of the filter.

Recursive Filters

The response of a recursive filter is a function of elements in the excitation as well as the response sequence. In the case of a linear, time-invariant, causal filter

$$y(nT) = \sum_{i=0}^{N} a_i x(nT - iT) - \sum_{i=1}^{N} b_i y(nT - iT) \tag{1.4}$$

i.e., if instant nT is taken to be the present, the present response is a function of the present and past N values of the excitation as well as the past N values of the response. Note that Eq. (1.4) simplifies to Eq. (1.3) if $b_i = 0$, and essentially the nonrecursive filter is a special case of the recursive one.

1.5 DIGITAL-FILTER NETWORKS

The basic digital-filter elements are the unit delay, the adder, and the multiplier. Their characterizations and symbols are given in Table 1.1. Their implementation can assume various forms, depending on the representation of the signals to be processed. If the signals are sequences of binary numbers, the unit delay will be in the form of a shift register, whereas the adder and multiplier will be combinational or sequential networks comprising NAND or NOR gates.

Digital-filter networks are collections of interconnected unit delays, adders, and multipliers. Their analysis is usually simple and can be carried out by using the element equations in Table 1.1.

Example 1.4 (*a*) Analyze the network of Fig. 1.5*a*. (*b*) Repeat part (*a*) for the network of Fig. 1.5*b*.

SOLUTION (*a*) The signals at nodes A and B are $y(nT - T)$ and $e^x y(nT - T)$, respectively. Thus

$$y(nT) = x(nT) + e^x y(nT - T)$$

Table 1.1 Digital-filter elements

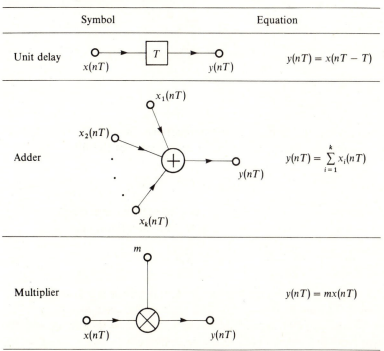

	Symbol	Equation
Unit delay	$x(nT)$ — T — $y(nT)$	$y(nT) = x(nT - T)$
Adder	$x_1(nT)$, $x_2(nT)$, $\ldots$, $x_k(nT)$ $\to$ $+$ $\to$ $y(nT)$	$y(nT) = \sum\limits_{i=1}^{k} x_i(nT)$
Multiplier	m, $x(nT)$ — $\otimes$ — $y(nT)$	$y(nT) = mx(nT)$

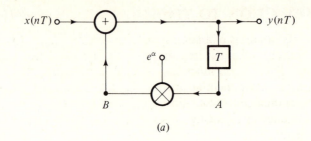

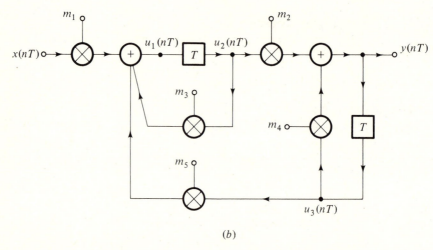

Figure 1.5 Digital-filter networks (Example 1.4): (a) first-order filter and (b) second-order filter.

(b) The analysis in this case is simplified by using the shift operator $\mathscr{E}$, which is defined by

$$\mathscr{E}^{-r}u(nT) = u(nT - rT)$$

From Fig. 1.5b

$$u_1(nT) = m_1 x(nT) + m_3 u_2(nT) + m_5 u_3(nT)$$
$$u_2(nT) = \mathscr{E}^{-1}u_1(nT) \qquad u_3(nT) = \mathscr{E}^{-1}y(nT)$$
$$y(nT) = m_2 u_2(nT) + m_4 u_3(nT)$$

and on eliminating $u_1(nT)$, $u_2(nT)$, and $u_3(nT)$, we have

$$y(nT) = a_1 x(nT - T) + b_1 y(nT - T) + b_2 y(nT - 2T)$$

where $\qquad a_1 = m_1 m_2 \qquad b_1 = m_3 + m_4 \qquad b_2 = m_2 m_5 - m_3 m_4$

Evidently, the analysis of digital-filter networks consists of solving a set of simultaneous equations. Hence signal flow-graph techniques can be used [2, 3].

1.6 INTRODUCTION TO TIME-DOMAIN ANALYSIS

The time-domain analysis of analog systems is facilitated by using several elementary functions such as the unit impulse, the unit step, etc. Similar functions can be used for digital filters. Such functions are defined in Table 1.2. The discrete-time unit step, unit ramp, exponential, and sinusoid are obtained by replacing t by nT in the corresponding continuous-time functions. The discrete-time unit impulse, however, is obtained by replacing a continuous-time impulse of strength unity (see Sec. 6.3) by unity.

Table 1.2 Discrete-time elementary functions

Function	Definition	Waveform
Unit impulse	$\delta(nT) = \begin{cases} 1 & n = 0 \\ 0 & n \neq 0 \end{cases}$	
Unit step	$u(nT) = \begin{cases} 1 & n \geq 0 \\ 0 & n < 0 \end{cases}$	
Unit ramp	$r(nT) = \begin{cases} nT & n \geq 0 \\ 0 & n < 0 \end{cases}$	
Exponential	$e^{\alpha nT}$	
Sinusoid	$\sin \omega nT$	

The time-domain response of simple digital filters can be determined by solving the difference equation directly using induction. Although this approach is somewhat primitive and inefficient, it merits consideration as it demonstrates the mode by which digital filters operate. The approach is best illustrated by examples.

Example 1.5 (*a*) Find the impulse response of the filter in Fig. 1.5*a*. The filter is initially relaxed; that is, $y(nT) = 0$ for $n < 0$. (*b*) Find the unit-step response of the filter.

SOLUTION (*a*) From Example 1.4*a*

$$y(nT) = x(nT) + e^\alpha y(nT - T) \tag{1.5}$$

With $x(nT) = \delta(nT)$, we can write

$$y(0) = 1 + e^\alpha y(-T) = 1$$
$$y(T) = 0 + e^\alpha y(0) = e^\alpha$$
$$y(2T) = 0 + e^\alpha y(T) = e^{2\alpha}$$
$$\dots\dots\dots\dots\dots\dots\dots\dots$$
$$y(nT) = e^{n\alpha}$$

The response is plotted in Fig. 1.6 for $\alpha < 0$, $\alpha = 0$, and $\alpha > 0$.
(*b*) With $x(nT) = u(nT)$

$$y(0) = 1 + e^\alpha y(-T) = 1$$
$$y(T) = 1 + e^\alpha y(0) = 1 + e^\alpha$$
$$y(2T) = 1 + e^\alpha y(T) = 1 + e^\alpha + e^{2\alpha}$$
$$\dots\dots\dots\dots\dots\dots\dots\dots\dots\dots$$
$$y(nT) = \sum_{k=0}^{n} e^{k\alpha}$$

This is a geometric progression, and hence

$$y(nT) - e^\alpha y(nT) = 1 - e^{(n+1)\alpha}$$

or

$$y(nT) = \frac{1 - e^{(n+1)\alpha}}{1 - e^\alpha}$$

For $\alpha < 0$, the steady-state value of the response is obtained by evaluating $y(nT)$ for $n \to \infty$, that is,

$$\lim_{n \to \infty} y(nT) = \frac{1}{1 - e^\alpha}$$

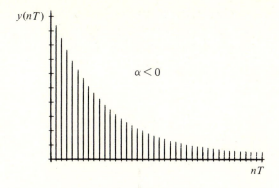

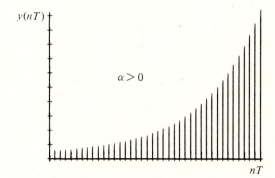

Figure 1.6 Impulse response of first-order filter (Example 1.5a).

For $\alpha = 0$, using l'Hospital's rule, we get

$$y(nT) = \lim_{\alpha \to 0} \frac{d(1 - e^{(n+1)\alpha})/d\alpha}{d(1 - e^{\alpha})/d\alpha} = n + 1$$

and hence $y(nT) \to \infty$ as $n \to \infty$. For $\alpha > 0$

$$\lim_{n \to \infty} y(nT) \approx \frac{e^{n\alpha}}{e^{\alpha} - 1} \to \infty$$

(see Fig. 1.7).

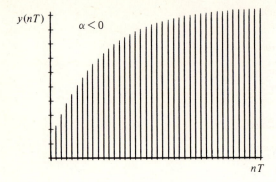

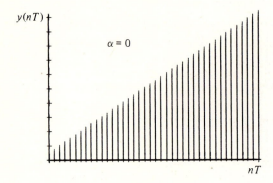

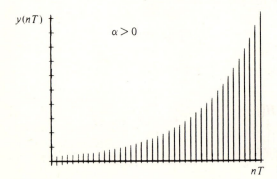

Figure 1.7 Unit-step response of first-order filter (Example 1.5b).

Example 1.6 Find the response of the filter in Fig. 1.5a if

$$x(nT) = \begin{cases} \sin \omega nT & \text{for } n \geq 0 \\ 0 & \text{otherwise} \end{cases}$$

Assume that $\alpha < 0$.

SOLUTION The filter is linear, and so

$$y(nT) = \mathcal{R} \sin \omega nT = \mathcal{R}\left(\frac{1}{2j} e^{j\omega nT} - \frac{1}{2j} e^{-j\omega nT}\right)$$

$$= \frac{1}{2j} \mathcal{R}e^{j\omega nT} - \frac{1}{2j} \mathcal{R}e^{-j\omega nT} = \frac{1}{2j} y_1(nT) - \frac{1}{2j} y_2(nT) \qquad (1.6)$$

With the filter initially relaxed, the use of Eq. (1.5) gives

$$y_1(nT) = \mathcal{R}e^{j\omega nT}$$

$$y_1(0) = e^0 + e^\alpha y_1(-T) = 1$$

$$y_1(T) = e^{j\omega T} + e^\alpha y_1(0) = e^\alpha + e^{j\omega T}$$

$$y_1(2T) = e^{j2\omega T} + e^\alpha y_1(T) = e^{2\alpha} + e^{\alpha + j\omega T} + e^{j2\omega T}$$

$$\cdots\cdots\cdots\cdots\cdots\cdots\cdots\cdots\cdots\cdots$$

$$y_1(nT) = e^{n\alpha} + e^{(n-1)\alpha + j\omega T} + \cdots + e^{j\omega nT}$$

$$= e^{j\omega nT}(1 + e^{\alpha - j\omega T} + \cdots + e^{n(\alpha - j\omega T)})$$

$$= e^{j\omega nT} \sum_{k=0}^{n} e^{k(\alpha - j\omega T)}$$

and, as in part (b) of Example 1.5,

$$y_1(nT) = \frac{e^{j\omega nT} - e^{(n+1)\alpha - j\omega T}}{1 - e^{\alpha - j\omega T}}$$

If we let

$$H(j\omega) = \frac{1}{1 - e^{\alpha - j\omega T}} = M(\omega)e^{j\theta(\omega)}$$

where

$$M(\omega) = |H(j\omega)| = \frac{1}{\sqrt{1 + e^{2\alpha} - 2e^\alpha \cos \omega T}} \qquad (1.7)$$

$$\theta(\omega) = \arg H(j\omega) = \omega T - \tan^{-1} \frac{\sin \omega T}{\cos \omega T - e^\alpha} \qquad (1.8)$$

$y_1(nT)$ becomes

$$y_1(nT) = M(\omega)(e^{j[\theta(\omega) + \omega nT]} - e^{(n+1)\alpha + j[\theta(\omega) - \omega T]}) \qquad (1.9)$$

By replacing ω by $-\omega$ in $y_1(nT)$ we get

$$y_2(nT) = M(-\omega)(e^{j[\theta(-\omega) - \omega nT]} - e^{(n+1)\alpha + j[\theta(-\omega) + \omega T]}) \qquad (1.10)$$

and since, according to Eqs. (1.7) and (1.8),

$$M(-\omega) = M(\omega) \qquad \theta(-\omega) = -\theta(\omega)$$

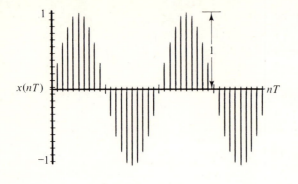

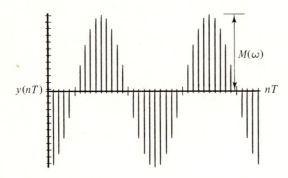

Figure 1.8 Steady-state sinusoidal response of first-order filter (Example 1.6).

Eqs. (1.6), (1.9), and (1.10) yield

$$y(nT) = \frac{M(\omega)}{2j} \left(e^{j[\theta(\omega) + \omega nT]} - e^{-j[\theta(\omega) + \omega nT]} \right)$$

$$- \frac{M(\omega)}{2j} e^{(n+1)\alpha} \left(e^{j[\theta(\omega) - \omega T]} - e^{-j[\theta(\omega) - \omega T]} \right)$$

$$= M(\omega) \sin \left[\omega nT + \theta(\omega) \right] - M(\omega) e^{(n+1)\alpha} \sin \left[\theta(\omega) - \omega T \right]$$

As $n \to \infty$, the second term reduces to zero since $\alpha < 0$, and therefore

$$\lim_{n \to \infty} y(nT) = M(\omega) \sin \left[\omega nT + \theta(\omega) \right]$$

In effect, the steady-state response of the filter to a sinusoid with amplitude of unity is a displaced sinusoid with amplitude $M(\omega)$, as illustrated in Fig. 1.8.

1.7 CONVOLUTION SUMMATION

The response of a digital filter to an arbitrary excitation can be expressed in terms of the impulse response of the filter.

An excitation $x(nT)$ can be written as

$$x(nT) = \sum_{k=-\infty}^{\infty} x_k(nT)$$

where

$$x_k(nT) = \begin{cases} x(kT) & \text{for } n = k \\ 0 & \text{otherwise} \end{cases}$$

Alternatively

$$x_k(nT) = x(kT)\delta(nT - kT)$$

and hence

$$x(nT) = \sum_{k=-\infty}^{\infty} x(kT)\delta(nT - kT) \tag{1.11}$$

Now consider a linear time-invariant filter in which

$$\mathscr{R}\delta(nT) = h(nT) \qquad \text{and} \qquad y(nT) = \mathscr{R}x(nT)$$

From Eq. (1.11)

$$y(nT) = \mathscr{R} \sum_{k=-\infty}^{\infty} x(kT)\delta(nT - kT) = \sum_{k=-\infty}^{\infty} x(kT)\mathscr{R}\delta(nT - kT)$$

$$= \sum_{k=-\infty}^{\infty} x(kT)h(nT - kT)$$

If the filter is also causal, $h(nT) = 0$ for $n < 0$ and

$$y(nT) = \sum_{k=-\infty}^{n} x(kT)h(nT - kT) = \sum_{k=0}^{\infty} h(kT)x(nT - kT) \tag{1.12}$$

Therefore, if $x(nT) = 0$ for $n < 0$, we have

$$y(nT) = \sum_{k=0}^{n} x(kT)h(nT - kT) = \sum_{k=0}^{n} h(kT)x(nT - kT) \tag{1.13}$$

This relation is of significant importance in the characterization as well as analysis of digital filters. It is known as the *convolution summation*.

Graphical Interpretation

The convolution summation is illustrated in Fig. 1.9. The impulse response $h(kT)$ is first folded about the y axis, as in Fig. 1.9c, and is then shifted to the right by a time interval nT, as in Fig. 1.9d, to yield $h(nT - kT)$. Then $x(kT)$ is multiplied by $h(nT - kT)$, as in Fig. 1.9e. The sum of all values in Fig. 1.9e is the response of the filter at instant nT.

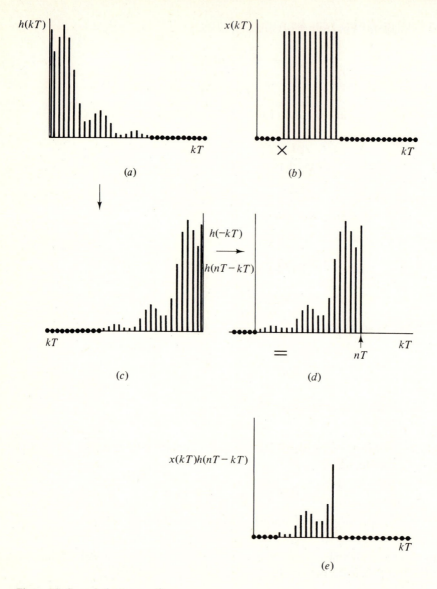

Figure 1.9 Convolution summation.

Example 1.7 (*a*) Using the convolution summation, find the unit-step response of the filter in Fig. 1.5*a*. (*b*) Hence find the response to an excitation

$$x(nT) = \begin{cases} 1 & \text{for } 0 \le n \le 4 \\ 0 & \text{otherwise} \end{cases}$$

SOLUTION (a) From part (a) of Example 1.5

$$h(nT) = e^{n\alpha}$$

Hence from Eq. (1.13)

$$y(nT) = \mathscr{R}u(nT) = \sum_{k=0}^{n} e^{k\alpha}u(nT - kT) = \sum_{k=0}^{n} e^{k\alpha}$$

$$= \begin{cases} \dfrac{1 - e^{(n+1)\alpha}}{1 - e^{\alpha}} & \text{for } n \geq 0 \\ 0 & \text{otherwise} \end{cases}$$

(b) For this part, we observe that

$$x(nT) = u(nT) - u(nT - 5T)$$

and so

$$\mathscr{R}x(nT) = \mathscr{R}u(nT) - \mathscr{R}u(nT - 5T)$$

Thus

$$y(nT) = \begin{cases} \dfrac{1 - e^{(n+1)\alpha}}{1 - e^{\alpha}} & \text{for } n \leq 4 \\ \dfrac{e^{(n-4)\alpha} - e^{(n+1)\alpha}}{1 - e^{\alpha}} & \text{for } n > 4 \end{cases}$$

1.8 STABILITY

A digital filter is said to be stable if and only if any bounded excitation results in a bounded response, i.e., if

$$|x(nT)| < \infty \qquad \text{for all } n$$

implies that

$$|y(nT)| < \infty \qquad \text{for all } n$$

For a linear, time-invariant, and causal filter, Eq. (1.12) gives

$$|y(nT)| \leq \sum_{k=0}^{\infty} |h(kT)| \cdot |x(nT - kT)|$$

and if

$$|x(nT)| \leq M < \infty \qquad \text{for all } n$$

we have

$$|y(nT)| \leq M \sum_{k=0}^{\infty} |h(kT)|$$

Clearly if

$$\sum_{k=0}^{\infty} |h(kT)| < \infty \tag{1.14}$$

then

$$|y(nT)| < \infty \qquad \text{for all } n$$

and, therefore, Eq. (1.14) constitutes a sufficient condition for stability.

The filter can be classified as stable only if the response is bounded for any excitation. Consider the specific excitation

$$x(nT - kT) = \begin{cases} M & \text{if } h(kT) > 0 \\ -M & \text{if } h(kT) < 0 \end{cases}$$

where M is a positive constant. From Eq. (1.12)

$$y(nT) = |y(nT)| = \sum_{k=0}^{\infty} M |h(kT)|$$

or

$$|y(nT)| = M \sum_{k=0}^{\infty} |h(kT)|$$

Therefore, for this excitation, the response will be bounded if and only if Eq. (1.14) holds, i.e., Eq. (1.14) constitutes a necessary and sufficient condition for stability.

Example 1.8 Check the filter of Fig. 1.5a for stability.

SOLUTION From part (a) of Example 1.5

$$\sum_{k=0}^{\infty} |h(kT)| = 1 + |e^{\alpha}| + \cdots + |e^{k\alpha}| + \cdots$$

The series converges if

$$\left| \frac{e^{(k+1)\alpha}}{e^{k\alpha}} \right| < 1 \qquad \text{or} \qquad e^{\alpha} < 1$$

Therefore, for stability

$$\alpha < 0$$

1.9 STATE-SPACE ANALYSIS

The analysis of analog systems is often simplified by using state-space techniques. Similar techniques can be developed for digital filters, as will now be shown.

Characterization

A state-space characterization can be obtained by reducing the Nth-order difference equation into a system of N first-order equations.

Consider a filter F_0 (Fig. 1.10a) in which

$$y(nT) = \sum_{i=0}^{N} a_i x(nT - iT) - \sum_{i=1}^{N} b_i y(nT - iT)$$

The filter can be decomposed into a pair of cascade filters as in Fig. 1.10b, where
F_1 and F_2 are characterized by

$$v(nT) = \mathscr{R}_1 x(nT) = \sum_{i=0}^{N} a_i x(nT - iT) \tag{1.15}$$

and $\quad\quad y(nT) = \mathscr{R}_2 v(nT) = v(nT) - \sum_{i=1}^{N} b_i y(nT - iT) \tag{1.16}$

respectively. Now assume that $x(nT)$ is applied directly at the input of F_2 and let

$$y'(nT) = \mathscr{R}_2 x(nT)$$

in which case

$$y'(nT) = x(nT) - \sum_{i=1}^{N} b_i y'(nT - iT)$$

With variables $q_1(nT)$, $q_2(nT)$, ..., $q_N(nT)$ defined as

$$q_1(nT) = y'(nT - NT)$$
$$q_2(nT) = y'(nT - NT + T)$$
$$\cdots\cdots\cdots\cdots\cdots\cdots\cdots \tag{1.17}$$
$$q_N(nT) = y'(nT - T)$$

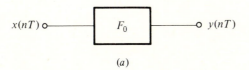

(a)

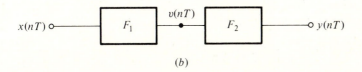

(b)

Figure 1.10 Representation of a digital filter by a cascade of two digital filters.

we can write

$$q_1(nT + T) = y'(nT - NT + T) = q_2(nT)$$

$$q_2(nT + T) = y'(nT - NT + 2T) = q_3(nT)$$

$$\cdots\cdots\cdots\cdots\cdots\cdots\cdots\cdots\cdots$$

$$q_{N-1}(nT + T) = y'(nT - T) = q_N(nT)$$

$$q_N(nT + T) = y'(nT) = x(nT) - b_1 y'(nT - T) - b_2 y'(nT - 2T)$$

$$- \cdots - b_N y'(nT - NT)$$

$$= x(nT) - b_1 q_N(nT) - b_2 q_{N-1}(nT) - \cdots - b_N q_1(nT)$$

$$(1.18)$$

These equations can be expressed in matrix form as

$$
\begin{bmatrix} q_1(nT + T) \\ q_2(nT + T) \\ \cdot \\ q_{N-1}(nT + T) \\ q_N(nT + T) \end{bmatrix}
=
\begin{bmatrix} 0 & 1 & 0 & \cdots & 0 \\ 0 & 0 & 1 & 0 & \cdots & 0 \\ \cdots\cdots\cdots\cdots\cdots\cdots\cdots \\ 0 & 0 & 0 & 0 & \cdots & 1 \\ -b_N & -b_{N-1} & \cdots & & -b_1 \end{bmatrix}
\begin{bmatrix} q_1(nT) \\ q_2(nT) \\ \cdot \\ q_{N-1}(nT) \\ q_N(nT) \end{bmatrix}
+
\begin{bmatrix} 0 \\ 0 \\ \cdot \\ 0 \\ 1 \end{bmatrix} x(nT)
$$

$$(1.19)$$

Now let us deduce the response of filter F_0 to an excitation $x(nT)$. From Eqs. (1.15) and (1.16)

$$y(nT) = \mathcal{R}_2 v(nT) = \mathcal{R}_2 \sum_{i=0}^{N} a_i x(nT - iT) = \sum_{i=0}^{N} a_i \mathcal{R}_2 x(nT - iT)$$

$$= \sum_{i=0}^{N} a_i y'(nT - iT)$$

since the filter is linear and time-invariant. By first eliminating $y'(nT)$ using Eq. (1.18) and then eliminating $y'(nT - iT)$ using Eq. (1.17) we have

$$y(nT) = a_0 x(nT) + c_1 q_1(nT) + \cdots + c_N q_N(nT)$$

where

$$c_1 = a_N - a_0 b_N$$

$$c_2 = a_{N-1} - a_0 b_{N-1}$$

$$\cdots\cdots\cdots\cdots\cdots\cdots$$

$$c_N = a_1 - a_0 b_1$$

Consequently, the response of F_0 can be expressed as

$$y(nT) = [c_1 \quad c_2 \quad \cdots \quad c_N] \begin{bmatrix} q_1(nT) \\ q_2(nT) \\ \cdot \\ q_N(nT) \end{bmatrix} + [a_0] x(nT) \qquad (1.20)$$

In summary, an Nth-order filter can be represented by the system

$$\mathbf{q}(nT + T) = \mathbf{A}\mathbf{q}(nT) + \mathbf{B}x(nT) \qquad (1.21)$$

$$y(nT) = \mathbf{C}\mathbf{q}(nT) + \mathbf{D}x(nT) \qquad (1.22)$$

where **A**, **B**, **C**, and **D** are the matrices in Eqs. (1.19) and (1.20). The N auxiliary variables $q_1(nT)$, $q_2(nT)$, $\dots$, $q_N(nT)$ are called *state variables*.

Example 1.9 A filter is characterized by

$$y(nT) = \tfrac{3}{2}x(nT) + 2x(nT - T) + \tfrac{1}{2}x(nT - 2T) - \tfrac{1}{2}y(nT - T) + \tfrac{1}{4}y(nT - 2T)$$

Deduce its state-space representation.

SOLUTION From Eqs. (1.19) and (1.20)

$$\mathbf{A} = \begin{bmatrix} 0 & 1 \\ \tfrac{1}{4} & -\tfrac{1}{2} \end{bmatrix} \qquad \mathbf{B} = \begin{bmatrix} 0 \\ 1 \end{bmatrix} \qquad \mathbf{C} = [\tfrac{7}{8} \quad \tfrac{5}{4}] \qquad \mathbf{D} = [\tfrac{3}{2}]$$

Time-Domain Analysis

The preceding state-space characterization leads directly to a relatively simple time-domain analysis.

For $n = 0, 1, 2, \dots$ Eq. (1.21) gives

$$\mathbf{q}(T) = \mathbf{A}\mathbf{q}(0) + \mathbf{B}x(0)$$

$$\mathbf{q}(2T) = \mathbf{A}\mathbf{q}(T) + \mathbf{B}x(T)$$

$$\mathbf{q}(3T) = \mathbf{A}\mathbf{q}(2T) + \mathbf{B}x(2T)$$

$$\dots\dots\dots\dots\dots\dots\dots\dots\dots\dots\dots$$

Hence

$$\mathbf{q}(2T) = \mathbf{A}^2\mathbf{q}(0) + \mathbf{A}\mathbf{B}x(0) + \mathbf{B}x(T)$$

$$\mathbf{q}(3T) = \mathbf{A}^3\mathbf{q}(0) + \mathbf{A}^2\mathbf{B}x(0) + \mathbf{A}\mathbf{B}x(T) + \mathbf{B}x(2T)$$

and in general

$$\mathbf{q}(nT) = \mathbf{A}^n\mathbf{q}(0) + \sum_{k=0}^{n-1} \mathbf{A}^{(n-1-k)}\mathbf{B}x(kT)$$

where $\mathbf{A}^0$ is the $N \times N$ unit matrix. Therefore, from Eq. (1.22)

$$y(nT) = \mathbf{C}\mathbf{A}^n\mathbf{q}(0) + \mathbf{C}\sum_{k=0}^{n-1} \mathbf{A}^{(n-1-k)}\mathbf{B}x(kT) + \mathbf{D}x(nT)$$

From Eq. (1.17)

$$\mathbf{q}(0) = \begin{bmatrix} q_1(0) \\ \vdots \\ q_N(0) \end{bmatrix} = \begin{bmatrix} y'(-NT) \\ \vdots \\ y'(-T) \end{bmatrix}$$

and if $x(nT) = 0$ for $n < 0$,

$$\mathbf{q}(0) = 0$$

Thus for an initially relaxed filter

$$y(nT) = \mathbf{C} \sum_{k=0}^{n-1} \mathbf{A}^{(n-1-k)} \mathbf{B} x(kT) + \mathbf{D} x(nT)$$

The impulse response $h(nT)$ of the filter is

$$h(nT) = \mathbf{C} \sum_{k=0}^{n-1} \mathbf{A}^{(n-1-k)} \mathbf{B} \delta(kT) + \mathbf{D} \delta(nT)$$

For $n = 0$

$$h(0) = \mathbf{D}\delta(0) = a_0$$

and for $n > 0$

$$h(nT) = \mathbf{C} \mathbf{A}^{(n-1)} \mathbf{B} \delta(0) + \mathbf{C} \mathbf{A}^{(n-2)} \mathbf{B} \delta(T) + \cdots + \mathbf{D}\delta(nT)$$

Therefore

$$h(nT) = \begin{cases} a_0 & \text{for } n = 0 \\ \mathbf{C} \mathbf{A}^{(n-1)} \mathbf{B} & \text{otherwise} \end{cases} \tag{1.23}$$

Similarly, the unit-step response of the filter is

$$y(nT) = \mathbf{C} \sum_{k=0}^{n-1} \mathbf{A}^{(n-1-k)} \mathbf{B} u(kT) + \mathbf{D} u(nT) = \mathbf{C} \sum_{k=0}^{n-1} \mathbf{A}^{(n-1-k)} \mathbf{B} + \mathbf{D}$$

Example 1.10 Assuming that the filter of Example 1.9 is initially relaxed, find $h(17T)$.

SOLUTION From Eq. (1.23)

$$h(17T) = \mathbf{C} \mathbf{A}^{16} \mathbf{B}$$

By forming $\mathbf{A}^2$, $\mathbf{A}^4$, $\mathbf{A}^8$, and then $\mathbf{A}^{16}$, we have

$$h(17T) = \begin{bmatrix} 7 & 5 \\ 8 & 4 \end{bmatrix} \begin{bmatrix} \dfrac{610}{65{,}536} & -\dfrac{987}{32{,}768} \\ -\dfrac{987}{131{,}072} & \dfrac{1597}{65{,}536} \end{bmatrix} \begin{bmatrix} 0 \\ 1 \end{bmatrix} = \dfrac{1076}{262{,}144}$$

REFERENCES

1. R. J. Schwarz and B. Friedland, "Linear Systems," McGraw-Hill, New York, 1965.
2. J. R. Abrahams and G. P. Coverley, "Signal Flow Analysis," Pergamon, New York, 1965.
3. B. C. Kuo, "Automatic Control Systems," Prentice-Hall, Englewood Cliffs, N.J., 1962.

PROBLEMS

1.1 By using appropriate tests, check the filters characterized by the following equations for time invariance, causality, and linearity:

(a) $\mathscr{R}x(nT) = 2x(nT - gT)$ where $g > 0$

(b) $\mathscr{R}x(nT) = \begin{cases} 6x(nT - 5T) & \text{for } x(nT) \leq 6 \\ 7x(nT - 5T) & \text{for } x(nT) > 6 \end{cases}$

(c) $\mathscr{R}x(nT) = (nT + 3T)x(nT - 3T)$

(d) $\mathscr{R}x(nT) = 5nTx^2(nT)$

(e) $\mathscr{R}x(nT) = 3x(nT + 3T)$

(f) $\mathscr{R}x(nT) = x(nT) \sin \omega nT$

(g) $\mathscr{R}x(nT) = K_1 \Delta x(nT)$ where $\Delta x(nT) = x(nT + T) - x(nT)$

(h) $\mathscr{R}x(nT) = K_2 \nabla x(nT)$ where $\nabla x(nT) = x(nT) - x(nT - T)$

(i) $\mathscr{R}x(nT) = x(nT + T)e^{-nT}$

(j) $\mathscr{R}x(nT) = x^2(nT + T)e^{-nT} \sin \omega nT$

1.2 Analyze the filter networks shown in Fig. P1.2.

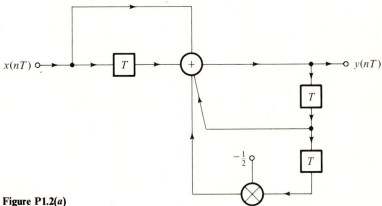

Figure P1.2(a)

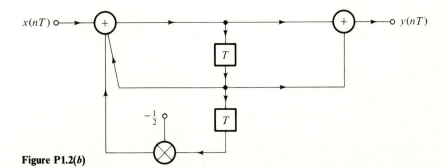

Figure P1.2(b)

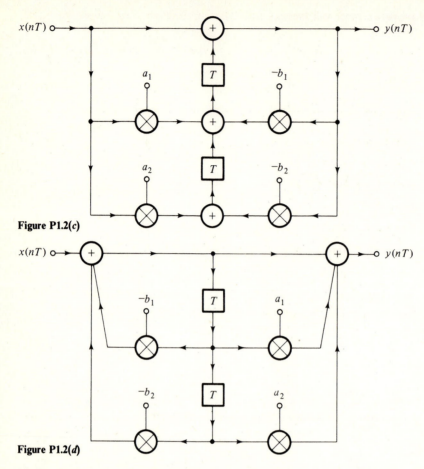

Figure P1.2(c)

Figure P1.2(d)

1.3 Two second-order filter sections of the type shown in Fig. P1.2c. are connected in cascade as in Fig. P1.3. The parameters of two sections are a_{11}, a_{21}, $-b_{11}$, $-b_{21}$ and a_{12}, a_{22}, $-b_{12}$, $-b_{22}$, respectively. Deduce the characterization of the combined filter.

Figure P1.3

1.4 The two sections in Prob. 1.3 are connected in parallel as in Fig. P1.4. Obtain the difference equation of the combined filter.

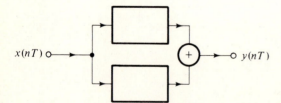

Figure P1.4

1.5 Fig. P1.5 shows a network with three inputs and three outputs. Derive a set of equations character-izing the network.

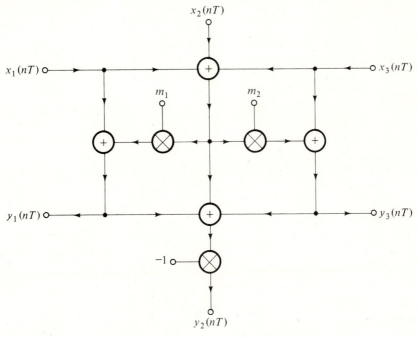

Figure P1.5

1.6 By using appropriate tests, check the filters of Fig. P1.6 for time invariance, linearity, and causality.

(a) The filter of Fig. P1.6a uses a device N whose response is given by

$$\mathscr{R}x(nT) = |x(nT)|$$

(b) The filter of Fig. P1.6b uses a multiplier M whose parameter is given by

$$m = 0.1x(nT)$$

(c) The filter of Fig. P1.6c uses a multiplier M whose parameter is given by

$$m = 0.1v(nT)$$

where $v(nT)$ is an independent signal.

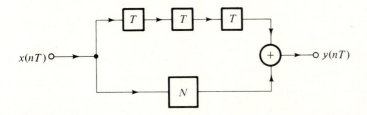

Figure P1.6(a)

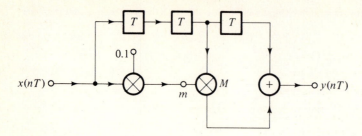

Figure P1.6(*b*)

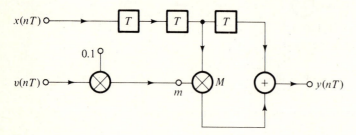

Figure P1.6(*c*)

1.7 Show that

$$r(nT) = \mu_2(nT - T) \qquad u(nT) = \mu_1(nT) \qquad \delta(nT) = T\mu_0(nT)$$

where

$$\mu_{i-1}(nT) = \frac{1}{T}\nabla\mu_i(nT)$$

and

$$\mu_2(nT) = \begin{cases} nT + T & \text{for } n \geq 0 \\ 0 & \text{otherwise} \end{cases}$$

1.8 The filter of Fig. P1.8 is initially relaxed. Find the time-domain response for $n = 0, 1, \ldots, 6$ if

$$x(nT) = \begin{cases} \sin \omega nT & \text{for } n \geq 0 \\ 0 & \text{otherwise} \end{cases}$$

where $\omega = \pi/6T$ and $T = 1$.

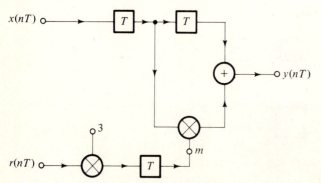

Figure P1.8

1.9 (*a*) Show that

$$r(nT) = \begin{cases} 0 & \text{for } n \le 0 \\ T \sum_{k=1}^{n} u(nT - kT) & \text{otherwise} \end{cases}$$

(*b*) By using this relation obtain the unit-ramp response of the filter shown in Fig. 1.5*a* in closed form. The filter is initially relaxed.

(*c*) Sketch the response for $\alpha > 0$, $\alpha = 0$, and $\alpha < 0$.

1.10 The excitation in Fig. 1.5*a* is

$$x(nT) = \begin{cases} 1 & \text{for } 0 \le n \le 4 \\ 2 & \text{for } n > 4 \\ 0 & \text{for } n < 0 \end{cases}$$

Find the response in closed form.

1.11 Repeat Prob. 1.10 for an excitation

$$x(nT) = \begin{cases} 1 & \text{for } n = 0 \\ 0 & \text{for } n < 0, n = 1, 2, 3, 4 \\ 1 & \text{for } n > 4 \end{cases}$$

1.12 Fig. P1.12 shows a second-order recursive filter. Compute the unit-step response for $0 \le n \le 15$ if

(*a*) $\alpha = 1$ $\beta = -\frac{1}{2}$

(*b*) $\alpha = \frac{1}{2}$ $\beta = -\frac{1}{8}$

(*c*) $\alpha = \frac{5}{4}$ $\beta = -\frac{25}{32}$

Compare the three responses and determine the frequency of ringing in terms of T, where possible.

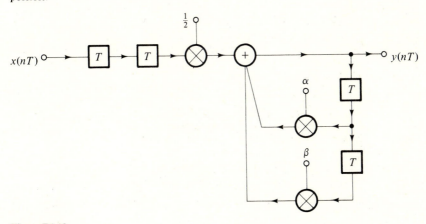

Figure P1.12

1.13 Fig. P1.13 shows a filter comprising a cascade of two first-order sections. The input signal is

$$x(nT) = \begin{cases} \sin \omega nT & \text{for } n \ge 0 \\ 0 & \text{otherwise} \end{cases}$$

and $T = 1$ ms. Compute the steady-state amplitude and phase angle of $y(nT)$ for a frequency $f = 10$ Hz. Repeat for $f = 100$ Hz.

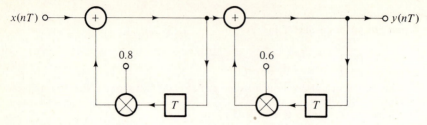

Figure P1.13

1.14 Fig. P1.14 shows a linear first-order filter.

(a) Assuming a sinusoidal excitation, derive an expression for the steady-state gain of the filter.

(b) Plot the gain in decibels (dB) versus $\log f$ for $f = 0$ to 1.0 kHz if $T = 1$ ms.

(c) Determine the lowest frequency at which the gain is reduced by 3 dB relative to the gain at zero frequency.

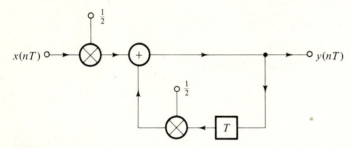

Figure P1.14

1.15 Two first-order filters of the type shown in Fig. 1.5a are connected in parallel as in Fig. P1.4. The multiplier constants for the two filters are $m_1 = e^{0.6}$ and $m_2 = e^{0.7}$. Find the unit-step response of the combined network in closed form.

1.16 The unit-step response of a filter is

$$y(nT) = \begin{cases} nT & \text{for } n \geq 0 \\ 0 & \text{for } n < 0 \end{cases}$$

(a) Using the convolution summation, find the unit-ramp response.

(b) Check the filter for stability.

1.17 An initially relaxed filter was tested with an input signal

$$x(nT) = 2u(nT)$$

and found to have a response

n	0	1	2	3	4	5	$\cdots$	100	$\cdots$
$y(nT)$	2	6	12	20	30	30	$\cdots$	30	$\cdots$

(a) Deduce the difference equation of the filter.

(b) Construct a possible network for the filter.

1.18 (*a*) Check the filter of Fig. P1.18*a* for stability.
(*b*) Repeat part (*a*) for the filter of Fig. P1.18*b*.

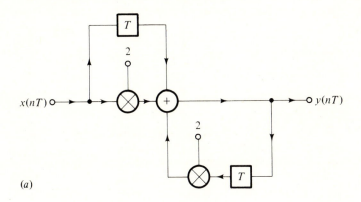

(*a*)

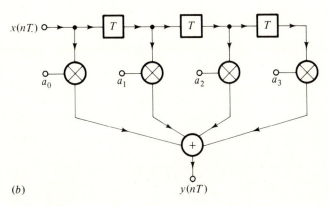

(*b*) $y(nT)$

Figure P1.18

1.19 Derive state-space representations for the filters of Figs. P1.2 to P1.4.

1.20 An initially relaxed filter is characterized by the matrices

$$\mathbf{A} = \begin{bmatrix} 0 & 1 \\ -\frac{5}{16} & -1 \end{bmatrix} \qquad \mathbf{B} = \begin{bmatrix} 0 \\ 1 \end{bmatrix} \qquad \mathbf{C} = [\tfrac{11}{8} \quad 2] \qquad \mathbf{D} = [2]$$

(*a*) Calculate the impulse response for $n = 0, 1, \ldots, 5$ and for $n = 17$ by using the state-space method.

(*b*) Deduce the difference equation of the filter.

(*c*) Repeat part (*a*) by using the difference equation.

1.21 Compute the unit-step response of the filter in Prob. 1.20 for $n = 5$ by using the state-space method.

1.22 A signal

$$x(nT) = 3u(nT) \cos \omega nT$$

is applied at the input of the filter in Prob. 1.20. Find the response at instant $5T$ if $\omega = 1/10T$ by using the convolution summation.

TWO

THE z TRANSFORM

2.1 INTRODUCTION

The analysis of linear, time-invariant digital filters is almost invariably carried out by using the z transform. The principal reason for this is that the difference equations characterizing such filters are transformed into algebraic equations which are usually much easier to manipulate.

We start here by defining the z transform and then proceed to discuss its salient properties. The application of the z transform to digital filters is postponed to Chap. 3.

2.2 DEFINITION

The (two-sided) z transform of a discrete-time function $f(nT)$ is defined as

$$F(z) = \sum_{n=-\infty}^{\infty} f(nT)z^{-n} \tag{2.1}$$

for all z for which $F(z)$ converges. The argument of $F(z)$, namely z, is a complex variable.

Equation (2.1) is usually represented by the simplified notation

$$F(z) = \mathscr{Z}f(nT)$$

2.3 THEOREMS

The properties of the z transform can conveniently be described by means of a number of theorems. A more detailed discussion of the subject can be found in the work of Jury [1].

Theorem 2.1: Laurent theorem [2] (a) If $G(z)$ is an analytic and single-valued function on two concentric circles C_1 and C_2 with center a and in the annulus between them, as depicted in Fig. 2.1, then $G(z)$ can be represented by the Laurent series

$$G(z) = \sum_{n=-\infty}^{\infty} A_n(z-a)^{-n} \tag{2.2}$$

Constant A_n is given by

$$A_n = \frac{1}{2\pi j} \oint_\Gamma G(z)(z-a)^{n-1}\, dz$$

where Γ is any simple closed contour in the counterclockwise sense lying in the annulus and encircling the inner circle.

(b) The Laurent series converges and represents $G(z)$ in the open annulus obtained by continuously increasing the radius of C_2 and decreasing the radius of C_1 until C_1 and C_2 reach points where $G(z)$ is singular.

(c) The Laurent series of $G(z)$ in its annulus of convergence is unique. However, $G(z)$ may have different Laurent series in different annuli about the same center.

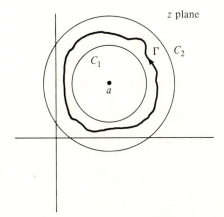

Figure 2.1 Domain of $G(z)$.

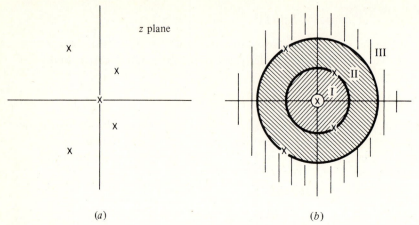

(a)   *(b)*

Figure 2.2 A function with three Laurent series about the origin: (*a*) zero-pole plot and (*b*) annuli of convergence.

A comparison of Eqs. (2.1) and (2.2) shows that the right-hand side in Eq. (2.1) is a Laurent series for $F(z)$ about the origin of the z plane and consequently Theorem 2.1 applies.

We shall be concerned exclusively with z transforms which are meromorphic functions, i.e., functions whose only singularities in the finite z plane are poles. According to Theorem 2.1*c*, such functions may have more than one Laurent series. For example, the function represented by the zero-pole plot of Fig. 2.2*a* has three Laurent series about the origin, one for each of the annuli identified in Fig. 2.2*b*. For convenience, we shall assume that the series in Eq. (2.1) is one that converges in the open annulus

$$R_1 \le |z| \le R_2 \tag{2.3}$$

where $R_2 \to \infty$ and $|z| = R_1$ is the circle passing through the most distant pole of $F(z)$ from the origin.

Theorem 2.2: Linearity If a and b are arbitrary constants and

$$\mathscr{L}f(nT) = F(z)$$
$$\mathscr{L}g(nT) = G(z)$$

where $f(nT)$ and $g(nT)$ are arbitrary discrete-time functions, then

$$\mathscr{L}[af(nT) + bg(nT)] = aF(z) + bG(z)$$

PROOF
$$\mathcal{Z}[af(nT) + bg(nT)] = \sum_{n=-\infty}^{\infty} [af(nT) + bg(nT)]z^{-n}$$

$$= a \sum_{n=-\infty}^{\infty} f(nT)z^{-n} + b \sum_{n=-\infty}^{\infty} g(nT)z^{-n}$$

$$= aF(z) + bG(z)$$

Theorem 2.3: Translation

(*a*)
$$\mathcal{Z}f(nT + mT) = z^m F(z)$$

[see page 38 for parts (*b*) and (*c*)]

PROOF (*a*)
$$\mathcal{Z}f(nT + mT) = \sum_{n=-\infty}^{\infty} f(nT + mT)z^{-n} = z^m \sum_{n=-\infty}^{\infty} f[(n+m)T]z^{-(n+m)}$$

$$= z^m \sum_{n=-\infty}^{\infty} f(nT)z^{-n} = z^m F(z)$$

Theorem 2.4: Complex scale change

$$\mathcal{Z}[w^{-n}f(nT)] = F(wz)$$

PROOF
$$\mathcal{Z}[w^{-n}f(nT)] = \sum_{n=-\infty}^{\infty} f(nT)w^{-n}z^{-n}$$

$$= \sum_{n=-\infty}^{\infty} f(nT)(wz)^{-n}$$

$$= F(wz)$$

Theorem 2.5: Complex differentiation

$$\mathcal{Z}[nTf(nT)] = -Tz \frac{dF(z)}{dz}$$

PROOF
$$\mathcal{Z}[nTf(nT)] = \sum_{n=-\infty}^{\infty} nTf(nT)z^{-n} = -Tz \sum_{n=-\infty}^{\infty} f(nT)(-n)z^{-n-1}$$

$$= -Tz \sum_{n=-\infty}^{\infty} f(nT) \frac{d}{dz}(z^{-n})$$

$$= -Tz \frac{d}{dz}\left[\sum_{n=-\infty}^{\infty} f(nT)z^{-n}\right] = -Tz \frac{dF(z)}{dz}$$

Theorem 2.6: Real convolution

(a)
$$\mathscr{L} \sum_{k=-\infty}^{\infty} f(kT)g(nT - kT) = F(z)G(z)$$

or
$$\mathscr{L} \sum_{k=-\infty}^{\infty} f(nT - kT)g(kT) = F(z)G(z)$$

[see page 39 for part (b)]

PROOF (a) $\mathscr{L} \displaystyle\sum_{k=-\infty}^{\infty} f(kT)g(nT - kT) = \displaystyle\sum_{n=-\infty}^{\infty} \displaystyle\sum_{k=-\infty}^{\infty} f(kT)g(nT - kT)z^{-n}$

$$= \sum_{k=-\infty}^{\infty} \sum_{n=-\infty}^{\infty} f(kT)g(nT - kT)z^{-n}$$

$$= \sum_{k=-\infty}^{\infty} f(kT)z^{-k} \sum_{n=-\infty}^{\infty} g(nT - kT)z^{-(n-k)}$$

$$= \sum_{n=-\infty}^{\infty} f(nT)z^{-n} \sum_{n=-\infty}^{\infty} g(nT)z^{-n}$$

$$= F(z)G(z)$$

2.4 ONE-SIDED z TRANSFORM

By analogy with the one-sided Laplace transform, the one-sided z transform is defined as

$$F_I(z) = \sum_{n=0}^{\infty} f(nT)z^{-n} = \mathscr{L}_I f(nT)$$

Obviously, this differs from the two-sided z transform only if $f(nT) \neq 0$ for $n < 0$. Furthermore, the preceding theorems apply except for Theorems 2.3 and 2.6, which are modified slightly, as follows.

Theorem 2.3 Translation

If
$$f(nT) = 0 \qquad \text{for } n < 0$$
and
$$\mathscr{L}_I f(nT) = F_I(z)$$

then for a positive integer m

(b)
$$\mathscr{L}_I f(nT + mT) = z^m \left[F_I(z) - \sum_{k=0}^{m-1} f(kT)z^{-k} \right]$$

(c)
$$\mathscr{L}_I f(nT - mT) = z^{-m} F_I(z)$$

Proof (b) $\mathcal{L}_I f(nT + mT) = \sum\limits_{n=0}^{\infty} f(nT + mT)z^{-n}$

$$= z^m \sum_{n=0}^{\infty} f[(n+m)T]z^{-(m+n)}$$

$$= z^m \sum_{k=m}^{\infty} f(kT)z^{-k}$$

$$= z^m \left[\sum_{k=0}^{\infty} f(kT)z^{-k} - \sum_{k=0}^{m-1} f(kT)z^{-k} \right]$$

$$= z^m \left[F_I(z) - \sum_{k=0}^{m-1} f(kT)z^{-k} \right]$$

(c) Since $f(nT) = f(nT - mT) = 0$ for $n < 0$, Theorem 2.3(a) applies.

Theorem 2.6: Real convolution If

$$F_I(z) = \mathcal{L}_I \ f(nT) \qquad \text{and} \qquad G_I(z) = \mathcal{L}_I \ g(nT)$$

then

(b) $\mathcal{L}_I \sum\limits_{k=0}^{n} f(kT)g(nT - kT) = \mathcal{L}_I \sum\limits_{k=0}^{n} f(nT - kT)g(kT) = F_I(z)G_I(z)$

Proof (b) The proof is the same as for Theorem 2.6(a).

We shall be concerned mostly with functions which are zero for $n < 0$. Hence the distinction between one-sided and two-sided transforms need not be made.

Example 2.1 Find the z transform of (a) $\delta(nT)$, (b) $u(nT)$, (c) $u(nT - T)K$, (d) $u(nT)Kw^n$, (e) $u(nT)e^{-anT}$, (f) $r(nT)$ (see Table 1.2), and (g) $u(nT) \sin \omega nT$.

Solution (a) $\mathcal{L}\delta(nT) = \delta(0) + \delta(T)z^{-1} + \cdots$

$$= 1$$

(b) $\mathcal{L}u(nT) = u(0) + u(T)z^{-1} + u(2T)z^{-2} + \cdots$

$$= 1 + z^{-1} + z^{-2} + \cdots$$

$$= (1 - z^{-1})^{-1} = \frac{z}{z-1}$$

(c) From Theorem 2.3

$$\mathcal{L}u(nT - T)K = Kz^{-1}\mathcal{L}u(nT) = \frac{K}{z-1}$$

(d) From Theorem 2.4

$$\mathscr{L}[u(nT)Kw^n] = K\mathscr{L}\left[\left(\frac{1}{w}\right)^{-n} u(nT)\right] = K\mathscr{L}u(nT)\Big|_{z\to z/w} = \frac{Kz}{z-w}$$

(e) By letting $K = 1$ and $w = e^{-\alpha T}$ in the preceding example we have

$$\mathscr{L}[u(nT)e^{-\alpha nT}] = \frac{z}{z - e^{-\alpha T}}$$

Table 2.1 Standard z transforms

$f(nT)$	$F(z)$
$\delta(nT)$	1
$u(nT)$	$\dfrac{z}{z-1}$
$u(nT-T)K$	$\dfrac{K}{z-1}$
$u(nT)Kw^n$	$\dfrac{Kz}{z-w}$
$u(nT-T)Kw^{n-1}$	$\dfrac{K}{z-w}$
$u(nT)e^{-\alpha nT}$	$\dfrac{z}{z-e^{-\alpha T}}$
$r(nT)$	$\dfrac{Tz}{(z-1)^2}$
$u(nT)\sin \omega nT$	$\dfrac{z \sin \omega T}{z^2 - 2z \cos \omega T + 1}$
$u(nT)\cos \omega nT$	$\dfrac{z(z - \cos \omega T)}{z^2 - 2z \cos \omega T + 1}$
$u(nT)e^{-\alpha nT}\sin \omega nT$	$\dfrac{ze^{-\alpha T} \sin \omega T}{z^2 - 2ze^{-\alpha T} \cos \omega T + e^{-2\alpha T}}$
$u(nT)e^{-\alpha nT}\cos \omega nT$	$\dfrac{z(z - e^{-\alpha T} \cos \omega T)}{z^2 - 2ze^{-\alpha T} \cos \omega T + e^{-2\alpha T}}$

(f) From Theorem 2.5

$$\mathscr{L}r(nT) = \mathscr{L}[nTu(nT)] = -Tz\frac{d}{dz}[\mathscr{L}u(nT)]$$

$$= \frac{Tz}{(z-1)^2}$$

(g) From part (e)

$$\mathscr{L}[u(nT)\sin \omega nT] = \mathscr{L}\left[\frac{u(nT)}{2j}\left(e^{j\omega nT} - e^{-j\omega nT}\right)\right]$$

$$= \frac{1}{2j}\mathscr{L}[u(nT)e^{j\omega nT}] - \frac{1}{2j}\mathscr{L}[u(nT)e^{-j\omega nT}]$$

$$= \frac{1}{2j}\left[\frac{z}{z - e^{j\omega T}} - \frac{z}{z - e^{-j\omega T}}\right]$$

$$= \frac{z\sin \omega T}{z^2 - 2z\cos \omega T + 1}$$

A list of the common z transforms is given in Table 2.1. An extensive list can be found in the work of Jury [1].

2.5 INVERSE z TRANSFORM

Function $f(nT)$ is said to be the *inverse z transform* of $F(z)$. Since the series in Eq. (2.1) has been assumed to converge in the open annulus described by Eq. (2.3), $f(nT)$ can be uniquely determined as

$$f(nT) = \frac{1}{2\pi j}\oint_{\Gamma} F(z)z^{n-1}\,dz = \mathscr{L}^{-1}F(z) \tag{2.4}$$

by virtue of Theorem 2.1a and c, where Γ is a contour in the counterclockwise sense enclosing all the singularities (poles) of $F(z)$. If

$$F(z)z^{n-1} = F_0(z) = \frac{N(z)}{\displaystyle\prod_{i=1}^{k}(z - p_i)^{m_i}}$$

where k and m_i are positive integers, then by using the residue theorem [2] we get

$$f(nT) = \sum_{i=1}^{k}\operatorname*{res}_{z=p_i}[F_0(z)]$$

where
$$\mathop{\mathrm{res}}_{z=p_i}[F_0(z)] = \frac{1}{(m-1)!} \lim_{z\to p_i} \frac{d^{m-1}}{dz^{m-1}}[(z-p_i)^m F_0(z)]$$

for a pole of order m, and

$$\mathop{\mathrm{res}}_{z=p_i}[F_0(z)] = \lim_{z\to p_i}[(z-p_i)F_0(z)]$$

for a simple pole.

Note that $F_0(z)$ may have a simple pole at the origin when $n = 0$ and possibly higher order poles for $n < 0$. This fact must be taken into account in the determination of $f(0), f(-T), f(-2T), \ldots$.

Example 2.2 Find the inverse z transforms of

$$(a) \quad F(z) = \frac{(2z-1)z}{2(z-1)(z+0.5)}$$

$$(b) \quad F(z) = \frac{1}{2(z-1)(z+0.5)}$$

SOLUTION (a) For $n \geq 0$

$$f(nT) = \mathop{\mathrm{res}}_{z=1}[F(z)z^{n-1}] + \mathop{\mathrm{res}}_{z=-0.5}[F(z)z^{n-1}]$$

$$= \frac{(2z-1)z^n}{2(z+0.5)}\bigg|_{z=1} + \frac{(2z-1)z^n}{2(z-1)}\bigg|_{z=-0.5}$$

$$= \tfrac{1}{3} + \tfrac{2}{3}(-\tfrac{1}{2})^n$$

For $n < 0$, one can show that $f(nT) = 0$, and hence for any value of n

$$f(nT) = u(nT)[\tfrac{1}{3} + \tfrac{2}{3}(-\tfrac{1}{2})^n]$$

(b) In this case $F(z)z^{n-1}$ has a pole at the origin if $n = 0$, and so $f(0)$ must be obtained individually, i.e.,

$$f(0) = \frac{1}{2(z-1)(z+0.5)}\bigg|_{z=0} + \frac{1}{2(z+0.5)z}\bigg|_{z=1} + \frac{1}{2(z-1)z}\bigg|_{z=-0.5}$$

$$= -1 + \tfrac{1}{3} + \tfrac{2}{3} = 0$$

Similarly, for $n < 0$ we have $f(nT) = 0$. On the other hand, for $n > 0$

$$f(nT) = \frac{z^{n-1}}{2(z+0.5)}\bigg|_{z=1} + \frac{z^{n-1}}{2(z-1)}\bigg|_{z=-0.5}$$

$$= \tfrac{1}{3} - \tfrac{1}{3}(-\tfrac{1}{2})^{n-1}$$

Alternatively, for any value of n

$$f(nT) = u(nT-T)[\tfrac{1}{3} - \tfrac{1}{3}(-\tfrac{1}{2})^{n-1}]$$

Because of the uniqueness of the Laurent series in a given annulus of convergence, the inverse z transform can also be obtained in other ways. For example, by equating coefficients, by performing long division, by expressing $F(z)$ in terms of binomial or partial fraction expansions, or by using the convolution theorem.

Use of Binomial Expansion

The binomial expansion can be used in simple one-pole z transforms.

Example 2.3 Find the inverse z transform of

$$F(z) = \frac{K}{z - w}$$

SOLUTION $\quad F(z) = \frac{K}{z}(1 - wz^{-1})^{-1} = \frac{K}{z}(1 + wz^{-1} + w^2z^{-2} + \cdots)$

$$= K(z^{-1} + wz^{-2} + w^2z^{-3} + \cdots)$$

$$= \sum_{n=0}^{\infty} [u(nT - T)Kw^{n-1}]z^{-n}$$

Hence $\qquad \mathscr{L}^{-1}\frac{K}{z - w} = f(nT) = u(nT - T)Kw^{n-1}$

Use of Partial Fractions

The z transform can be expressed in terms of partial fractions as

$$F(z) = \sum_{i=1}^{k} F_i(z)$$

where $F_1(z)$, $F_2(z)$, ... are simple one-pole z transforms. Thus

$$\mathscr{L}^{-1}F(z) = \sum_{i=1}^{k} \mathscr{L}^{-1}F_i(z)$$

where each inverse z transform on the right-hand side can be obtained from Table 2.1.

Example 2.4 Find the inverse z transforms of

$$(a) \quad F(z) = \frac{z}{(z - \frac{1}{2})(z - \frac{1}{4})}$$

$$(b) \quad F(z) = \frac{z}{z^2 + z + \frac{1}{2}}$$

SOLUTION (a) $F(z)$ can be expressed as

$$F(z) = \frac{z}{(z - \frac{1}{2})(z - \frac{1}{4})} = \frac{2}{z - \frac{1}{2}} - \frac{1}{z - \frac{1}{4}}$$

and from Example 2.3

$$f(nT) = 2u(nT - T)(\tfrac{1}{2})^{n-1} - u(nT - T)(\tfrac{1}{4})^{n-1}$$

$$= 4u(nT - T)[(\tfrac{1}{2})^n - (\tfrac{1}{4})^n]$$

(b) By expanding $F(z)/z$ we get

$$\frac{F(z)}{z} = \frac{1}{z^2 + z + \frac{1}{2}} = \frac{A}{z - p_1} + \frac{B}{z - p_2}$$

where $\qquad p_1 = \dfrac{e^{j3\pi/4}}{\sqrt{2}} \qquad p_2 = \dfrac{e^{-j3\pi/4}}{\sqrt{2}} \qquad A = -j \qquad B = j$

and so $\qquad F(z) = \dfrac{-jz}{z - p_1} + \dfrac{jz}{z - p_2}$

From Table 2.1

$$f(nT) = u(nT)(-jp_1^n + jp_2^n)$$

$$= (\tfrac{1}{2})^{n/2}u(nT)\frac{1}{j}(e^{j3\pi n/4} - e^{-j3\pi n/4})$$

$$= 2(\tfrac{1}{2})^{n/2}u(nT)\sin\frac{3\pi n}{4}$$

Alternatively, one could expand $F(z)$ into partial fractions, as in part (a).

Use of Convolution Theorem

From Theorem 2.6(a)

$$\mathscr{Z}^{-1}[F(z)G(z)] = \sum_{k=-\infty}^{\infty} f(kT)g(nT - kT)$$

Thus, if a z transform can be split into a product of two z transforms whose inverses are available, performing the convolution summation will yield the desired inverse.

Example 2.5 Find the inverse z transforms of

$$(a) \quad Y(z) = \frac{z}{(z - 1)^2}$$

$$(b) \quad Y(z) = \frac{z}{(z - 1)^3}$$

SOLUTION (a) Let

$$F(z) = \frac{z}{z-1} \quad \text{and} \quad G(z) = \frac{1}{z-1}$$

From Table 2.1

$$f(nT) = u(nT) \quad \text{and} \quad g(nT) = u(nT - T)$$

and hence for $n < 0$, the convolution summation yields $y(nT) = 0$. For $n \geq 0$

$$y(nT) = \sum_{k=0}^{n} u(kT)u(nT - T - kT)$$

$$= u(0)u(nT - T) + u(T)u(nT - 2T) + \cdots$$

$$+ u(nT - T)u(0) + u(nT)u(-T)$$

$$= 1 + 1 + \cdots + 1 + 0 = n$$

Therefore

$$y(nT) = nu(nT)$$

(b) For this example we can write

$$F(z) = \frac{z}{(z-1)^2} \quad \text{and} \quad G(z) = \frac{1}{z-1}$$

in which case

$$f(nT) = nu(nT)$$

$$g(nT) = u(nT - T)$$

Thus for $n < 0$, $y(nT) = 0$. For $n \geq 0$

$$y(nT) = \sum_{k=0}^{n} ku(kT)u(nT - T - kT)$$

$$= 0[u(nT - T)] + 1[u(nT - 2T)] + \cdots + (n-1)u(0) + nu(-T)$$

$$= 0 + 1 + 2 + \cdots + n - 1 + 0$$

$$= \sum_{k=1}^{n-1} k$$

Now by using the following simple manipulation

$$\sum_{k=1}^{n-1} k: 1 + 2 + \cdots + n - 1$$

$$+ \sum_{k=1}^{n-1} k: n - 1 + n - 2 + \cdots + 1$$

$$\overline{\qquad\qquad\qquad\qquad\qquad\qquad\qquad}$$

$$2\sum_{k=1}^{n-1} k: n + n + \cdots + n$$

we obtain

$$y(nT) = \sum_{k=1}^{n-1} k = \tfrac{1}{2}n(n-1)$$

Therefore $y(nT) = \tfrac{1}{2}n(n-1)u(nT)$

2.6 COMPLEX CONVOLUTION

The inverse z transform of a product of z transforms can be formed by using the real convolution. By analogy, the z transform of a product of two time-domain functions can be formed by using the complex convolution.

Theorem 2.7: Complex convolution If

$$\mathscr{L}f(nT) = F(z) = \sum_{n=-\infty}^{\infty} f(nT)z^{-n}$$

$$\mathscr{L}g(nT) = G(z) = \sum_{n=-\infty}^{\infty} g(nT)z^{-n}$$

then

$$Y(z) = \mathscr{L}[f(nT)g(nT)] = \frac{1}{2\pi j}\oint_{\Gamma_1} F(v)G\left(\frac{z}{v}\right)v^{-1}\,dv$$

$$= \frac{1}{2\pi j}\oint_{\Gamma_2} F\left(\frac{z}{v}\right)G(v)v^{-1}\,dv$$

where Γ_1 (or Γ_2) is a contour in the common region of convergence of $F(v)$ and $G(z/v)$ [or $F(z/v)$ and $G(v)$].

PROOF From Eq. (2.4)

$$Y(z) = \sum_{n=-\infty}^{\infty} [f(nT)g(nT)]z^{-n}$$

$$= \sum_{n=-\infty}^{\infty} f(nT)\left[\frac{1}{2\pi j}\oint_{\Gamma_2} G(v)v^{n-1}\,dv\right]z^{-n}$$

$$= \frac{1}{2\pi j}\oint_{\Gamma_2}\left[\sum_{n=-\infty}^{\infty} f(nT)\left(\frac{z}{v}\right)^{-n}\right]G(v)v^{-1}\,dv$$

$$= \frac{1}{2\pi j}\oint_{\Gamma_2} F\left(\frac{z}{v}\right)G(v)v^{-1}\,dv$$

If Γ_2 is a circle in the region of convergence and $v = \rho e^{j\theta}$ and $z = re^{j\phi}$, the above integral can be put in the form

$$Y(re^{j\phi}) = \frac{1}{2\pi}\int_0^{2\pi} F\left[\frac{r}{\rho}e^{j(\phi-\theta)}\right]G(\rho e^{j\theta})\,d\theta$$

which is recognized as a convolution integral.

The complex convolution will be used later in the design of nonrecursive filters.

Example 2.6 Find the z transform of

$$y(nT) = u(nT)e^{-\alpha nT} \sin \omega nT$$

SOLUTION Let

$$f(nT) = u(nT)e^{-\alpha nT} \quad \text{and} \quad g(nT) = u(nT) \sin \omega nT$$

From Table 2.1

$$F(z) = \frac{z}{z - e^{-\alpha T}} \quad \text{and} \quad G(z) = \frac{z \sin \omega T}{(z - e^{j\omega T})(z - e^{-j\omega T})}$$

and hence

$$Y(z) = \frac{1}{2\pi j} \oint_{\Gamma_2} F\left(\frac{z}{v}\right) G(v) v^{-1} \, dv$$

$$= \frac{1}{2\pi j} \oint_{\Gamma_2} \frac{-z e^{\alpha T} \sin \omega T}{(v - z e^{\alpha T})(v - e^{j\omega T})(v - e^{-j\omega T})} \, dv$$

$F(z/v)$ and $G(v)$ converge in the regions

$$|v| < |z e^{\alpha T}| \quad \text{and} \quad |v| > 1$$

respectively, as illustrated in Fig. 2.3. By evaluating the residues at $v = e^{+j\omega T}$ and $e^{-j\omega T}$ we have

$$Y(z) = \frac{z e^{-\alpha T} \sin \omega T}{z^2 - 2z e^{-\alpha T} \cos \omega T + e^{-2\alpha T}}$$

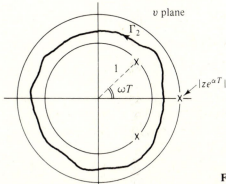

Figure 2.3 Region of convergence of $F(z/v)$ and $G(v)$ (Example 2.6).

REFERENCES

1. E. I. Jury, "Theory and Application of the z-Transform Method," Wiley, New York, 1964.
2. E. Kreyszig, "Advanced Engineering Mathematics," Wiley, New York, 1968.

PROBLEMS

2.1 (a) Find all Laurent series of

$$F(z) = \frac{1}{1 - z^2}$$

with center $z = 1$.

(b) Find all Laurent series of

$$F(z) = \frac{7z^2 + 9z - 18}{z^3 - 9z}$$

with center $z = 0$.

(c) Illustrate the region of convergence for each of the above series.

2.2 Find the z transforms of the following functions:

(a) $u(nT)(2 + 3e^{-2nT})$

(b) $u(nT) - \delta(nT)$

(c) $\varepsilon^{-k}u(nT)$

(d) $\nabla u(nT)$

(e) $u(nT)[1 + (-1)^n]e^{-nT}$

(f) $u(nT) \sin (\omega nT + \psi)$

(g) $u(nT) \cosh \alpha nT$

(h) $u(nT)nT$

(i) $u(nT)(nT)^2$

(j) $u(nT)(nT)^3$

(k) $u(nT)nTe^{-4nT}$

Identify the region of convergence in each case.

2.3 Find the z transforms of the following discrete-time signals:

(a) $x(nT) = \begin{cases} 1 & \text{for } 0 \leq n \leq k \\ 0 & \text{otherwise} \end{cases}$

(b) $x(nT) = \begin{cases} nT & \text{for } 0 \leq n \leq 5 \\ 0 & \text{otherwise} \end{cases}$

(c) $x(nT) = \begin{cases} 0 & \text{for} & n < 0 \\ 1 & \text{for } 0 \leq n \leq 5 \\ 2 & \text{for } 5 < n \leq 10 \\ 3 & \text{for} & n > 10 \end{cases}$

(d) $x(nT) = \begin{cases} 0 & \text{for} & n < 0 \\ nT & \text{for } 0 \leq n < 5 \\ nT - 5T & \text{for } 5 \leq n < 10 \\ nT - 10T & \text{for} & n \geq 10 \end{cases}$

2.4 By using Theorem 2.6(b), form the z transforms of

(a) $y(nT) = \sum\limits_{k=0}^{n} r(nT - kT)u(kT)$

(b) $y(nT) = \sum\limits_{k=0}^{n} u(nT - kT)u(kT)e^{-\alpha kT}$

2.5 Prove that

$$\mathscr{Z} \sum\limits_{k=0}^{n} f(kT) = \frac{z}{z-1} \mathscr{Z} f(nT)$$

2.6 Show that the initial and final values of $f(nT)$ are given by

(a) $f(0) = \lim\limits_{z \to \infty} F(z)$

(b) $f(\infty) = \lim\limits_{z \to 1}(z - 1)F(z)$

where $F(z)$ is the one-sided z transform of $f(nT)$.

2.7 Find $f(0)$ and $f(\infty)$ for the following one-sided z transforms:

(a) $F(z) = \dfrac{2z - 1}{z - 1}$

(b) $F(z) = \dfrac{(e^{-\alpha T} - 1)z}{z^2 - (1 + e^{-\alpha T})z + e^{-\alpha T}}$

(c) $F(z) = \dfrac{Tze^{-4T}}{(z - e^{-4T})^2}$

2.8 Find the z transform of $y(nT)$.

(a) $y(nT) = \sum\limits_{i=0}^{N} a_i x(nT - iT)$

(b) $y(nT) = \sum\limits_{i=0}^{N} a_i x(nT - iT) - \sum\limits_{i=1}^{N} b_i y(nT - iT)$

2.9 Form the z transform of

$$x(nT) = [u(nT) - u(nT - NT)]W^{kn}$$

2.10 Find the inverse z transforms of

(a) $F(z) = \dfrac{2}{2z - 1}$ (b) $F(z) = \dfrac{5}{z - e^{-T}}$

(c) $F(z) = \dfrac{3z}{3z + 2}$ (d) $F(z) = \dfrac{2z}{z^2 - 2z + 1}$

2.11 The z transform

$$F(z) = \frac{N(z)}{D(z)}$$

has a simple pole at $z = p_i$. Show that

$$\operatorname*{res}_{z = p_i} F(z) = \lim\limits_{z \to p_i} (z - p_i)F(z) = \frac{N(p_i)}{D'(p_i)} \qquad \text{where } D'(z) = \frac{dD(z)}{dz}$$

2.12 Find the inverse z transforms of the following by using Eq. (2.4):

(a) $F(z) = \dfrac{z^2}{z^2 + 1}$ 　　(b) $F(z) = \dfrac{2z^2}{2z^2 - 2z + 1}$

(c) $F(z) = \dfrac{1}{(z - 0.8)^4}$ 　　(d) $F(z) = \dfrac{6z}{(2z^2 + 2z + 1)(3z - 1)}$

2.13 Find the inverse z transforms of the following by using the partial-fraction method:

(a) $F(z) = \dfrac{(z - 1)^2}{z^2 - 0.1z - 0.56}$ 　　(b) $F(z) = \dfrac{4z^3}{(2z + 1)(2z^2 - 2z + 1)}$

2.14 Find the inverse z transforms of the following by using Theorem 2.6(a):

(a) $F(z) = \dfrac{z^2}{z^2 - 2z + 1}$ 　　(b) $F(z) = \dfrac{z^2}{(z - e^{-T})(z - 1)}$

2.15 Find the inverse z transform of the following by equating coefficients of equal powers of z:

$$F(z) = \frac{z(z + 1)}{(z - 1)^3}$$

2.16 Find the inverse z transform of the following by means of long division:

$$F(z) = \frac{z(z^2 + 4z + 1)}{(z - 1)^4}$$

2.17 Find the z transform of

$$x(nT) = u(nT)e^{-anT} \sin (\omega nT + \psi)$$

by using the complex-convolution theorem.

THREE

THE APPLICATION OF THE z TRANSFORM

3.1 INTRODUCTION

Through the use of the z transform, a digital filter can be characterized by a discrete-time transfer function, which plays the same key role as the continuous-time transfer function in an analog filter. In this chapter, the discrete-time transfer function is first defined, its properties are then examined, and finally its application in time-domain and frequency-domain analysis is described.

3.2 THE DISCRETE-TIME TRANSFER FUNCTION

The transfer function of a digital filter is defined as the ratio of the z transform of the response to the z transform of the excitation.

Consider a linear, time-invariant digital filter, and let $x(nT)$, $y(nT)$, and $h(nT)$ be the excitation, response, and impulse response, respectively. By using the convolution summation described in Sec. 1.7 we have

$$y(nT) = \sum_{k=-\infty}^{\infty} x(kT)h(nT - kT)$$

and therefore, from Theorem 2.6(a),

$$\mathscr{L}y(nT) = \mathscr{L}h(nT)\mathscr{L}x(nT)$$

or
$$Y(z) = H(z)X(z) \qquad (3.1)$$

51

In effect, the transfer function of a digital filter is the z transform of the impulse response.

In later chapters we shall be dealing with analog and digital filters at the same time. To avoid confusion, we describe the transfer function of an analog filter as continuous-time and that of a digital filter as discrete-time.

Derivation of $H(z)$

The exact form of $H(z)$ can readily be derived from the difference equation characterizing the filter, from the filter network, or from a state-space characterization if one is available.

For causal, recursive filters

$$y(nT) = \sum_{i=0}^{N} a_i x(nT - iT) - \sum_{i=1}^{N} b_i y(nT - iT)$$

and hence

$$\mathscr{L} y(nT) = \sum_{i=0}^{N} a_i z^{-i} \mathscr{L} x(nT) - \sum_{i=1}^{N} b_i z^{-i} \mathscr{L} y(nT)$$

or

$$\frac{Y(z)}{X(z)} = H(z) = \frac{\displaystyle\sum_{i=0}^{N} a_i z^{N-i}}{\displaystyle z^{N} + \sum_{i=1}^{N} b_i z^{N-i}} \tag{3.2}$$

By factoring the numerator and denominator polynomials $H(z)$ can be put in the form

$$H(z) = \frac{H_0 \displaystyle\prod_{i=1}^{N}(z - z_i)}{\displaystyle\prod_{i=1}^{N}(z - p_i)} \tag{3.3}$$

where $z_1, z_2, \ldots, z_N$ are the zeros and $p_1, p_2, \ldots, p_N$ are the poles of $H(z)$. Thus digital filters, like analog filters, can be represented by zero-pole plots like the one in Fig. 3.1.

In nonrecursive filters, $b_i = 0$ for $i = 1, 2, \ldots, N$, and so the poles in these filters are all located at the origin of the z plane.

The z-domain characterizations of the unit delay, the adder, and the multiplier are obtained from Table 1.1 as

$$Y(z) = z^{-1}X(z) \qquad Y(z) = \sum_{i=1}^{k} X_i(z) \qquad Y(z) = mX(z)$$

By using these relations $H(z)$ can be derived directly from the filter network.

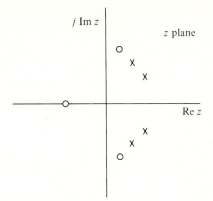

Figure 3.1 Typical zero-pole plot for $H(z)$.

Example 3.1 Find the transfer function of the filter shown in Fig. 3.2.

SOLUTION We can write

$$U(z) = X(z) + \tfrac{1}{2}z^{-1}U(z) - \tfrac{1}{4}z^{-2}U(z)$$

$$Y(z) = U(z) + z^{-1}U(z)$$

Hence $U(z) = \dfrac{X(z)}{1 - \tfrac{1}{2}z^{-1} + \tfrac{1}{4}z^{-2}}$ $Y(z) = (1 + z^{-1})U(z)$

and $$H(z) = \dfrac{z(z + 1)}{z^2 - \tfrac{1}{2}z + \tfrac{1}{4}}$$

Alternatively, $H(z)$ can be deduced from the state-space equations (see Sec. 1.9)

$$\mathbf{q}(nT + T) = \mathbf{A}\mathbf{q}(nT) + \mathbf{B}x(nT)$$

$$y(nT) = \mathbf{C}\mathbf{q}(nT) + \mathbf{D}x(nT) \tag{3.4}$$

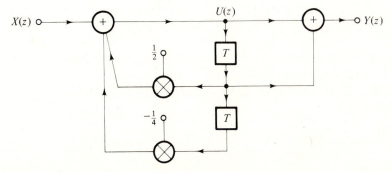

Figure 3.2 Second-order recursive filter (Example 3.1).

By applying the two-sided z transform

$$\mathscr{L}\mathbf{q}(nT + T) = \mathbf{A}\mathscr{L}\mathbf{q}(nT) + \mathbf{B}\mathscr{L}x(nT) = \mathbf{A}\mathbf{Q}(z) + \mathbf{B}X(z) \qquad (3.5)$$

Also

$$\mathscr{L}\mathbf{q}(nT + T) = z\mathscr{L}\mathbf{q}(nT) = z\mathbf{Q}(z) \qquad (3.6)$$

and from Eqs. (3.5) and (3.6)

$$z\mathbf{Q}(z) = \mathbf{A}\mathbf{Q}(z) + \mathbf{B}X(z)$$

or

$$\mathbf{Q}(z) = (z\mathbf{I} - \mathbf{A})^{-1}\mathbf{B}X(z)$$

where $\mathbf{I}$ is the $N \times N$ unit matrix. Now from Eq. (3.4)

$$Y(z) = \mathbf{C}\mathbf{Q}(z) + \mathbf{D}X(z)$$

and on eliminating $\mathbf{Q}(z)$ we have

$$H(z) = \mathbf{C}(z\mathbf{I} - \mathbf{A})^{-1}\mathbf{B} + \mathbf{D}$$

3.3 STABILITY

As can be seen in Eq. (3.2), the discrete-time transfer function is a rational function of z with real coefficients, and for causal filters the degree of the numerator polynomial is equal to or less than that of the denominator polynomial. We shall now show that the poles of the transfer function determine whether the filter is stable or not, as in analog systems.

Consider an Nth-order causal filter characterized by $H(z)$ and for the sake of convenience assume that the poles of $H(z)$ are simple. The impulse response of such a filter is given by Eq. (2.4) as

$$h(nT) = \mathscr{L}^{-1}H(z) = \frac{1}{2\pi j}\oint_\Gamma H(z)z^{n-1}\,dz$$

For $n = 0$

$$h(0) = \operatorname*{res}_{z=0}\frac{H(z)}{z} + \sum_{i=1}^{N}\operatorname*{res}_{z=p_i}\frac{H(z)}{z}$$

and for $n > 0$

$$h(nT) = \sum_{i=1}^{N}\operatorname*{res}_{z=p_i}[H(z)z^{n-1}] = \sum_{i=1}^{N}\operatorname*{res}_{z=p_i}[H(z)]p_i^{n-1}$$

With

$$p_i = r_i e^{j\psi_i}$$

such that

$$r_i \leq r_m < 1 \qquad \text{for } i = 1, 2, \ldots, N \qquad (3.7)$$

we can write

$$\sum_{n=0}^{\infty} |h(nT)| = |h(0)| + \sum_{n=1}^{\infty} \left| \sum_{i=1}^{N} \operatorname*{res}_{z=p_i} [H(z)] r_i^{n-1} e^{j(n-1)\psi i} \right|$$

$$\leq |h(0)| + \sum_{n=1}^{\infty} \sum_{i=1}^{N} \left| \operatorname*{res}_{z=p_i} H(z) \right| r_i^{n-1} \tag{3.8}$$

Since $H(z)$ is analytic in the neighborhood of point $z = 0$ and each point $z = p_i$, $h(0)$ as well as the residues of $H(z)$ are finite, and so

$$\left| \operatorname*{res}_{z=p_i} H(z) \right| \leq R_m \qquad \text{for } i = 1, 2, \ldots, N$$

Thus from Eqs. (3.7) and (3.8)

$$\sum_{n=0}^{\infty} |h(nT)| \leq |h(0)| + NR_m \sum_{n=1}^{\infty} r_m^{n-1} < \infty$$

since the summation on the right-hand side converges. Therefore, Eq. (3.7) constitutes a sufficient condition for stability (see Sec. 1.8).

If a single pole of $H(z)$ is located on or outside the unit circle $|z| = 1$, say pole p_k, then as $n \to \infty$

$$h(nT) \approx \operatorname*{res}_{z=p_k} [H(z)] r_k^{n-1} e^{j(n-1)\psi k}$$

since $r_i^{n-1} \to 0$ for $i \neq k$. Hence

$$\sum_{n=0}^{\infty} |h(nT)| \approx \lim_{Q \to \infty} \sum_{n=0}^{Q} \left| \operatorname*{res}_{z=p_k} H(z) \right| r_k^{n-1} \to \infty$$

i.e., in this case $H(z)$ represents an unstable filter. Therefore, a necessary and sufficient condition for stability is

$$|p_i| < 1 \qquad \text{for } i = 1, 2, \ldots, N$$

The permissible region for the location of poles is illustrated in Fig. 3.3.

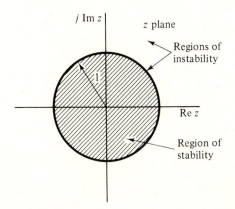

Figure 3.3 Permissible z-plane region for the location of the poles of $H(z)$.

In nonrecursive filters the poles are located at the origin of the z plane; i.e., these filters are always stable.

Example 3.2 Check the filter of Fig. 3.4 for stability.

SOLUTION The transfer function of the filter is

$$H(z) = \frac{z^2 - z + 1}{z^2 - z + \frac{1}{2}} = \frac{z^2 - z + 1}{(z - p_1)(z - p_2)}$$

where

$$p_1, p_2 = \frac{1}{2} \pm j\frac{1}{2}$$

Since

$$|p_1|, |p_2| < 1$$

the filter is stable.

Stability Criterion

Any digital filter can be checked for stability by finding the poles of the transfer function. An alternative and quicker approach, however, is to use a stability criterion due to Jury [1]. This is analogous to the Routh-Hurwitz criterion in analog systems [2].

Consider a filter characterized by

$$H(z) = \frac{N(z)}{D(z)} \tag{3.9}$$

where

$$D(z) = \sum_{i=0}^{N} b_i z^{N-i}$$

and assume that $b_0 > 0$. The coefficients of $D(z)$ can be used to construct an array of numbers, as in Table 3.1. The first two rows of the array are formed by entering

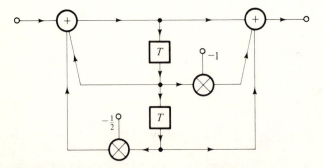

Figure 3.4 Second-order recursive filter (Example 3.2).

Table 3.1 Jury's array

Row	Coefficients			
1	b_0	b_1	$\cdots$	b_N
2	b_N	b_{N-1}	$\cdots$	b_0
3	c_0	c_1	$\cdots$	c_{N-1}
4	c_{N-1}	c_{N-2}	$\cdots$	c_0
5	d_0	d_1	$\cdots$	d_{N-2}
6	d_{N-2}	d_{N-3}	$\cdots$	d_0
$\cdots$	....			
$2N-3$	r_0	r_1	r_2	

the coefficients of $D(z)$ directly. The elements of the third and fourth rows are computed as

$$c_i = \begin{vmatrix} b_0 & b_{N-i} \\ b_N & b_i \end{vmatrix} \qquad \text{for } i = 0, 1, \ldots, N-1$$

those of the fifth and sixth rows as

$$d_i = \begin{vmatrix} c_0 & c_{N-1-i} \\ c_{N-1} & c_i \end{vmatrix} \qquad \text{for } i = 0, 1, \ldots, N-2$$

and so on until $2N-3$ rows are obtained. The last row will comprise three elements, say r_0, r_1, and r_2.

Jury's stability criterion states that the filter characterized by Eq. (3.9) is stable if and only if the following conditions are satisfied:

(i) $D(1) > 0$

(ii) $(-1)^N D(-1) > 0$

(iii) $b_0 > |b_N|$

$|c_0| > |c_{N-1}|$

$|d_0| > |d_{N-2}|$

....

$|r_0| > |r_2|$

Example 3.3 (*a*) A digital filter is characterized by

$$H(z) = \frac{z^4}{4z^4 + 3z^3 + 2z^2 + z + 1}$$

Check the filter for stability. (*b*) Repeat part (*a*) if

$$H(z) = \frac{z^2 + 2z + 1}{z^4 + 6z^3 + 3z^2 + 4z + 5}$$

SOLUTION (*a*) Jury's algorithm gives the following table:

1	4	3	2	1	1
2	1	1	2	3	4
3	15	11	6	1	
4	1	6	11	15	
5	224	159	79		

Since

$$D(1) = 11 \qquad (-1)^4 D(-1) = 3$$

and $b_0 > |b_4|$, $|c_0| > |c_3|$, $|d_0| > |d_2|$, the filter is stable.
(*b*) In this case

$$(-1)^4 D(-1) = -1$$

i.e., condition (ii) is violated and the filter is unstable.

3.4 TIME-DOMAIN ANALYSIS

The time-domain response of a digital filter to any excitation $x(nT)$ can be readily obtained from Eq. (3.1) as

$$y(nT) = \mathscr{Z}^{-1}[H(z)X(z)]$$

Any one of the inversion techniques described in Sec. 2.5 can be used.

Example 3.4 Find the unit-step response of the filter shown in Fig. 3.4.

SOLUTION From Example 3.2

$$H(z) = \frac{z^2 - z + 1}{(z - p_1)(z - p_2)}$$

where $\quad p_1 = \tfrac{1}{2} - j\dfrac{1}{2} = \dfrac{e^{-j\pi/4}}{\sqrt{2}} \qquad p_2 = \tfrac{1}{2} + j\dfrac{1}{2} = \dfrac{e^{j\pi/4}}{\sqrt{2}}$

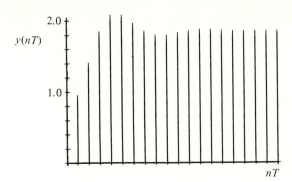

Figure 3.5 Unit-step response (Example 3.4).

and from Table 2.1

$$X(z) = \frac{z}{z-1}$$

On expanding $H(z)X(z)/z$ into partial fractions, we have

$$H(z)X(z) = \frac{Az}{z-1} + \frac{B_1 z}{z - p_1} + \frac{B_2 z}{z - p_2}$$

where
$$A = 2 \qquad B_1 = \frac{e^{j5\pi/4}}{\sqrt{2}} \qquad B_2 = B_1^* = \frac{e^{-j5\pi/4}}{\sqrt{2}}$$

Hence

$$y(nT) = \mathscr{Z}^{-1}[H(z)X(z)]$$

$$= 2u(nT) + \frac{1}{(\sqrt{2})^{n+1}} u(nT)(e^{j(n-5)\pi/4} + e^{-j(n-5)\pi/4})$$

$$= 2u(nT) + \frac{1}{(\sqrt{2})^{n-1}} u(nT) \cos\left[(n-5)\frac{\pi}{4}\right]$$

The response of the filter is plotted in Fig. 3.5.

3.5 FREQUENCY-DOMAIN ANALYSIS

An analog filter characterized by a continuous-time transfer function $H(s)$ has a steady-state sinusoidal response of the form

$$\lim_{t \to \infty} y(t) = \lim_{t \to \infty} \mathscr{R}u(t) \sin \omega t = M(\omega) \sin\left[\omega t + \theta(\omega)\right]$$

where
$$M(\omega) = |H(j\omega)| \qquad \theta(\omega) = \arg H(j\omega)$$

$M(\omega)$ is the gain and $\theta(\omega)$ is the phase shift of the filter at frequency ω. The plots of $M(\omega)$ and $\theta(\omega)$ versus ω, which are the amplitude response and phase response, respectively, constitute the basic frequency-domain characterization of the filter. If these plots are available, one can determine the steady-state response not only to any sinusoidal waveform but also to any other waveform that can be expressed as a linear combination of sinusoids.

Let us consider an Nth-order digital filter characterized by $H(z)$, and for the sake of simplicity let us exclude multiple poles. The sinusoidal response of such a filter is

$$y(nT) = \mathscr{L}^{-1}[H(z)X(z)]$$

where

$$X(z) = \mathscr{L}u(nT) \sin \omega nT = \frac{z \sin \omega T}{(z - e^{j\omega T})(z - e^{-j\omega T})} \tag{3.10}$$

or

$$y(nT) = \frac{1}{2\pi j} \oint_\Gamma H(z)X(z)z^{n-1} \, dz$$

$$= \sum \text{res} \, [H(z)X(z)z^{n-1}] \tag{3.11}$$

For $n > 0$, Eqs. (3.10) and (3.11) yield

$$y(nT) = \sum_{i=1}^{N} \mathop{\text{res}}_{z=p_i} [H(z)]X(p_i)p_i^{n-1} + \frac{1}{2j}[H(e^{j\omega T})e^{j\omega nT} - H(e^{-j\omega T})e^{-j\omega nT}]$$

In a stable filter $|p_i| < 1$ for $i = 1, 2, \ldots, N$, and hence as $n \to \infty$, the summation part in the above equation tends to zero. Therefore, at steady state

$$y(nT) \approx \frac{1}{2j}[H(e^{j\omega T})e^{j\omega nT} - H(e^{-j\omega T})e^{-j\omega nT}]$$

Now

$$H(e^{-j\omega T}) = \sum_{n=-\infty}^{\infty} h(nT)e^{j\omega nT} = \left[\sum_{n=-\infty}^{\infty} h(nT)e^{-j\omega nT}\right]^* = H^*(e^{j\omega T})$$

and if

$$H(e^{j\omega T}) = M(\omega)e^{j\theta(\omega)}$$

where

$$M(\omega) = |H(e^{j\omega T})| \qquad \theta(\omega) = \arg H(e^{j\omega T})$$

the response of the filter can be expressed as

$$y(nT) = M(\omega) \sin [\omega nT + \theta(\omega)]$$

Clearly, the effect of a digital filter on a sinusoidal excitation, like that of an analog filter, is to introduce a gain $M(\omega)$ and a phase shift $\theta(\omega)$. Therefore, a digital filter, like an analog filter, can be represented in the frequency domain by an amplitude response and a phase response. The main difference between analog and digital filters is that in the first case the transfer function is evaluated on the imaginary axis of the s plane whereas in the second case this is evaluated on the unit circle $|z| = 1$ of the z plane.

For a transfer function expressed in terms of its zeros and poles as in Eq. (3.3)

$$H(e^{j\omega T}) = M(\omega)e^{j\theta(\omega)} = \frac{H_0 \prod_{i=1}^{N}(e^{j\omega T} - z_i)}{\prod_{i=1}^{N}(e^{j\omega T} - p_i)}$$

and by letting

$$e^{j\omega T} - z_i = M_{zi}e^{j\psi_{zi}} \tag{3.12}$$

$$e^{j\omega T} - p_i = M_{pi}e^{j\psi_{pi}} \tag{3.13}$$

we obtain

$$M(\omega) = \frac{H_0 \prod_{i=1}^{N} M_{zi}}{\prod_{i=1}^{N} M_{pi}}$$

$$\theta(\omega) = \sum_{i=1}^{N} \psi_{zi} - \sum_{i=1}^{N} \psi_{pi}$$

Thus $M(\omega)$ and $\theta(\omega)$ can be determined by drawing the phasors of Eqs. (3.12) and (3.13) in the z plane and then measuring their magnitudes and angles. This procedure is illustrated in Fig. 3.6 for a second-order filter.

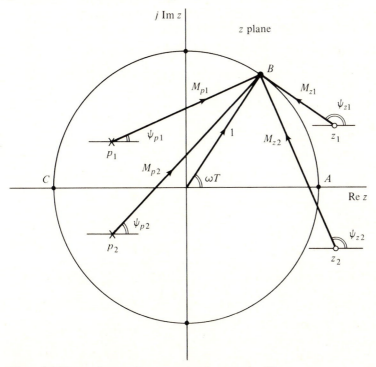

Figure 3.6 Phasor diagram for a second-order recursive filter.

3.4 Show that

$$H(z) = \frac{\displaystyle\sum_{i=0}^{M} a_i z^{M-i}}{z^N + \displaystyle\sum_{i=1}^{N} b_i z^{N-i}}$$

represents a causal filter only if $M \leq N$.

3.5 (a) Find the impulse response of the filter shown in Fig. 3.2. (b) Repeat part (a) for the filter of Fig. 3.4.

3.6 Obtain the impulse response of the filter in Prob. 3.2. Sketch the response.

3.7 Starting from first principles, show that

$$H(z) = \frac{z}{(z - \frac{1}{4})^4}$$

represents a stable filter.

3.8 (a) A recursive filter is represented by

$$H(z) = \frac{z^6}{6z^6 + 5z^5 + 4z^4 + 3z^3 + 2z^2 + z + 1}$$

Check the filter for stability.

(b) Repeat part (a) if

$$H(z) = \frac{(z + 2)^2}{6z^6 + 5z^5 - 4z^4 + 3z^3 + 2z^2 + z + 1}$$

3.9 Obtain (a) the transfer function, (b) the impulse response, and (c) the necessary condition for stability for the filter of Fig. P3.9. The constants m_1 and m_2 are given by

$$m_1 = 2r \cos \theta \qquad \text{and} \qquad m_2 = -r^2$$

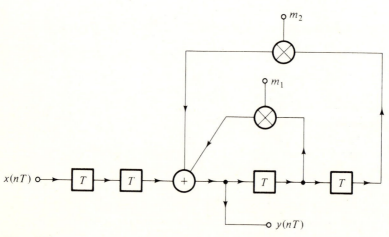

Figure P3.9

3.10 Find the permissible range for m in Fig. P3.10 if the filter is to be stable.

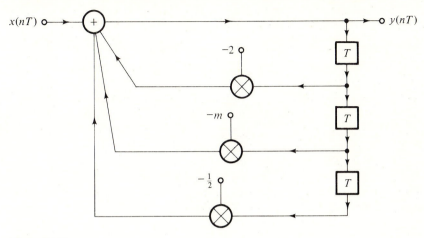

Figure P3.10

3.11 Find the unit-step response of the filter shown in Fig. 3.2.

3.12 Find the unit-ramp response of the filter shown in Fig. 3.4 if $T = 1$.

3.13 The input excitation in Fig. 3.2 is

$$x(nT) = \begin{cases} n & \text{for } 0 \le n \le 2 \\ 4 - n & \text{for } 2 < n \le 4 \\ 0 & \text{for } n > 4 \end{cases}$$

Determine the response for $0 \le n \le 5$ by using the z transform.

3.14 Repeat Prob. 1.12 by using the z transform. For each of the three cases deduce the exact frequency of ringing if $T = 1$ and also the steady-state value of the response.

3.15 A filter has a transfer function

$$H(z) = \frac{1}{z^2 + \frac{1}{4}}$$

(a) Find the response if

$$x(nT) = u(nT) \sin \omega nT$$

(b) Deduce the steady-state sinusoidal response.

3.16 A filter is characterized by

$$H(z) = \frac{1}{(z - r)^2}$$

where $|r| < 1$. Show that the steady-state sinusoidal response is given by

$$y(nT) = M(\omega) \sin [\omega nT + \theta(\omega)] \qquad \text{where} \quad \begin{aligned} M(\omega) &= |H(e^{j\omega T})| \\ \theta(\omega) &= \arg H(e^{j\omega T}) \end{aligned}$$

3.17 (a) Show that

$$M^2(\omega) = H(z)H(z^{-1})\Big|_{z=e^{j\omega T}}$$

(b) By using this relation show that

$$H(z) = \frac{1 - az + bz^2}{b - az + z^2}$$

represents an allpass filter, i.e., one with constant gain at all frequencies.

3.18 Figure P3.18 shows a nonrecursive filter.

(a) Derive expressions for the gain and phase shift.

(b) Determine the transmission zeros of the filter, i.e., zero-gain frequencies.

(c) Sketch the amplitude and phase responses.

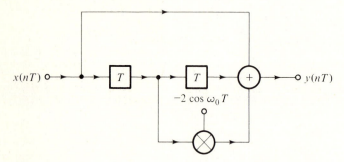

Figure P3.18

3.19 The group delay in digital as in analog filters is defined by

$$\tau = -\frac{d\theta(\omega)}{d\omega}$$

Show that the equation

$$y(nT) = x(nT) + 2x(nT - T) + 3x(nT - 2T) + 4x(nT - 3T) + 3x(nT - 4T)$$
$$+ 2x(nT - 5T) + x(nT - 6T)$$

represents a constant-delay filter.

3.20 Derive expressions for the gain and phase shift of the filter shown in Fig. 3.4.

3.21 Table P3.21 gives the transfer function coefficients of four filters labeled A to D. By plotting $20 \log M(\omega)$ versus ω for values of ω in the range 0 to 5.0 rad/s, identify a lowpass, a highpass, a bandpass, and a bandstop filter. Each filter has a transfer function of the form

$$H(z) = H \prod_{i=1}^{2} \frac{A_{0i} + A_{1i}z + A_{0i}z^2}{B_{0i} + B_{1i}z + z^2}$$

and the sampling frequency is 10 rad/s in each case.

3.22 Show that the gain and phase shift in a digital filter satisfy the relations

$$M(\omega_s - \omega) = M(\omega) \qquad \text{and} \qquad \theta(\omega_s - \omega) = -\theta(\omega)$$

Table P3.21

Filter	i	A_{0i}	A_{1i}	B_{0i}	B_{1i}
A	1	2.222545×10^{-1}	-4.445091×10^{-1}	4.520149×10^{-2}	1.561833×10^{-1}
	2	3.085386×10^{-1}	-6.170772×10^{-1}	4.509715×10^{-1}	2.168171×10^{-1}
		$H = 1.0$			
B	1	5.490566	9.752955	7.226400×10^{-1}	4.944635×10^{-1}
	2	5.871082×10^{-1}	-1.042887	7.226400×10^{-1}	-4.944634×10^{-1}
		$H = 2.816456 \times 10^{-2}$			
C	1	1.747744×10^{-1}	1.517270×10^{-8}	5.741567×10^{-1}	1.224608
	2	1.399382	1.214846×10^{-7}	5.741567×10^{-1}	-1.224608
		$H = 8.912509 \times 10^{-1}$			
D	1	9.208915	1.561801×10	5.087094×10^{-1}	-1.291110
	2	2.300089	1.721670	8.092186×10^{-1}	-1.069291
		$H = 6.669086 \times 10^{-4}$			

FOUR

REALIZATION

4.1 INTRODUCTION

The design of digital filters comprises four general steps:

1. Approximation
2. Realization
3. Study of arithmetic errors
4. Implementation

The approximation step is the process of generating a transfer function satisfying a set of desired specifications, which may concern the amplitude, phase, and possibly the time-domain response of the filter. The realization step is the process of converting the transfer function into a filter network. The approximation and realization steps are carried out on the assumption that the arithmetic devices to be employed are of infinite precision. Since practical devices are of finite precision, it becomes necessary to study the effects of arithmetic errors on the performance of the filter.

The implementation of a digital filter can assume two forms, software and hardware. In the first case, implementation involves the simulation of the filter network on a digital computer. In the second case, on the other hand, this involves the conversion of the filter network into a dedicated piece of hardware.

We consider here the second step in the design process, namely the realization. In addition, we discuss some important topological properties of digital-filter networks in general.

Realization can be accomplished by using the following methods:

1. Direct
2. Direct canonic
3. Cascade
4. Parallel
5. Ladder
6. Wave

The first five of these methods will be considered in the following sections. The wave method as well as the merits and demerits of the various methods will be considered later, in Chap. 12.

4.2 DIRECT REALIZATION

Let us consider an Nth-order transfer function such as

$$H(z) = \frac{N(z)}{D(z)} = \frac{N(z)}{1 + D'(z)} = \frac{Y(z)}{X(z)} \tag{4.1}$$

where

$$N(z) = \sum_{i=0}^{N} a_i z^{-i} \tag{4.2}$$

and

$$D'(z) = \sum_{i=1}^{N} b_i z^{-i}$$

We can write

$$Y(z) = U_1(z) + U_2(z)$$

where

$$U_1(z) = N(z)X(z) \tag{4.3}$$

and

$$U_2(z) = -D'(z)Y(z)$$

and hence the realization of $H(z)$ can be broken down into the realization of two simpler transfer functions, $N(z)$ and $-D'(z)$, as illustrated in Fig. 4.1.

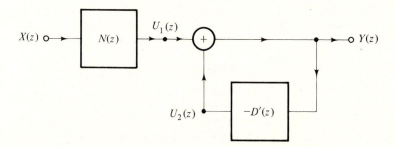

Figure 4.1 Decomposition of $H(z)$ into two simpler transfer functions.

Consider the realization of $N(z)$. From Eqs. (4.2) and (4.3)

$$U_1(z) = [a_0 + z^{-1}N_1(z)]X(z)$$

where

$$N_1(z) = \sum_{i=1}^{N} a_i z^{-i+1}$$

and thus $N(z)$ can be realized by using a multiplier with parameter a_0 in parallel with a network characterized by $z^{-1}N_1(z)$. In turn, $z^{-1}N_1(z)$ can be realized by using a unit delay in cascade with a network characterized by $N_1(z)$. Since the unit delay may precede or follow the realization of $N_1(z)$, two possibilities exist for $N(z)$, as depicted in Fig. 4.2.

The above procedure can now be applied to $N_1(z)$. That is, $N_1(z)$ can be expressed as

$$N_1(z) = a_1 + z^{-1}N_2(z) \qquad \text{where } N_2(z) = \sum_{i=2}^{N} a_i z^{-i+2}$$

and as before two networks can be obtained for $N_1(z)$. Clearly, there are four networks for $N(z)$. Two of them are shown in Fig. 4.3.

This cycle of events can be repeated N times, whereupon $N_N(z)$ will reduce to a single multiplier. In each cycle of the procedure there are two possibilities, and since there are N cycles, a total of 2^N distinct networks can be deduced for $N(z)$. Three of them are shown in Fig. 4.4.

$-D'(z)$ can be realized in exactly the same way. Networks for $-D'(z)$ can be obtained by replacing $a_0, a_1, a_2, \ldots$ in Fig. 4.4 by $0, -b_1, -b_2, \ldots$.

Finally, the realization of $H(z)$ can be accomplished by interconnecting realizations of $N(z)$ and $-D'(z)$ according to Fig. 4.1.

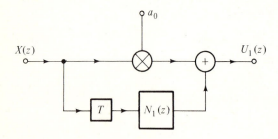

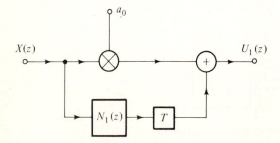

Figure 4.2 Two realizations of $N(z)$.

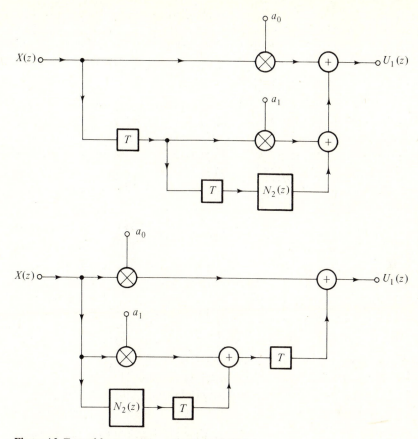

Figure 4.3 Two of four possible realizations of $N(z)$.

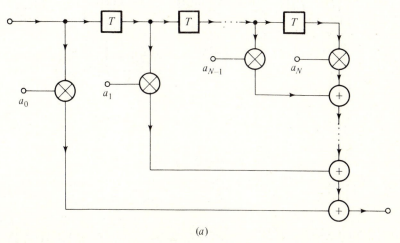

(a)

Figure 4.4 Three of 2^N possible realizations of $N(z)$.

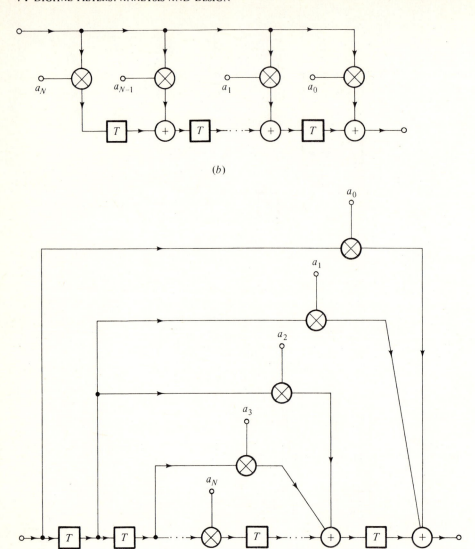

(b)

(c)

Figure 4.4 (*continued*)

Example 4.1 Realize the transfer function

$$H(z) = \frac{a_0 + a_1 z^{-1} + a_2 z^{-2}}{1 + b_1 z^{-1} + b_2 z^{-2}}$$

SOLUTION Two realizations of $H(z)$ can be readily obtained from Fig. 4.4a and b, as shown in Fig. 4.5a and b.

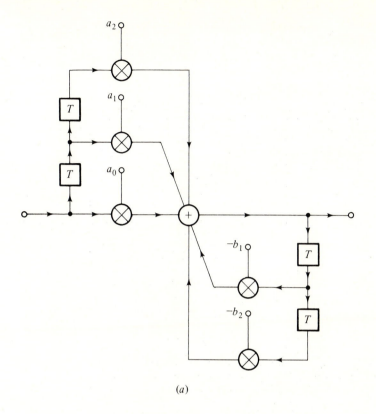

(a)

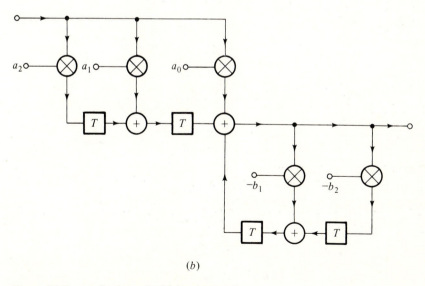

(b)

Figure 4.5 Two realizations of $H(z)$ (Example 4.1).

4.3 DIRECT CANONIC REALIZATION

A digital network is said to be *canonic* if the number of unit delays employed is equal to the order of the transfer function.

Equation (4.1) can be expressed as

$$Y(z) = N(z)Y'(z) \quad \text{where } Y'(z) = X(z) - D'(z)Y'(z)$$

in which case $H(z)$ can be realized as in Fig. 4.6 by using the network of Fig. 4.4a for $N(z)$ as well as $-D'(z)$. Evidently, the signals at nodes A', B', ... are equal to the respective signals at nodes A, B, ... Therefore, the unit delays in path $A'B'$ can be eliminated to yield a canonic realization for $H(z)$.

4.4 CASCADE REALIZATION

The transfer function can be factored into a product of second-order transfer functions as

$$H(z) = \prod_{i=1}^{M} H_i(z)$$

where

$$H_i(z) = \frac{a_{0i} + a_{1i}z^{-1} + a_{2i}z^{-2}}{1 + b_{1i}z^{-1} + b_{2i}z^{-2}}$$

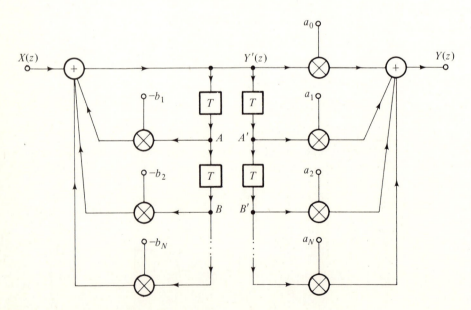

Figure 4.6 Derivation of the canonic realization of $H(z)$.

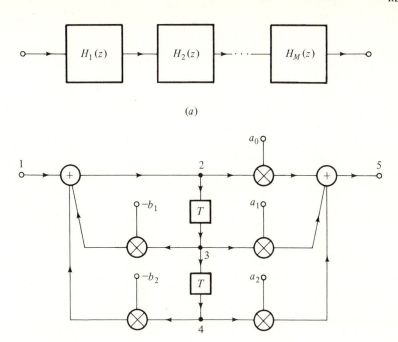

(a)

(b)

Figure 4.7 (a) Cascade realization of $H(z)$; (b) canonic second-order section.

Hence
$$Y(z) = [H_1(z)X(z)]H_2(z) \cdots H_M(z)$$
$$= [H_2(z)Y_1(z)]H_3(z) \cdots H_M(z)$$
$$\dots\dots\dots\dots\dots\dots\dots\dots\dots\dots\dots$$
$$= H_M(z)Y_{M-1}(z)$$

where
$$Y_1(z) = H_1(z)X(z)$$
$$Y_i(z) = H_i(z)Y_{i-1}(z) \qquad \text{for } i = 2, 3, \dots, M-1$$

In this way, $H(z)$ can be realized by using the cascade configuration of Fig. 4.7a. The individual sections can be realized by employing the canonic second-order network of Fig. 4.7b.

4.5 PARALLEL REALIZATION

Alternatively, the transfer function can be expanded into partial fractions as

$$H(z) = \sum_{i=1}^{M} H_i(z)$$

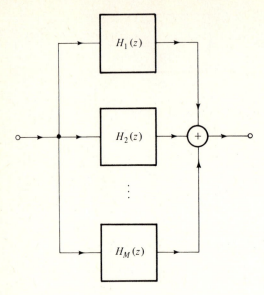

Figure 4.8 Parallel realization of $H(z)$.

where
$$H_i(z) = \frac{a_{0i} + a_{1i}z^{-1}}{1 + b_{1i}z^{-1} + b_{2i}z^{-2}}$$

In this case

$$Y(z) = \sum_{i=1}^{M} H_i(z)X(z)$$

and so the parallel configuration of Fig. 4.8 is obtained.

An alternative parallel realization can be readily obtained by expanding $H(z)/z$ into partial fractions.

4.6 LADDER REALIZATION

The last realization method to be considered here is the ladder realization, due to Mitra and Sherwood [1]. This is based on the configuration of Fig. 4.9a.

With $N = 4$ in Fig. 4.9a, straightforward analysis yields

$$Y(z) = H_1(z)Y'(z) + H_2(z)X(z) \tag{4.4}$$

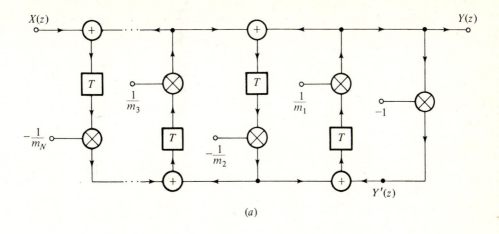

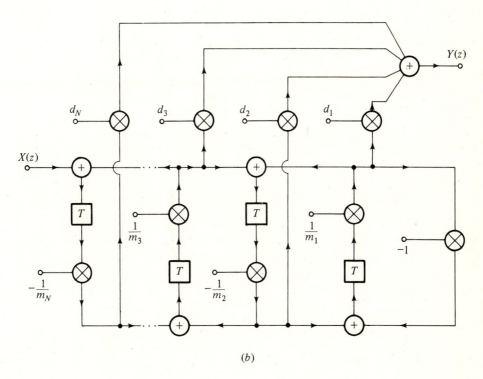

Figure 4.9 Ladder realization of (a) $H(z) = (-1)^k/D(z)$ and (b) $H(z) = N(z)/D(z)$.

where $H_1(z) = \dfrac{N_1(z)}{D_1(z)} = \dfrac{m_4 m_3 m_2 z^3 + (m_4 + m_2)z}{m_4 m_3 m_2 m_1 z^4 + (m_4 m_3 + m_4 m_1 + m_2 m_1)z^2 + 1}$

$$= \cfrac{1}{m_1 z + \cfrac{1}{m_2 z + \cfrac{1}{m_3 z + \cfrac{1}{m_4 z}}}}$$

$$H_2(z) = \frac{1}{D_1(z)}$$

and since

$$Y'(z) = -Y(z)$$

Eq. (4.4) gives

$$\frac{Y(z)}{X(z)} = H(z) = \frac{H_2(z)}{1 + H_1(z)} \qquad (4.5)$$

Similarly, for any value of N it can be shown that

$$H_1(z) = \frac{N_1(z)}{D_1(z)} = \cfrac{1}{m_1 z + \cfrac{1}{m_2 z + \cfrac{1}{\cdots\cdots\cdots\cdots}}}$$

$$\cfrac{1}{m_N z}$$

$$H_2(z) = \frac{(-1)^k}{D_1(z)}$$

where k is the largest integer equal to or less than $N/2$; that is,

$$(-1)^k = \begin{cases} +1 & \text{for } k = 1, 4, 5, 8, 9, \ldots \\ -1 & \text{for } k = 2, 3, 6, 7, \ldots \end{cases}$$

Let us consider the transfer function

$$H(z) = \frac{(-1)^k}{D(z)} = \frac{(-1)^k}{1 + D'(z)}$$

where

$$D'(z) = \sum_{i=1}^{N} b_i z^i$$

We can write

$$H(z) = \frac{(-1)^k}{\text{Od } D(z) + \text{Ev } D(z)} \tag{4.6}$$

where Od $D(z)$ and Ev $D(z)$ denote the odd part and even part of $D(z)$, respectively. By comparing Eqs. (4.5) and (4.6) the following identifications can be made:

$$H_1(z) = \begin{cases} \dfrac{\text{Od } D(z)}{\text{Ev } D(z)} & \text{for even } N \\[2mm] \dfrac{\text{Ev } D(z)}{\text{Od } D(z)} & \text{for odd } N \end{cases}$$

$$H_2(z) = \begin{cases} \dfrac{(-1)^k}{\text{Ev } D(z)} & \text{for even } N \\[2mm] \dfrac{(-1)^k}{\text{Od } D(z)} & \text{for odd } N \end{cases}$$

Now by expressing $H_1(z)$ as a continued-fraction expansion about infinity we have

$$H_1(z) = \cfrac{1}{c_1 z + \cfrac{1}{c_2 z + \cfrac{1}{\begin{array}{c} \cdots\cdots\cdots \\ \cfrac{1}{c_N z} \end{array}}}} \tag{4.7}$$

and therefore if

$$m_i = c_i$$

the configuration of Fig. 4.9a becomes a realization of $H(z)$.

The synthesis can be extended to any transfer functions of the form

$$H(z) = \frac{N(z)}{D(z)} \qquad \text{where } N(z) = \sum_{i=0}^{N} a_i z^i$$

by modifying the basic configuration as illustrated in Fig. 4.9b. For $N = 4$ and $m_i = c_i$, analysis yields

$$\frac{Y(z)}{X(z)} = H(z) = \frac{\displaystyle\sum_{i=1}^{4} d_i n_i(z)}{D(z)}$$

where
$$n_1(z) = 1$$
$$n_2(z) = c_1 z + 1$$
$$n_3(z) = -(c_1 c_2 z^2 + c_2 z + 1)$$
$$n_4(z) = -[c_1 c_2 c_3 z^3 + c_2 c_3 z^2 + (c_1 + c_3)z + 1]$$

By assigning

$$N(z) = \sum_{i=1}^{4} d_i n_i(z)$$

and then equating coefficients of like powers of z, the matrix equation

$$\begin{bmatrix} -c_1 c_2 c_3 & 0 & 0 & 0 \\ -c_2 c_3 & -c_1 c_2 & 0 & 0 \\ -(c_1 + c_3) & -c_2 & c_1 & 0 \\ -1 & -1 & 1 & 1 \end{bmatrix} \begin{bmatrix} d_4 \\ d_3 \\ d_2 \\ d_1 \end{bmatrix} = \begin{bmatrix} a_3 \\ a_2 \\ a_1 \\ a_0 \end{bmatrix} \tag{4.8}$$

can be formed. The solution of this equation yields the necessary values of d_1, d_2, ...

The realization method relies on the existence of the continued-fraction expansion in Eq. (4.7), and as a consequence it sometimes breaks down, e.g., if $H(z)$ has poles on the imaginary axis of the z plane [1].

Example 4.2 Realize the transfer function

$$H(z) = \frac{10^{-2}(-3.517 + 0.665z + 0.665z^2 - 3.517z^3)}{1 - 3.266z + 3.739z^2 - 1.53z^3}$$

using the ladder method.

SOLUTION $H(z)$ can be expressed as

$$H(z) = 0.02299 + \frac{-0.0582 + 0.0817z - 0.0793z^2}{1 - 3.266z + 3.739z^2 - 1.53z^3}$$

$$= 0.02299 + H'(z)$$

and hence $H(z)$ can be realized by using a multiplier in parallel with a network characterized by $H'(z)$. Since the order of $H'(z)$ is odd, we can write

$$H_1(z) = -\frac{3.739z^2 + 1}{1.53z^3 + 3.266z}$$

$$= \cfrac{1}{-0.4092z + \cfrac{1}{-1.309z + \cfrac{1}{-2.856z}}}$$

and thus

$$m_1 = c_1 = -0.4092 \qquad m_2 = c_2 = -1.309 \qquad m_3 = c_3 = -2.856$$

For $N = 3$, analysis yields

$$n_1(z) = -1$$

$$n_2(z) = -(c_1 z + 1)$$

$$n_3(z) = c_1 c_2 z^2 + c_2 z + 1$$

and so d_1, d_2, and d_3 are given by

$$\begin{bmatrix} c_1 c_2 & 0 & 0 \\ c_2 & -c_1 & 0 \\ 1 & -1 & -1 \end{bmatrix} \begin{bmatrix} d_3 \\ d_2 \\ d_1 \end{bmatrix} = \begin{bmatrix} a_2 \\ a_1 \\ a_0 \end{bmatrix}$$

or

$$d_3 = \frac{a_2}{c_1 c_2} = -0.148 \qquad d_2 = \frac{d_3 c_2 - a_1}{c_1} = -0.274$$

$$d_1 = d_3 - d_2 - a_0 = 0.184$$

4.7 TOPOLOGICAL PROPERTIES

Digital-filter networks can be represented in terms of flow graphs which have certain topological properties of theoretical as well as of practical interest. We discuss some of these properties here.

Signal Flow-Graph Representation

Consider the canonic network of Fig. 4.7b, and let $y_1(n)$, $y_2(n)$, ... be the signals flowing out of nodes 1, 2, ... We can write

$$y_1(n) = i(n)$$

$$y_2(n) = i(n) - b_1 y_3(n) - b_2 y_4(n)$$

$$y_3(n) = \mathscr{E}^{-1} y_2(n)$$

$$y_4(n) = \mathscr{E}^{-1} y_3(n)$$

$$y_5(n) = a_0 y_2(n) + a_1 y_3(n) + a_2 y_4(n)$$

$$o(n) = y_5(n)$$

where $i(n)$ and $o(n)$ are the input and output, respectively. Thus a conventional flow graph can be constructed for the network comprising a number of nodes, a number of directed branches, a source $i(n)$, and a sink $o(n)$, as illustrated in Fig. 4.10a. The source and sink can be treated as branches by using the self-loop of

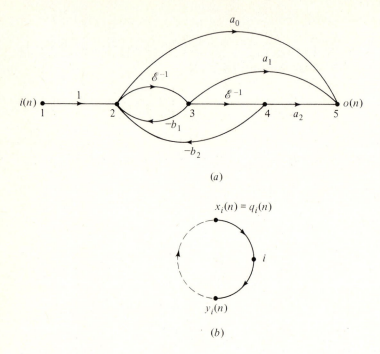

(a)

(b)

Figure 4.10 (a) Flow-graph representation of the filter depicted in Fig. 4.7b. (b) Representation of a source or a sink by means of a self-loop.

Fig. 4.10b, where $q_i(n)$ is an independent quantity for a source and $q_i(n) = 0$ for a sink. By designating the signal entering node j of the network via path ij as x_{ij} and the signal flowing out of node j as y_j the preceding equations can be put in the concise form

$$y_j = \sum_{i=1}^{5} x_{ij} \qquad \text{for } j = 1, 2, \dots, 5$$

where $x_{uv} = 0$ if no direct path exists between nodes u and v.

Similarly, for a digital network with N nodes

$$y_j = \sum_{i=1}^{N} x_{ij} \qquad \text{for } j = 1, 2, \dots, N$$

Tellegen's theorem Consider two distinct sets of signals **S** and **S'** defined by

$$\mathbf{S} = \{x_{11}, \dots, x_{NN}, y_1, \dots, y_N\} \qquad \mathbf{S'} = \{x'_{11}, \dots, x'_{NN}, y'_1, \dots, y'_N\}$$

These may pertain to one and the same network, in which case their differences may be due to differences in the input signals or possibly to variations in one or more multiplier constants. Alternatively, **S** and **S'** may pertain to distinct but topologically compatible networks (networks which have the same topology).

Tellegen's theorem as applied to digital networks [2, 3] states that the elements of **S** and **S′** satisfy the general relation

$$\sum_{i=1}^{N}\sum_{j=1}^{N}(y_j x'_{ij} - y'_i x_{ji}) = 0$$

PROOF The validity of this theorem can be established by writing the left-hand side of the relation as

$$\sum_{j=1}^{N} y_j \sum_{i=1}^{N} x'_{ij} - \sum_{i=1}^{N} y'_i \sum_{j=1}^{N} x_{ji}$$

and then replacing

$$\sum_{i=1}^{N} x'_{ij} \qquad \text{and} \qquad \sum_{j=1}^{N} x_{ji}$$

by y'_j and y_i, respectively.

Similarly, if

$$\mathscr{L} x_{ij} = X_{ij} \qquad \mathscr{L} y_j = Y_j$$

the preceding steps yield

$$\sum_{i=1}^{N}\sum_{j=1}^{N}(Y_j X'_{ij} - Y'_i X_{ji}) = 0 \tag{4.9}$$

i.e., Tellegen's theorem holds in both the time domain and z domain.

Reciprocity

A flow graph with M accessible nodes is said to be a *multipole* (or M-pole). Such a flow graph can be represented as depicted in Fig. 4.11, where F is a source-free flow graph.

Consider an M-pole, and let

$$\mathbf{S} = \{X_1, \ldots, X_M, Y_1, \ldots, Y_M\} \qquad \mathbf{S'} = \{X'_1, \ldots, X'_M, Y'_1, \ldots, Y'_M\}$$

be possible sets of signals. The M-pole is said to be reciprocal [2] if

$$\sum_{i=1}^{M}(X_i Y'_i - X'_i Y_i) = 0 \tag{4.10}$$

for all possible pairs of **S** and **S′**.

The M-pole can be represented by the set of equations

$$Y_i = \sum_{j=1}^{M} H_{ji} X_j \qquad \text{for } i = 1, 2, \ldots, M \tag{4.11}$$

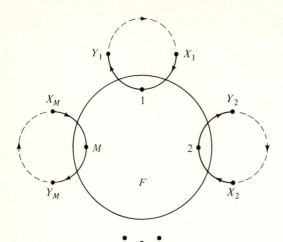

Figure 4.11 Multipole flow graph.

where H_{ji} is the transfer function from node j to node i. From Eqs. (4.10) and (4.11)

$$\sum_{i=1}^{M} \left(X_i \sum_{j=1}^{M} H_{ji} X'_j - X'_i \sum_{j=1}^{M} H_{ji} X_j \right) = 0$$

or

$$\sum_{i=1}^{M} \sum_{j=1}^{M} H_{ji} X_i X'_j - \sum_{i=1}^{M} \sum_{j=1}^{M} H_{ji} X_j X'_i = 0$$

and on interchanging subscripts in the second summation we have

$$\sum_{i=1}^{M} \sum_{j=1}^{M} (H_{ji} - H_{ij}) X_i X'_j = 0$$

Therefore, an M-pole is a reciprocal flow graph if and only if

$$H_{ij} = H_{ji}$$

for all values of i and j.

Interreciprocity

Now consider two distinct M-poles G and G' having the same number of accessible nodes, and let $\mathbf{S}$ and $\mathbf{S}'$ as defined above be sets of signals pertaining to G and G', respectively. The two M-poles are said to be interreciprocal [2] if

$$\sum_{i=1}^{M} (X_i Y'_i - X'_i Y_i) = 0 \qquad (4.12)$$

for all possible pairs of $\mathbf{S}$ and $\mathbf{S}'$.

G and G' can be described by

$$Y_i = \sum_{j=1}^{M} H_{ji} X_j \qquad \text{for } i = 1, 2, \ldots, M \qquad (4.13)$$

and

$$Y_i' = \sum_{j=1}^{M} H_{ji}' X_j' \qquad \text{for } i = 1, 2, \ldots, M \qquad (4.14)$$

respectively. From Eqs. (4.12) to (4.14)

$$\sum_{i=1}^{M} \sum_{j=1}^{M} (H_{ji}' - H_{ij}) X_i X_j' = 0$$

and therefore G and G' are interreciprocal flow graphs if and only if

$$H_{ij} = H_{ji}' \qquad (4.15)$$

for all values of i and j.

Transposition

Given a flow graph G, a corresponding flow graph G' can be derived by reversing the direction in each and every branch in G. The flow graph so derived is said to be the *transpose* (or *adjoint*) of G [2–4]. If the transmittance of branch ij in G is designated as B_{ij} and that of branch ji in G' as B_{ji}', then by definition

$$B_{ij} = B_{ji}' \qquad (4.16)$$

Let

$$\mathbf{S} = \{X_{11}, \ldots, X_{NN}, Y_1, \ldots, Y_N\} \qquad \mathbf{S}' = \{X_{11}', \ldots, X_{NN}', Y_1', \ldots, Y_N'\}$$

be possible sets of signals pertaining to G and G', respectively, and assume that nodes 1 to M are accessible. From Eq. (4.9)

$$\sum_{i=1}^{M} (Y_i X_{ii}' - Y_i' X_{ii}) + \sum_{i=M+1}^{N} (Y_i X_{ii}' - Y_i' X_{ii}) + \sum_{i=1}^{N} \sum_{\substack{j=1 \\ j \neq i}}^{N} (Y_j X_{ij}' - Y_i' X_{ji}) = 0$$

$$(4.17)$$

The first summation represents contributions due to external self-loops, i.e., sources and sinks, whereas the second and third summations represent contributions due to internal self-loops and internal branches, respectively. For both internal self-loops and branches we can write

$$X_{ji} = B_{ji} Y_j \qquad \text{and} \qquad X_{ij}' = B_{ij}' Y_i'$$

and hence

$$\sum_{i=1}^{N} \sum_{\substack{j=1 \\ j \neq i}}^{N} (Y_j X_{ij}' - Y_i' X_{ji}) = \sum_{i=1}^{N} \sum_{\substack{j=1 \\ j \neq i}}^{N} (B_{ij}' - B_{ji}) Y_j Y_i' = 0$$

$$\sum_{i=M+1}^{N} (Y_i X_{ii}' - Y_i' X_{ii}) = \sum_{i=M+1}^{N} (B_{ii}' - B_{ii}) Y_i Y_i' = 0$$

according Eq. (4.16). For external self-loops

$$X'_{ii} = X'_i \qquad X_{ii} = X_i$$

according to Fig. 4.11, and as a result Eq. (4.17) simplifies to Eq. (4.12). Therefore G and its transpose G' are interreciprocal flow graphs. Consequently, if H_{ij} is the transfer function from node i to node j in G and H'_{ji} is the transfer function from node j to node i in G', then

$$H_{ij} = H'_{ji}$$

Because of this property, transposition can serve as a means of deriving alternative digital-filter networks.

Example 4.3 Form the transpose of the canonic network of Fig. 4.7b.

SOLUTION The transpose flow graph is readily obtained, as shown in Fig. 4.12a, by using Fig. 4.10a. The corresponding network is shown in Fig. 4.12b.

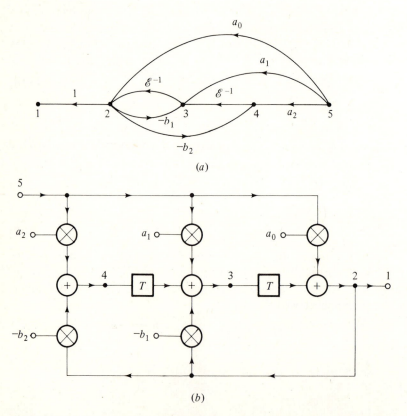

(a)

(b)

Figure 4.12 (a) Transpose flow graph of the network depicted in Fig. 4.7b (Example 4.3). (b) Corresponding transpose network.

Sensitivity Analysis

In the study of arithmetic errors (see Chap. 11) we shall be concerned with the sensitivities of digital-filter networks to variations in the multiplier constants. In a network characterized by

$$H(z) = f(z, m_1, m_2, \ldots)$$

where $m_1, m_2, \ldots$ are multiplier constants, the sensitivities are given by

$$S_{m_i}^H(z) = \frac{\partial H(z)}{\partial m_i} \qquad \text{for } i = 1, 2, \ldots$$

Although differentiation of the transfer function will ultimately yield the sensitivities, much manipulation is often necessary, especially in complicated networks. Fortunately, however, differentiation can be avoided through the concept of transposition [2, 3], as we shall now show.

The network of Fig. 4.13a, the transpose of this network, and the network with m changed to $m + \Delta m$ can be represented by the 2-poles of Fig. 4.13b, c, d, respectively. By virtue of Tellegen's theorem

$$\sum_{i=1}^{N} W_i = 0 \tag{4.18}$$

where

$$W_i = \sum_{j=1}^{N} (Y_j'' X_{ij}' - Y_i' X_{ji}'')$$

For $i = 1$

$$W_1 = Y_1'' X_{11}' - Y_1' X_{11}'' + \sum_{j=2}^{N} (B_{1j}' - B_{j1}) Y_j'' Y_1' = -Q Y_1' \tag{4.19}$$

since

$$X_{11}' = X_1' = 0 \qquad X_{11}'' = X_1'' = Q \qquad B_{1j}' = B_{j1}$$

For $i = 2$

$$W_2 = Y_2'' X_{22}' - Y_2' X_{22}'' + \sum_{\substack{j=1 \\ j \neq 2}}^{N} (B_{2j}' - B_{j2}) Y_j'' Y_2' = Q Y_2'' \tag{4.20}$$

since

$$X_{22}' = X_2' = Q \qquad X_{22}'' = X_2'' = 0 \qquad B_{2j}' = B_{j2}$$

For $i = 3$

$$W_3 = Y_4'' X_{34}' - Y_3' X_{43}'' + \sum_{\substack{j=1 \\ j \neq 4}}^{N} (B_{3j}' - B_{j3}) Y_j'' Y_3' = -\Delta m Y_3' Y_4'' \tag{4.21}$$

since

$$X_{34}' = m Y_3' \qquad X_{43}'' = (m + \Delta m) Y_4'' \qquad B_{3j}' = B_{j3}$$

Finally, for $i \geq 4$

$$W_i = \sum_{j=1}^{N} (B_{ij}' - B_{ji}) Y_j'' Y_i' = 0 \tag{4.22}$$

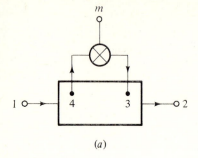

(a)

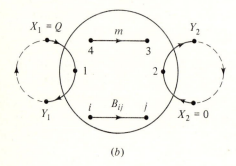

(b)

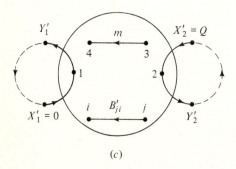

(c)

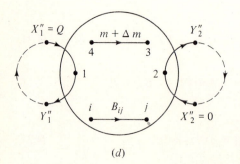

(d)

Figure 4.13 Derivation of sensitivities: (a) digital-filter network, (b) two-pole representation, (c) two-pole representation of the transpose network, (d) two-pole representation of the network with m changed to $m + \Delta m$.

since $$B'_{ij} = B_{ji} \quad \text{for } i \geq 4$$

Consequently, from Eqs. (4.18) to (4.22)

$$W_1 + W_2 + W_3 = 0$$

or $$Y''_2 - Y'_1 = \frac{\Delta m \, Y'_3 \, Y''_4}{Q} \tag{4.23}$$

Now

$$\Delta H_{12} = H''_{12} - H_{12} = H''_{12} - H'_{21} = \frac{Y''_2 - Y'_1}{Q} \tag{4.24}$$

and from Eqs. (4.23) and (4.24)

$$\frac{\Delta H_{12}}{\Delta m} = \frac{Y'_3 \, Y''_4}{Q^2} = H'_{23} H''_{14} = H_{32} H''_{14}$$

As $\Delta m \to 0$, $H''_{14} \to H_{14}$, and therefore

$$S_m^H(z) = \lim_{\Delta m \to 0} \frac{\Delta H_{12}}{\Delta m} = H_{14}(z) H_{32}(z) \tag{4.25}$$

i.e., the sensitivity to variations in multiplier constant m can be formed by multiplying the transfer function from the input of the network to the input of the multiplier by the transfer function from the output of the multiplier to the output of the network. This technique can eliminate considerable manipulation in complicated networks.

Example 4.4 Find the sensitivities for the network of Fig. 4.14.

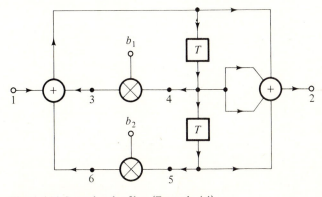

Figure 4.14 Second-order filter (Example 4.4).

SOLUTION Straightforward analysis yields

$$H_{14}(z) = \frac{z}{z^2 - b_1 z - b_2} \qquad H_{15}(z) = \frac{1}{z^2 - b_1 z - b_2}$$

$$H_{32}(z) = H_{62}(z) = \frac{(z+1)^2}{z^2 - b_1 z - b_2}$$

Hence from Eq. (4.25)

$$S_{b1}^H(z) = \frac{z(z+1)^2}{(z^2 - b_1 z - b_2)^2} \qquad \text{and} \qquad S_{b2}^H(z) = \frac{(z+1)^2}{(z^2 - b_1 z - b_2)^2}$$

REFERENCES

1. S. K. Mitra and R. J. Sherwood, Digital Ladder Networks, *IEEE Trans. Audio Electroacoust.*, vol. AU-21, pp. 30–36, February 1973.
2. A. Fettweis, A General Theorem for Signal-Flow Networks, with Applications, *Arch. Elektron. Uebertrag.*, vol. 25, pp. 557–561, 1971.
3. R. E. Seviora and M. Sablatash, A Tellegen's Theorem for Digital Filters, *IEEE Trans. Circuit Theory*, vol. CT-18, pp. 201–203, January 1971.
4. L. B. Jackson, On the Interaction of Roundoff Noise and Dynamic Range in Digital Filters, *Bell Syst. Tech. J.*, vol. 49, pp. 159–184, February 1970.

ADDITIONAL REFERENCES

Antoniou, A.: Realization of Digital Filters, *IEEE Trans. Audio Electroacoust.*, vol. AU-20, pp. 95–97, March 1972.

Crochiere, R. E.: Computational Methods for Sensitivity Analysis of Digital Filters, *M.I.T. Res. Lab. Electron. Q. Prog. Rep.* 109, pp. 113–123, April 15, 1973.

Director, S. W., and R. A. Rohrer: The Generalized Adjoint Network and Network Sensitivities, *IEEE Trans. Circuit Theory*, vol. CT-16, pp. 318–323, August 1969.

Szczupak, J., and S. K. Mitra: Digital Filter Realization Using Successive Multiplier-Extraction Approach, *IEEE Trans. Acoust., Speech, Signal Process.*, vol. ASSP-23, pp. 235–239, April 1975.

—— and ——: Detection, Location, and Removal of Delay-Free Loops in Digital Filter Configurations, *IEEE Trans. Acoust., Speech, Signal Process.*, vol. ASSP-23, pp. 558–562, December 1975.

PROBLEMS

4.1 By using first the direct and then the canonic method, realize the following transfer functions:

(a) $\quad H(z) = \dfrac{4(z-1)^4}{4z^4 + 3z^3 + 2z^2 + z + 1}$ (b) $\quad H(z) = \dfrac{(z+1)^2}{4z^3 - 2z^2 + 1}$

4.2 By using first the cascade and then the parallel method, realize the following transfer functions:

(a) $\quad H(z) = \dfrac{16(z+1)z^2}{(4z^2 - 2z + 1)(4z + 3)}$ (b) $\quad H(z) = \dfrac{6z}{6z^3 + 4z^2 + z - 1}$

4.3 A filter is characterized by the equations

$$\mathbf{q}(nT + T) = \mathbf{A}\mathbf{q}(nT) + \mathbf{B}x(nT) \qquad y(nT) = \mathbf{C}\mathbf{q}(nT) + \mathbf{D}x(nT)$$

where
$$\mathbf{A} = \begin{bmatrix} 0 & 1 & 0 \\ 0 & 0 & 1 \\ \frac{25}{64} & -\frac{29}{32} & \frac{3}{4} \end{bmatrix} \qquad \mathbf{B} = \begin{bmatrix} 0 \\ 0 \\ 1 \end{bmatrix} \qquad \mathbf{C} = \begin{bmatrix} \frac{25}{64} & \frac{3}{32} & \frac{11}{4} \end{bmatrix} \qquad \mathbf{D} = \begin{bmatrix} 1 \end{bmatrix}$$

Obtain a canonic realization for the filter.

4.4 Realize the transfer function

$$H(z) = \frac{0.0154z^3 + 0.0462z^2 + 0.0462z + 0.0154}{z^3 - 1.990z^2 + 1.572z - 0.4583}$$

by using the ladder method.

4.5 (*a*) Construct a flow chart for the software implementation of an N-section parallel filter.

(*b*) Repeat part (*a*) for an N-section cascade filter.

4.6 Given a continuous-time transfer function $H_A(s)$, a corresponding discrete-time transfer function $H_D(z)$ can be formed as

$$H_D(z) = H_A(s) \Big|_{s = \frac{2}{T}\frac{z-1}{z+1}}$$

(see Sec. 7.6).

(*a*) Obtain a digital-filter network by using the transfer function

$$H_A(s) = \frac{s^2}{s^2 + \sqrt{2}s + 1}$$

(*b*) Evaluate the gain of the filter for $\omega = 0$ and $\omega = \pi/T$.

4.7 (*a*) The flow graph of Fig. P4.7*a* represents a recursive filter. Deduce the transfer function of the filter.

(*b*) Repeat part (*a*) for the flow graph of Fig. P4.7*b*.

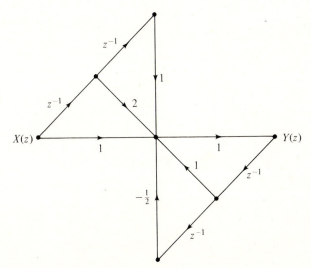

Figure P4.7 (*a*)

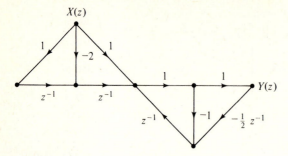

Figure P4.7 (b)

4.8 (a) Convert the flow graph of Fig. P4.8 into a topologically equivalent network.
(b) Obtain an alternative realization by using the canonic method.

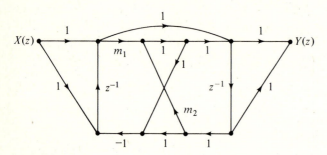

Figure P4.8

4.9 Derive flow-graph representations for the filters of Figs. 4.6 and 4.9a.

4.10 A flow graph is said to be *computable* if there are no closed delay-free loops. Check the flow graphs of Figs. P4.10a and P4.10b for computability.

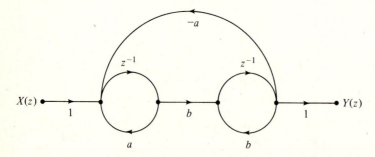

Figure P4.10 (a)

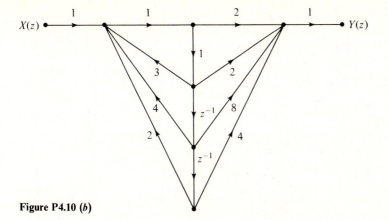

Figure P4.10 (*b*)

4.11 The network in Fig. P4.11 is excited first with a unit step and then with a unit impulse. Show that the signal distributions in the two cases satisfy Tellegen's theorem.

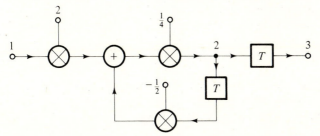

Figure P4.11

4.12 Figure P4.12 depicts two topologically compatible networks. Show that the signal distributions in the two networks satisfy Tellegen's theorem.

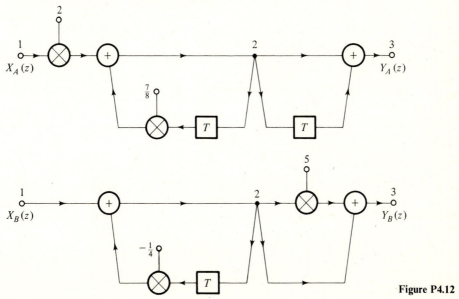

Figure P4.12

4.13 The 2-pole of Fig. P4.13a can be represented by the equations

$$Y_1(z) = H_{11}(z)X_1(z) + H_{21}(z)X_2(z)$$

$$Y_2(z) = H_{12}(z)X_1(z) + H_{22}(z)X_2(z)$$

(a) Find $H_{11}(z)$, $H_{12}(z)$, ... Hence check the flow graph for reciprocity.
(b) Repeat part (a) for the 2-pole of Fig. P4.13b.

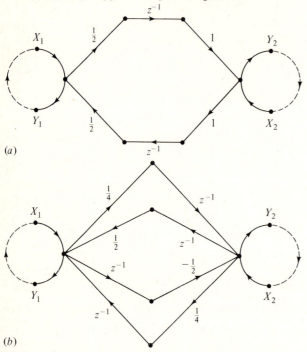

(a)

(b)

Figure P4.13

4.14 (a) Figure P4.14a shows a pair of 2-poles. Check the pair for interreciprocity.
(b) Repeat part (a) for the 2-poles of Fig. P4.14b.

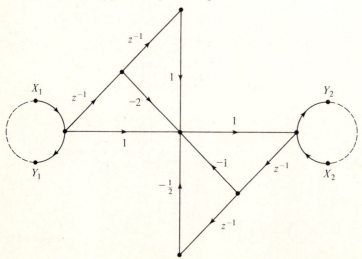

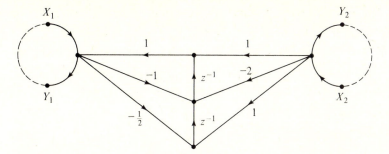

Figure P4.14 (a)

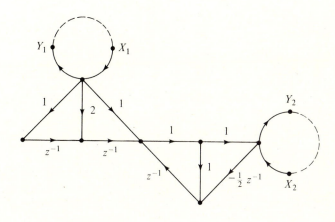

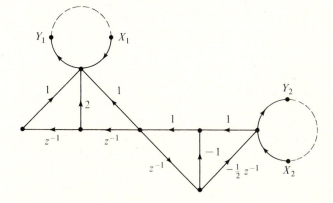

Figure P4.14 (b)

4.15 (*a*) Form the transpose of the network shown in Fig. 1.5*b*.

(*b*) Obtain an alternative ladder structure by applying the transpose approach to the network of Fig. 4.9*b*.

4.16 Figure P4.16 depicts an allpass network. Obtain an alternative allpass network by using the transpose method.

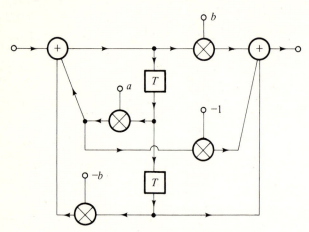

Figure P4.16

4.17 (*a*) Find the sensitivities of the network shown in Fig. P4.17*a*.

(*b*) Repeat part (*a*) for the network of Fig. P4.17*b*.

4.18 The gain and phase-shift sensitivities of a digital filter are defined as

$$S_{m_i}^M = \frac{\partial M(\omega)}{\partial m_i} \qquad S_{m_i}^\theta = \frac{\partial \theta(\omega)}{\partial m_i}$$

Assuming that

$$\left. \frac{\partial H(z)}{\partial m_i} \right|_{z = e^{j\omega T}} = \text{Re } S_{m_i}^H(e^{j\omega T}) + j \text{ Im } S_{m_i}^H(e^{j\omega T})$$

derive expressions for $S_{m_i}^M$ and $S_{m_i}^\theta$.

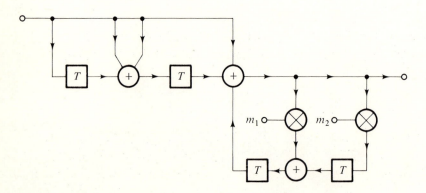

Figure P4.17 (*a*)

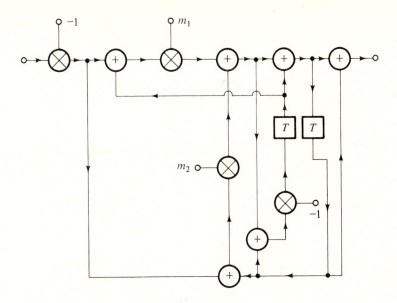

Figure P4.17 (*b*)

FIVE

ANALOG-FILTER APPROXIMATIONS

5.1 INTRODUCTION

The approximation problem in the design of recursive digital filters is usually solved by using the following analog-filter approximations [1–5]:

1. Butterworth
2. Tschebyscheff
3. Elliptic
4. Bessel

The derivation and properties of these approximations form the subject of this chapter.

The derivation of the elliptic approximation is somewhat complicated, as it entails a basic understanding of elliptic functions. Fortunately, however, the ultimate results can be put in a simple form which is easy to apply, even for the uninitiated. The elliptic approximation is treated in detail here because this is the most efficient of the above four if prescribed loss specifications are to be met.

The chapter begins with an introductory discussion of analog-filter terminology.

5.2 BASIC CONCEPTS

An analog filter like that in Fig. 5.1 can be represented by the equation

$$\frac{V_o(s)}{V_i(s)} = H(s) = \frac{N(s)}{D(s)}$$

where $V_i(s)$ and $V_o(s)$ are the Laplace transforms of the input and output voltages $v_i(t)$ and $v_o(t)$, respectively, $H(s)$ is the transfer function, and $N(s)$ and $D(s)$ are polynomials in s ($=\sigma + j\omega$). The loss (or attenuation) of the filter in decibels is defined by

$$A(\omega) = 20 \log \left| \frac{V_i(j\omega)}{V_o(j\omega)} \right| = 20 \log \frac{1}{|H(j\omega)|} = 10 \log L(\omega^2)$$

where

$$L(\omega^2) = \frac{1}{H(j\omega)H(-j\omega)} \tag{5.1}$$

The plot of $A(\omega)$ versus ω is the *loss characteristic.*

The phase shift and group delay of the filter are given by

$$\theta(\omega) = \arg H(j\omega) \quad \text{and} \quad \tau = -\frac{d\theta(\omega)}{d\omega}$$

respectively. Their plots versus ω are the *phase* and *delay characteristics.*

With $\omega = s/j$ in Eq. (5.1) the function

$$L(-s^2) = \frac{D(s)D(-s)}{N(s)N(-s)}$$

can be formed. This is called the *loss function* of the filter, and, as is evident, its zeros are the poles of $H(s)$ and their negatives, whereas its poles are the zeros of $H(s)$ and their negatives. Typical zero-pole plots for $H(s)$ and $L(-s^2)$ are shown in Fig. 5.2.

An ideal lowpass filter is one that will pass only low-frequency components. Its loss characteristic is of the form depicted in Fig. 5.3a. The frequency ranges 0 to ω_c and ω_c to ∞ are the *passband* and *stopband*, respectively. The boundary between the passband and stopband, namely ω_c, is the *cutoff frequency.*

Similarly, ideal highpass bandpass and bandstop filters can be identified having loss characteristics like those depicted in Fig. 5.3b to d.

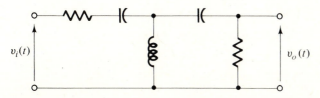

Figure 5.1 Analog filter.

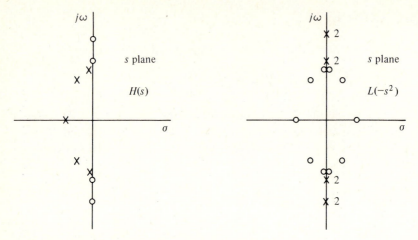

Figure 5.2 Zero-pole plots for $H(s)$ and $L(-s^2)$.

A practical lowpass filter differs from an ideal one in that the passband loss is not zero, the stopband loss is not infinite, and the transition between passband and stopband is gradual. The loss characteristic might assume the form shown in Fig. 5.4a, where ω_p is the passband edge, ω_a is the stopband edge, A_p is the maximum passband loss, and A_a is the minimum stopband loss. The cutoff frequency ω_c is a hypothetical boundary between passband and stopband, which may be the 3-dB frequency or possibly the square root of $\omega_p \omega_a$ (in elliptic filters).

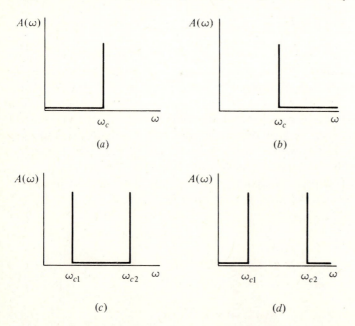

Figure 5.3 Ideal loss characteristics: (a) lowpass, (b) highpass, (c) bandpass, (d) bandstop.

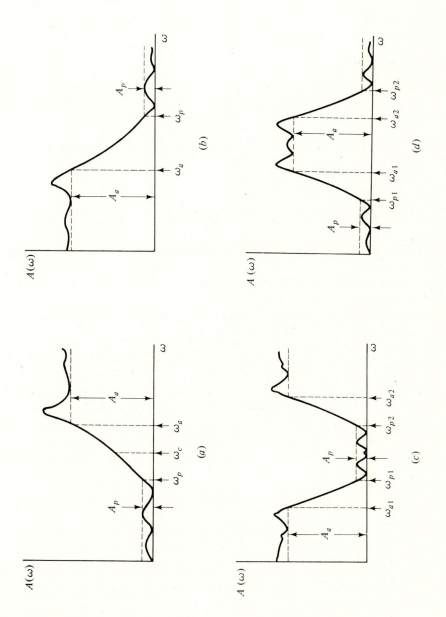

Figure 5.4 Nonideal loss characteristics: (*a*) lowpass, (*b*) highpass, (*c*) bandpass, (*d*) bandstop.

Typical characteristics for practical highpass, bandpass, and bandstop filters are shown in Fig. 5.4b to d.

A filter approximation is a realizable transfer function such that the plot of $A(\omega)$ versus ω approaches one of the ideal characteristics in Fig. 5.3. A transfer function is realizable if it characterizes a stable and causal network. Such a transfer function must satisfy the following constraints:

1. It must be a rational function of s with real coefficients.
2. Its poles must lie in the left-half s plane.
3. The degree of the numerator polynomial must be equal to or less than that of the denominator polynomial.

In the following four sections we focus our attention on normalized lowpass approximations, in which $\omega_c \approx 1$ rad/s. The derivation of denormalized lowpass, highpass, bandpass, and bandstop approximations is almost invariably accomplished through transformations of normalized lowpass approximations. The appropriate transformations are described in Sec. 5.7.

5.3 BUTTERWORTH APPROXIMATION

The simplest lowpass approximation, the Butterworth approximation, is derived by assuming that $L(\omega^2)$ is a polynomial of the form

$$L(\omega^2) = B_0 + B_1 \omega^2 + \cdots + B_n \omega^{2n} \tag{5.2}$$

such that

$$\lim_{\omega^2 \to 0} L(\omega^2) = 1$$

in a maximally flat sense.

Derivation

The Taylor series of $L(x + h)$, where $x = \omega^2$, is

$$L(x + h) = L(x) + h \frac{dL(x)}{dx} + \cdots + \frac{h^k}{k!} \frac{d^k L(x)}{dx^k}$$

Since it is required that as many derivatives of $L(x)$ as possible be zero for $x = 0$, we may assign

$$L(0) = 1$$

$$\frac{d^k L(x)}{dx^k} \bigg|_{x=0} = 0 \qquad \text{for } k \leq n - 1$$

5.4 TS

In the l
ω, and
A more
approx
prescrit

Deriva

The los
is of th

where

and

$F(\omega)$, l
transfer

where H

$A(\omega)$

Thus from Eq. (5.2)

$$B_0 = 1 \quad \text{and} \quad B_1 = B_2 = \cdots = B_{n-1} = 0$$

or

$$L(\omega^2) = 1 + B_n \omega^{2n}$$

Now for a normalized approximation in which

$$L(1) = 2$$

that is, $A(\omega) \approx 3$ dB at $\omega = 1$ rad/s, $B_n = 1$ and

$$L(\omega^2) = 1 + \omega^{2n} \tag{5.3}$$

Hence the loss in a normalized lowpass Butterworth approximation is

$$A(\omega) = 10 \log (1 + \omega^{2n})$$

This is plotted in Fig. 5.5 for $n = 3, 6, 9$.

Normalized Transfer Function

With $\omega = s/j$ in Eq. (5.3) we have

$$L(-s^2) = 1 + (-s^2)^n = \prod_{k=1}^{2n} (s - s_k)$$

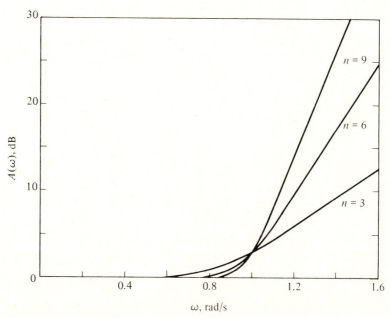

Figure 5.5 Typical Butterworth loss characteristics ($n = 3, 6, 9$).

Figure 5.7

The derivation of $H_N(s)$ involves three general steps:

1. The exact form of $F(\omega)$ is deduced such that the desired loss characteristic is achieved.
2. The exact form of $L(\omega^2)$ is obtained.
3. The zeros of $L(-s^2)$ and in turn the poles of $H_N(s)$ are found.

Close examination of the Tschebyscheff loss characteristic depicted in Fig. 5.7 reveals that $F(\omega)$ and $L(\omega^2)$ must have the following properties:

Property 1: $F(\omega) = 0$ if $\omega = \pm\Omega_{01},\ \pm\Omega_{02}$

Property 2: $F^2(\omega) = 1$ if $\omega = 0,\ \pm\hat{\Omega}_1,\ \pm1$

Property 3: $\dfrac{dL(\omega^2)}{d\omega} = 0$ if $\omega = 0,\ \pm\Omega_{01},\ \pm\hat{\Omega}_1,\ \pm\Omega_{02}$

From property 1, $F(\omega)$ must be a polynomial of the form

$$F(\omega) = M_1(\omega^2 - \Omega_{01}^2)(\omega^2 - \Omega_{02}^2)$$

($M_1,\ M_2,\ \ldots,\ M_6$ represent constants in this analysis.) From property 2, $1 - F^2(\omega)$ has zeros at $\omega = 0,\ \pm\hat{\Omega}_1,\ \pm1$. Furthermore, the derivative of $1 - F^2(\omega)$ with respect to ω, namely

$$\frac{d}{d\omega}[1 - F^2(\omega)] = -2F(\omega)\frac{dF(\omega)}{d\omega} = -\frac{1}{\varepsilon^2}\frac{dL(\omega^2)}{d\omega} \tag{5.7}$$

has zeros at $\omega = 0,\ \pm\Omega_{01},\ \pm\hat{\Omega}_1,\ \pm\Omega_{02}$, according to property 3. Consequently, $1 - F^2(\omega)$ must have at least double zeros at $\omega = 0,\ \pm\hat{\Omega}_1$. Therefore, we can write

$$1 - F^2(\omega) = M_2\omega^2(\omega^2 - \hat{\Omega}_1^2)^2(\omega^2 - 1)$$

Now from Eq. (5.7) and properties 1 and 3

$$\frac{dF(\omega)}{d\omega} = \frac{1}{2\varepsilon^2 F(\omega)}\frac{dL(\omega^2)}{d\omega} = M_3\omega(\omega^2 - \hat{\Omega}_1^2)$$

By combining the above results we can form the differential equation

$$\left[\frac{dF(\omega)}{d\omega}\right]^2 = \frac{M_4[1 - F^2(\omega)]}{1 - \omega^2}$$

which can be expressed in terms of definite integrals as

$$M_5\int_0^F \frac{dx}{\sqrt{1 - x^2}} + M_6 = \int_0^\omega \frac{dy}{\sqrt{1 - y^2}}$$

Hence F and ω are interrelated by the equation

$$M_5\cos^{-1}F + M_6 = \cos^{-1}\omega = \theta \tag{5.8}$$

i.e., for a given value of θ

$$\omega = \cos\theta \quad \text{and} \quad F = \cos\left(\frac{\theta}{M_5} - \frac{M_6}{M_5}\right)$$

What remains to be done is to determine constants M_5 and M_6. If $\omega = 0$, $\theta = \pi/2$; and if $\omega = 1$, $\theta = 0$, as depicted in Fig. 5.8. Now F will correspond to $F(\omega)$ only if it has two zeros in the range $0 \le \theta \le \pi/2$ (property 1), and its magnitude is unity if $\theta = 0, \pi/2$ (property 2). F must thus be of the form illustrated in Fig. 5.8. As can be seen, for $\theta = 0$

$$F = \cos\left(-\frac{M_6}{M_5}\right) = 1$$

or $M_6 = 0$. In addition, one period of F must be equal to one-quarter period of ω, that is,

$$2\pi M_5 = \frac{\pi}{2} \quad \text{or} \quad M_5 = \tfrac{1}{4}$$

Therefore, the exact form of $F(\omega)$ can be obtained from Eq. (5.8) as

$$F(\omega) = \cos\left(4\cos^{-1}\omega\right)$$

Alternatively, by expressing $\cos 4\theta$ in terms of $\cos\theta$, $F(\omega)$ can be put in the form

$$F(\omega) = 1 - 8\omega^2 + 8\omega^4$$

This polynomial is a fourth-order Tschebyscheff polynomial and is often designated as $T_4(\omega)$.

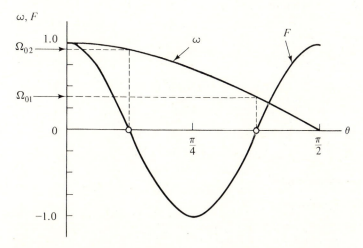

Figure 5.8 Plots of ω and F versus θ.

Similarly, for an nth-order Tschebyscheff approximation, one can show that

$$F(\omega) = T_n(\omega) = \cos (n \cos^{-1} \omega)$$

and hence from Eq. (5.5)

$$L(\omega^2) = 1 + \varepsilon^2 [\cos (n \cos^{-1} \omega)]^2 \qquad (5.9)$$

This relation gives the loss characteristic for $|\omega| \leq 1$. For $|\omega| > 1$, the quantity $\cos^{-1} \omega$ becomes complex, i.e.,

$$\cos^{-1} \omega = j\theta \qquad (5.10)$$

and since

$$\omega = \cos j\theta = \tfrac{1}{2}(e^{j(j\theta)} + e^{-j(j\theta)}) = \cosh \theta$$

we have

$$\theta = \cosh^{-1} \omega$$

Now from Eq. (5.10)

$$\cos^{-1} \omega = j \cosh^{-1} \omega$$

and

$$\cos (n \cos^{-1} \omega) = \cos (jn \cosh^{-1} \omega) = \cosh (n \cosh^{-1} \omega)$$

Thus for $|\omega| > 1$, Eq. (5.9) becomes

$$L(\omega^2) = 1 + \varepsilon^2 [\cosh (n \cosh^{-1} \omega)]^2 \qquad (5.11)$$

In summary, the loss in a normalized lowpass Tschebyscheff approximation is given by

$$A(\omega) = 10 \log [1 + \varepsilon^2 T_n^2(\omega)]$$

where

$$T_n(\omega) = \begin{cases} \cos (n \cos^{-1} \omega) & |\omega| \leq 1 \\ \cosh (n \cosh^{-1} \omega) & |\omega| > 1 \end{cases}$$

The loss characteristics for $n = 4$, $A_p = 1$ dB and $n = 7$, $A_p = 0.5$ dB are plotted in Fig. 5.9. As can be seen,

$$A(0) = \begin{cases} A_p & \text{for even } n \\ 0 & \text{for odd } n \end{cases}$$

as is generally the case in the Tschebyscheff approximation.

Normalized Transfer Function

With $\omega = s/j$, Eq. (5.11) becomes

$$L(-s^2) = 1 + \varepsilon^2 \left[\cosh \left(n \cosh^{-1} \frac{s}{j}\right)\right]^2$$

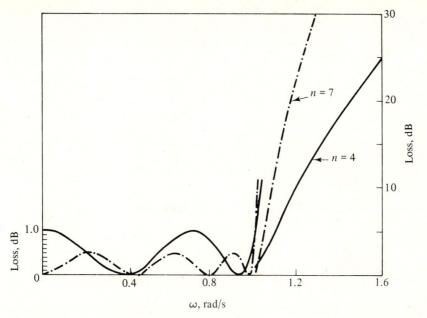

Figure 5.9 Typical Tschebyscheff loss characteristics ($n = 4$, $A_p = 1$ dB and $n = 7$, $A_p = 0.5$ dB).

and if $s = \sigma + j\omega$ is a zero of $L(-s^2)$, we can write

$$u + jv = \cosh^{-1}(-j\sigma + \omega)$$

$$\cosh[n(u + jv)] = \pm\frac{j}{\varepsilon} \tag{5.12}$$

Hence

$$-j\sigma + \omega = \cosh(u + jv) = \cosh u \cos v + j \sinh u \sin v$$

or

$$\sigma = -\sinh u \sin v \tag{5.13}$$

$$\omega = \cosh u \cos v \tag{5.14}$$

Similarly, from Eq. (5.12)

$$\cosh nv \cos nv + j \sinh nu \sin nv = \pm\frac{j}{\varepsilon}$$

or

$$\cosh nu \cos nv = 0$$

and

$$\sinh nu \sin nv = \pm\frac{1}{\varepsilon} \tag{5.15}$$

The solution of the first equation is

$$v = \frac{(2k - 1)\pi}{2n} \qquad \text{for } k = 1, 2, \dots \tag{5.16}$$

and since $\sin nv = \pm 1$, Eq. (5.15) yields

$$u = \pm \frac{1}{n} \sinh^{-1} \frac{1}{\varepsilon}$$

Therefore, from Eqs. (5.13), (5.14), and (5.16)

$$\sigma_k = \pm \sinh\left(\frac{1}{n}\sinh^{-1}\frac{1}{\varepsilon}\right)\sin\frac{(2k-1)\pi}{2n}$$

$$\omega_k = \cosh\left(\frac{1}{n}\sinh^{-1}\frac{1}{\varepsilon}\right)\cos\frac{(2k-1)\pi}{2n}$$

for $k = 1, 2, \ldots, 2n$. Evidently,

$$\frac{\sigma_k^2}{\sinh^2 u} + \frac{\omega_k^2}{\cosh^2 u} = 1$$

i.e., the zeros of $L(-s^2)$ are located on an ellipse, as depicted in Fig. 5.10. The normalized transfer function can now be formed as

$$H_N(s) = \frac{H_0}{\displaystyle\prod_{i=1}^{n}(s - p_i)}$$

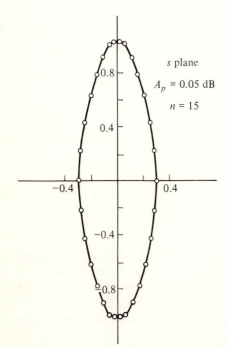

s plane

$A_p = 0.05$ dB

$n = 15$

Figure 5.10 Zero-pole plot of $L(-s^2)$ for a fifteenth-order Tschebyscheff filter.

where p_i for $i = 1, 2, \ldots, n$ are the left-half s-plane zeros of $L(-s^2)$. Constant H_0 can be chosen as

$$H_0 = \begin{cases} 10^{-0.05 A_p} \displaystyle\prod_{i=1}^{n} (-p_i) & \text{for even } n \\[2em] \displaystyle\prod_{i=1}^{n} (-p_i) & \text{for odd } n \end{cases}$$

so as to achieve zero minimum passband loss.

Example 5.2 Form $H_N(s)$ if $n = 4$ and $A_p = 1.0$ dB.

SOLUTION From Eq. (5.6)

$$x = \frac{1}{\varepsilon} = \frac{1}{\sqrt{10^{0.1} - 1}} = 1.965227$$

and

$$\sinh^{-1} \frac{1}{\varepsilon} = \ln \left(x + \sqrt{x^2 + 1} \right) = 1.427975$$

Hence

$$\sigma_k = \pm 0.364625 \sin \frac{(2k - 1)\pi}{8}$$

$$\omega_k = 1.064402 \cos \frac{(2k - 1)\pi}{8}$$

Thus

$$H_N(s) = \frac{H_0}{\displaystyle\prod_{i=1}^{2} (s - p_i)(s - p_i^*)}$$

where

$$p_1, p_1^* = -0.139536 \pm j0.983379$$

$$p_2, p_2^* = -0.336870 \pm j0.407329$$

$$H_0 = 0.245653$$

(see Program B.7).

5.5 ELLIPTIC APPROXIMATION

Although the Tschebyscheff approximation leads to a much better passband characteristic than the Butterworth approximation, the corresponding stopband characteristics are similar. They are both very good at high frequencies but tend to deteriorate progressively as the frequency is decreased. An improved stopband characteristic can be achieved by using the elliptic approximation. In this the passband loss oscillates between zero and a prescribed maximum A_p, as in the Tschebyscheff approximation, and the stopband loss oscillates between infinity and a prescribed minimum A_a.

The elliptic approximation is more efficient than the preceding two in that the transition between passband and stopband is steeper for a given approximation order.

Our approach to this approximation follows the formulation of Grossman [6], which, although involved, is probably the simplest.

Fifth-Order Approximation

The loss characteristic in a fifth-order normalized elliptic approximation is of the form depicted in Fig. 5.11, where

$$\omega_p = \sqrt{k} \qquad \omega_a = \frac{1}{\sqrt{k}} \qquad \omega_c = \sqrt{\omega_a \omega_p} = 1$$

The constants k and k_1 given by

$$k = \frac{\omega_p}{\omega_a}$$

and

$$k_1 = \left(\frac{10^{0.1 A_p} - 1}{10^{0.1 A_a} - 1} \right)^{1/2} \tag{5.17}$$

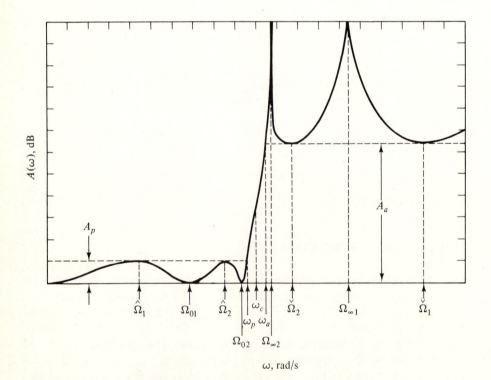

Figure 5.11 Loss characteristic of a fifth-order elliptic filter.

are the *selectivity factor* and *discrimination factor*, respectively. The loss is given by

$$A(\omega) = 10 \log L(\omega^2)$$

where

$$L(\omega^2) = 1 + \varepsilon^2 F^2(\omega) \tag{5.18}$$

and

$$\varepsilon^2 = 10^{0.1A_p} - 1 \tag{5.19}$$

Function $F(\omega)$ and in turn $L(\omega^2)$, $L(-s^2)$, and $H(s)$ are ratios of polynomials in this case.

According to the elliptic loss characteristic of Fig. 5.11, the prerequisite properties of $F(\omega)$ and $L(\omega^2)$ are as follows:

Property 1: $F(\omega) = 0$ if $\omega = 0, \pm\Omega_{01}, \pm\Omega_{02}$

Property 2: $F(\omega) = \infty$ if $\omega = \infty, \pm\Omega_{\infty 1}, \pm\Omega_{\infty 2}$

Property 3: $F^2(\omega) = 1$ if $\omega = \pm\hat{\Omega}_1, \pm\hat{\Omega}_2, \pm\sqrt{k}$

Property 4: $F^2(\omega) = \dfrac{1}{k_1^2}$ if $\omega = \pm\check{\Omega}_1, \pm\check{\Omega}_2, \pm\dfrac{1}{\sqrt{k}}$

Property 5: $\dfrac{dL(\omega^2)}{d\omega} = 0$ if $\omega = \pm\hat{\Omega}_1, \pm\hat{\Omega}_2, \pm\check{\Omega}_1, \pm\check{\Omega}_2$

By using these properties we shall attempt to derive the exact form of $F(\omega)$. The approach is analogous to that used earlier in the Tschebyscheff approximation.

From properties 1 and 2

$$F(\omega) = \frac{M_1\omega(\omega^2 - \Omega_{01}^2)(\omega^2 - \Omega_{02}^2)}{(\omega^2 - \Omega_{\infty 1}^2)(\omega^2 - \Omega_{\infty 2}^2)} \tag{5.20}$$

(M_1 to M_7 represent constants), and from properties 2 and 3

$$1 - F^2(\omega) = \frac{M_2(\omega^2 - \hat{\Omega}_1^2)^2(\omega^2 - \hat{\Omega}_2^2)^2(\omega^2 - k)}{(\omega^2 - \Omega_{\infty 1}^2)^2(\omega^2 - \Omega_{\infty 2}^2)^2}$$

where the double zeros at $\omega = \pm\hat{\Omega}_1, \pm\hat{\Omega}_2$ are due to property 5 (see Sec. 5.4). Similarly, from properties 2, 4, and 5

$$1 - k_1^2 F^2(\omega) = \frac{M_3(\omega^2 - \check{\Omega}_1^2)^2(\omega^2 - \check{\Omega}_2^2)^2(\omega^2 - 1/k)}{(\omega^2 - \Omega_{\infty 1}^2)^2(\omega^2 - \Omega_{\infty 2}^2)^2}$$

and from property 5

$$\frac{dF(\omega)}{d\omega} = \frac{M_4(\omega^2 - \hat{\Omega}_1^2)(\omega^2 - \hat{\Omega}_2^2)(\omega^2 - \check{\Omega}_1^2)(\omega^2 - \check{\Omega}_2^2)}{(\omega^2 - \Omega_{\infty 1}^2)^2(\omega^2 - \Omega_{\infty 2}^2)^2}$$

By combining the above results we can form the important relation

$$\left[\frac{dF(\omega)}{d\omega}\right]^2 = \frac{M_5[1 - F^2(\omega)][1 - k_1^2 F^2(\omega)]}{(1 - \omega^2/k)(1 - k\omega^2)} \tag{5.21}$$

Alternatively, we can write

$$\int_0^F \frac{dx}{\sqrt{(1-x^2)(1-k_1^2 x^2)}} = \sqrt{M_5} \int_0^\omega \frac{dy}{\sqrt{(1-y^2/k)(1-ky^2)}} + M_7$$

and if $y = \sqrt{k}\, y'$, $y' = y$

$$\int_0^F \frac{dx}{\sqrt{(1-x^2)(1-k_1^2 x^2)}} = M_6 \int_0^{\omega/\sqrt{k}} \frac{dy}{\sqrt{(1-y^2)(1-k^2 y^2)}} + M_7$$

These are elliptic integrals of the first kind, and they can be put in the more convenient form

$$\int_0^{\phi_1} \frac{d\theta_1}{\sqrt{1 - k_1^2 \sin^2 \theta_1}} = M_6 \int_0^\phi \frac{d\theta}{\sqrt{1 - k^2 \sin^2 \theta}} + M_7$$

by using the transformations

$$x = \sin \theta_1 \qquad F = \sin \phi_1 \qquad y = \sin \theta \qquad \frac{\omega}{\sqrt{k}} = \sin \phi$$

The above two integrals can assume complex values if complex values are allowed for ϕ_1 and ϕ. By letting

$$\int_0^\phi \frac{d\theta}{\sqrt{1 - k^2 \sin^2 \theta}} = z \qquad \text{where } z = u + jv$$

the solution of Eq. (5.21) can be expressed in terms of a pair of simultaneous equations as

$$\frac{\omega}{\sqrt{k}} = \sin \phi = \text{sn}\,(z, k) \tag{5.22}$$

$$F = \sin \phi_1 = \text{sn}\,(M_6 z + M_7, k_1) \tag{5.23}$$

The entities at the right-hand side are elliptic functions.

Further progress in this analysis can be made by using the properties of elliptic functions (see Appendix A).

As demonstrated in Sec. A.7, Eq. (5.22) is a transformation which maps trajectory $ABCD$ in Fig. 5.12a onto the positive real axis of the ω plane, as depicted in Fig. 5.12b. Since the behavior of $F(\omega)$ is known for all real values of ω, constants M_6 and M_7 can be determined. In turn, the exact form of $F(\omega)$ can be derived.

If $z = u$ and $0 \le u \le K$ (domain 1 in Sec. A.7), Eqs. (5.22) and (5.23) become

$$\omega = \sqrt{k}\, \text{sn}\,(u, k) \tag{5.24}$$

$$F = \text{sn}\,(M_6 u + M_7, k_1) \tag{5.25}$$

where ω and F have real periods of $4K$ and $4K_1/M_6$, respectively (see Sec. A.6). If $\omega = 0$, $u = 0$; and if $\omega = \sqrt{k}$, $u = K$, as illustrated in Fig. 5.13. Now F will correspond to $F(\omega)$ if it has zeros at $u = 0$ and at two other points in the range

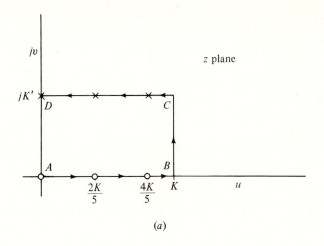

(a)

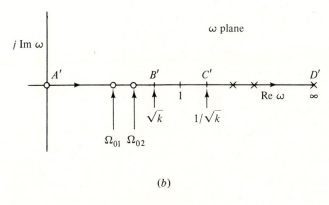

(b)

Figure 5.12 Mapping properties of Eq. (5.22).

$0 < u \le K$ (property 1), and its magnitude is unity at $u = K$ (property 3). Consequently, F must be of the form illustrated in Fig. 5.13. Clearly, for $u = 0$

$$F = \text{sn}\,(M_7, k_1) = 0$$

or $M_7 = 0$. Furthermore, five quarter-periods of F must be equal to one quarter-period of ω, that is,

$$M_6 = \frac{5K_1}{K}$$

and so from Eq. (5.25)

$$F = \text{sn}\left(\frac{5K_1 u}{K}, k_1\right)$$

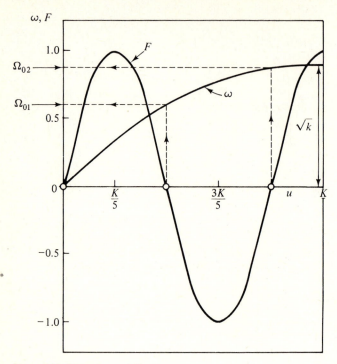

Figure 5.13 Plots of ω and F versus u.

Now F has z-plane zeros at

$$u = \frac{2Ki}{5} \qquad \text{for } i = 0, 1, 2$$

and, therefore, $F(\omega)$ must have ω-plane zeros (zero-loss frequencies) at

$$\Omega_{0i} = \sqrt{k} \, \text{sn}\left(\frac{2Ki}{5}, k\right) \qquad \text{for } i = 0, 1, 2$$

according to Eq. (5.24) (see Fig. 5.12).

If $z = u + jK'$ and $0 \le u \le K$ (domain 3 in Sec. A.7), Eqs. (5.22) and (5.23) assume the form

$$\omega = \frac{1}{\sqrt{k} \, \text{sn} \, (u, k)} \tag{5.26}$$

$$F = \text{sn}\left[\frac{5K_1(u + jK')}{K}, k_1\right] \tag{5.27}$$

If $\omega = \infty$, $u = 0$ and F must be infinite (property 2), i.e.,

$$F = \text{sn}\left(\frac{j5K_1K'}{K}, k_1\right) = \infty$$

and from Eq. (A.19)

$$F = \frac{j \text{ sn } (5K_1 K'/K, k_1')}{\text{cn } (5K_1 K'/K, k_1')} = \infty \qquad \text{where } k_1' = \sqrt{1 - k_1^2}$$

Hence it is necessary that

$$\text{cn} \left(\frac{5K_1 K'}{K}, k_1' \right) = 0$$

and therefore the relation

$$\frac{5K'}{K} = \frac{K_1'}{K_1} \tag{5.28}$$

must hold. The quantities K, K' are functions of k, and similarly K_1, K_1' are functions of k_1; in turn, k_1 is a function of A_p, A_a by definition. In effect, Eq. (5.28) constitutes an implicit constraint among filter specifications. We shall assume here that Eq. (5.28) holds. The implications of this assumption will be examined at a later point.

With Eq. (5.28) satisfied, Eq. (5.27) becomes

$$F = \text{sn} \left(\frac{5K_1}{K} u + jK_1', k_1 \right)$$

and after some manipulation

$$F = \frac{1}{k_1 \text{ sn } (5K_1 u/K, k_1)}$$

Evidently, $F = \infty$ if

$$u = 2Ki/5 \qquad \text{for } i = 0, 1, 2 \tag{5.29}$$

that is, F has poles at

$$z = \frac{2Ki}{5} + jK' \qquad \text{for } i = 0, 1, 2$$

as depicted in Fig. 5.12, and since line CD maps onto line $C'D'$, F corresponds to $F(\omega)$. That is, $F(\omega)$ has two poles in the range $1/\sqrt{k} \leq \omega < \infty$ and one at $\omega = \infty$ (property 2). The poles of $F(\omega)$ (infinite-loss frequencies) can be obtained from Eqs. (5.26) and (5.29) as

$$\Omega_{\infty i} = \frac{1}{\sqrt{k} \text{ sn } (2Ki/5, k)} \qquad \text{for } i = 0, 1, 2$$

Therefore, the infinite-loss frequencies are the reciprocals of the zero-loss frequencies, i.e.,

$$\Omega_{\infty i} = \frac{1}{\Omega_{0i}}$$

and by eliminating $\Omega_{\infty i}$ in Eq. (5.20), we have

$$F(\omega) = \frac{M_1' \omega (\omega^2 - \Omega_{01}^2)(\omega^2 - \Omega_{02}^2)}{(1 - \omega^2 \Omega_{01}^2)(1 - \omega^2 \Omega_{02}^2)} \tag{5.30}$$

The only unknown at this point is constant M_1'. With $z = K + jv$ and $0 \le v \le K'$ (domain 2 in Sec. A.7), Eqs. (5.22) and (5.23) can be put in the form

$$\omega = \frac{\sqrt{k}}{\text{dn}\,(v, k')} \qquad F = \text{sn}\left[\frac{5K_1(K + jv)}{K}, k_1\right]$$

If $\omega = 1$, $v = K'/2$ and $F(1) = M_1'$, according to Eq. (5.30). Hence

$$M_1' = \text{sn}\left(5K_1 + j\frac{5K'K_1}{2K}, k_1\right)$$

or

$$M_1' = \text{sn}\left(K_1 + \frac{jK_1'}{2}, k_1\right)$$

according to Eqs. (5.28) and (A.8), and after some manipulation

$$M_1' = \frac{1}{\text{dn}\,(K_1'/2, k_1')} = \frac{1}{\sqrt{k_1}}$$

Nth-Order Approximation (n odd)

For an nth-order approximation with n odd, constant M_7 in Eq. (5.23) is zero, and n quarter-periods of F must correspond to one quarter-period of ω, that is,

$$M_6 = \frac{nK_1}{K}$$

Therefore, Eq. (5.23) assumes the form

$$F = \text{sn}\left(\frac{nK_1 z}{K}, k_1\right) \tag{5.31}$$

where the relation

$$\frac{nK'}{K} = \frac{K_1'}{K_1}$$

must hold. The expression for $F(\omega)$ can be shown to be

$$F(\omega) = \frac{(-1)^r \omega}{\sqrt{k_1}} \prod_{i=1}^{r} \frac{\omega^2 - \Omega_i^2}{1 - \omega^2 \Omega_i^2}$$

where

$$r = \frac{n-1}{2}$$

and

$$\Omega_i = \sqrt{k}\,\text{sn}\left(\frac{2Ki}{n}, k\right) \qquad \text{for } i = 1, 2, \ldots, r$$

Zeros and Poles of $L(-s^2)$

The next task is to determine the zeros and poles of $L(-s^2)$. From Eqs. (5.18) and (5.31), the z-domain representation of the loss function can be expressed as

$$L(z) = 1 + \varepsilon^2 \ \mathrm{sn}^2 \left(\frac{nK_1 z}{K}, k_1 \right)$$

and by factoring

$$L(z) = \left[1 + j\varepsilon \ \mathrm{sn} \left(\frac{nK_1 z}{K}, k_1 \right) \right] \left[1 - j\varepsilon \ \mathrm{sn} \left(\frac{nK_1 z}{K}, k_1 \right) \right]$$

If z_1 is a root of the first factor, $-z_1$ must be a root of the second factor since the elliptic sine is an odd function of z. Consequently, the zeros of $L(z)$ can be determined by solving the equation

$$\mathrm{sn} \left(\frac{nK_1 z}{K}, k_1 \right) = \frac{j}{\varepsilon}$$

In practice, the value of k_1 is very small. For example, $k_1 \leq 0.0161$ if $A_p \leq 1$ dB and $A_a \geq 30$ dB and decreases further if A_p is reduced or A_a is increased. We can thus assume that $k_1 = 0$, in which case

$$\mathrm{sn} \left(\frac{nK_1 z}{K}, 0 \right) = \sin \frac{nK_1 z}{K} = \frac{j}{\varepsilon}$$

where $K_1 = \pi/2$, according to Eq. (A.2). Alternatively,

$$\frac{-jn\pi z}{2K} = \sinh^{-1} \frac{1}{\varepsilon}$$

and on using the identity

$$\sinh^{-1} x = \ln \left(x + \sqrt{x^2 + 1} \right)$$

and Eq. (5.19), we obtain one zero of $L(z)$ as

$$z_0 = jv_0$$

where

$$v_0 = \frac{K}{n\pi} \ln \frac{10^{0.05 A_p} + 1}{10^{0.05 A_p} - 1}$$

Now $\mathrm{sn} \ (nK_1 z/K, k_1)$ has a real period of $4K/n$, and as a result all z_i given by

$$z_i = z_0 + \frac{4Ki}{n} \qquad \text{for } i = 0, 1, 2, \ldots$$

must also be zeros of $L(z)$.

The zeros of $L(\omega^2)$ can be deduced by using the transformation between the z and ω planes, namely Eq. (5.22). In turn, the zeros of $L(-s^2)$ can be obtained by letting $\omega = s/j$. For $i = 0$, there is a real zero of $L(-s^2)$ at $s = \sigma_0$, where

$$\sigma_0 = j\sqrt{k} \ \mathrm{sn} \ (jv_0, k) \tag{5.32}$$

and for $i = 1, 2, \ldots, n - 1$ there are $n - 1$ distinct complex zeros at $s = \sigma_i + j\omega_i$, where

$$\sigma_i + j\omega_i = j\sqrt{k}\ \mathrm{sn}\left(jv_0 + \frac{4Ki}{n}, k\right) \tag{5.33}$$

The remaining n zeros are negatives of zeros already determined.

For $n = 5$, the required values of the elliptic sine are

$$\mathrm{sn}\left(jv_0 + \frac{4K}{5}\right)$$

$$\mathrm{sn}\left(jv_0 + \frac{8K}{5}\right) = \mathrm{sn}\left(jv_0 + 2K - \frac{2K}{5}\right) = -\mathrm{sn}\left(jv_0 - \frac{2K}{5}\right)$$

$$\mathrm{sn}\left(jv_0 + \frac{12K}{5}\right) = \mathrm{sn}\left(jv_0 + 2K + \frac{2K}{5}\right) = -\mathrm{sn}\left(jv_0 + \frac{2K}{5}\right)$$

$$\mathrm{sn}\left(jv_0 + \frac{16K}{5}\right) = \mathrm{sn}\left(jv_0 + 4K - \frac{4K}{5}\right) = \mathrm{sn}\left(jv_0 - \frac{4K}{5}\right)$$

Hence Eq. (5.33) can be put in the form

$$\sigma_i + j\omega_i = j\sqrt{k}\,(-1)^i\,\mathrm{sn}\left(jv_0 \pm \frac{2Ki}{5}, k\right) \qquad \text{for } i = 1, 2$$

Similarly, for any odd value of n

$$\sigma_i + j\omega_i = j\sqrt{k}\,(-1)^i\,\mathrm{sn}\left(jv_0 \pm \frac{2Ki}{n}, k\right) \qquad \text{for } i = 1, 2, \ldots, \frac{n-1}{2}$$

Now with the aid of the addition formula we can show that

$$\sigma_i + j\omega_i = \frac{(-1)^i\sigma_0 V_i \pm j\Omega_i W}{1 + \sigma_0^2\Omega_i^2} \qquad \text{for } i = 1, 2, \ldots, \frac{n-1}{2}$$

where

$$W = \sqrt{(1 + k\sigma_0^2)\left(1 + \frac{\sigma_0^2}{k}\right)} \tag{5.34}$$

$$V_i = \sqrt{(1 - k\Omega_i^2)\left(1 - \frac{\Omega_i^2}{k}\right)} \tag{5.35}$$

$$\Omega_i = \sqrt{k}\,\mathrm{sn}\left(\frac{2Ki}{n}, k\right) \tag{5.36}$$

A complete description of $L(-s^2)$ is at this point available. It has zeros at $s = \pm\sigma_0$, $\pm(\sigma_i + j\omega_i)$ and double poles at $s = \pm j/\Omega_i$, which can be evaluated by using the series representation of elliptic functions given in Sec. A.8. From Eqs. (5.32) and (A.30).

$$\sigma_0 = \frac{-2q^{1/4} \sum_{m=0}^{\infty} (-1)^m q^{m(m+1)} \sinh [(2m+1)\Lambda]}{1 + 2 \sum_{m=1}^{\infty} (-1)^m q^{m^2} \cosh 2m\Lambda} \tag{5.37}$$

where

$$\Lambda = \frac{1}{2n} \ln \frac{10^{0.05A_p} + 1}{10^{0.05A_p} - 1}$$

The parameter q, known as the *modular constant*, is given by

$$q = e^{-\pi K'/K} \tag{5.38}$$

Similarly, from Eqs. (5.36) and (A.30)

$$\Omega_i = \frac{2q^{1/4} \sum_{m=0}^{\infty} (-1)^m q^{m(m+1)} \sin \dfrac{(2m+1)\pi i}{n}}{1 + 2 \sum_{m=1}^{\infty} (-1)^m q^{m^2} \cos \dfrac{2m\pi i}{n}} \quad \text{for } i = 1, 2, \ldots, \frac{n-1}{2}. \tag{5.39}$$

The modular constant q can be determined by evaluating K and K' numerically. A quicker method, however, is to use the following procedure.

Since dn $(0, k) = 1$, Eq. (A.32) gives

$$\sqrt{k'} = \frac{1 - 2q + 2q^4 - 2q^9 + \cdots}{1 + 2q + 2q^4 + 2q^9 + \cdots} \tag{5.40}$$

Now $q < 1$ since $K, K' > 0$, and hence a first approximation for q is

$$q_0 = \frac{1}{2} \frac{1 - \sqrt{k'}}{1 + \sqrt{k'}}$$

By eliminating $\sqrt{k'}$ using Eq. (5.40), rationalizing, and then performing long division we have

$$q \approx q_0 + 2q^5 - 5q^9 + 10q^{13}$$

Thus, if q_{m-1} is an approximation for q

$$q_m \approx q_0 + 2q_{m-1}^5 - 5q_{m-1}^9 + 10q_{m-1}^{13}$$

is a better approximation. By using this recursive relation repeatedly we can show that

$$q \approx q_0 + 2q_0^5 + 15q_0^9 + 150q_0^{13}$$

Since k is known, the quantities k', q_0, q, σ_0, Ω_i, σ_i, and ω_i can be evaluated. Subsequently, the normalized transfer function $H_N(s)$ can be formed.

Nth-Order Approximation (*n* even)

So far we have been concerned with odd-order approximations. However, the results can be easily extended to the case of even *n*.

Function *F* is of the form

$$F = \text{sn}\left(\frac{nK_1}{K}z + K_1, k_1\right)$$

where the relation

$$\frac{nK'}{K} = \frac{K_1'}{K_1}$$

must again hold. The expression for $F(\omega)$ in this case is given by

$$F(\omega) = \frac{(-1)^r}{\sqrt{k_1}}\prod_{i=1}^{r}\frac{\omega^2 - \Omega_i^2}{1 - \omega^2\Omega_i^2}$$

where $\quad r = \dfrac{n}{2} \quad \Omega_i = \sqrt{k}\,\text{sn}\left[\dfrac{(2i-1)K}{n}, k\right] \quad$ for $i = 1, 2, \ldots, r$

The zeros of $L(-s^2)$ are

$$s_i = \pm(\sigma_i + j\omega_i)$$

where $\qquad \sigma_i + j\omega_i = \dfrac{\pm[\sigma_0 V_i + j(-1)^i\Omega_i W]}{1 + \sigma_0^2\Omega_i^2}$

The parameters W, V_i, and σ_0 are given by Eqs. (5.34), (5.35), and (5.37), as in the case of odd *n*, and the values of Ω_i can be computed by replacing i by $i - \frac{1}{2}$ in the right-hand side of Eq. (5.39).

Specification Constraint

The results of the preceding sections are based on the assumption that the relation

$$\frac{nK'}{K} = \frac{K_1'}{K_1} \tag{5.41}$$

holds. As pointed out earlier, this equation constitutes a constraint among filter specifications of the form

$$f(n, k) = f(A_p, A_c)$$

Consequently, if three of the four parameters are specified, the fourth is automatically fixed. It is thus of interest to put Eq. (5.41) in a more useful form which can be used to evaluate the corresponding fourth parameter.

From the definition of the elliptic sine sn $(K_1, k_1) = 1$ and from Eq. (A.30)

$$k_1 = 4\sqrt{q_1}\left(\frac{1 + q_1^2 + q_1^6 + \cdots}{1 + 2q_1 + 2q_1^4 + \cdots}\right)^2 \qquad \text{where } q_1 = e^{-\pi K_1'/K_1}$$

In practice, k_1 is close to zero, k_1' is close to unity, K_1'/K_1 is large, and, as a result, $q_1 \ll 1$. Hence we can assume that

$$k_1 \approx 4\sqrt{q_1} \quad \text{or} \quad k_1^2 = 16q_1 = 16e^{-\pi K_1'/K_1}$$

By eliminating K_1'/K_1, using Eq. (5.41), we have

$$k_1^2 = 16e^{-\pi n K'/K}$$

and from Eq. (5.38)

$$k_1^2 = 16q^n$$

Therefore, from Eq. (5.17) the desired formula is

$$\frac{10^{0.1A_p} - 1}{10^{0.1A_a} - 1} = 16q^n \tag{5.42}$$

If n, k, and A_p are specified, the resulting minimum stopband loss is given by

$$A_a = 10 \log \left(\frac{10^{0.1A_p} - 1}{16q^n} + 1 \right) \tag{5.43}$$

This is plotted against k in Fig. 5.14 for various values of n and A_p.

Alternatively, if k, A_a, and A_p are specified, the required approximation order must satisfy the inequality

$$n \geq \frac{\log 16D}{\log (1/q)} \quad \text{where } D = \frac{10^{0.1A_a} - 1}{10^{0.1A_p} - 1}$$

Normalized Transfer Function

An elliptic normalized lowpass filter with a selectivity factor k, a maximum passband loss of A_p dB, and a minimum stopband loss equal to or in excess of A_a dB has a transfer function of the form

$$H_N(s) = \frac{H_0}{D_0(s)} \prod_{i=1}^{r} \frac{s^2 + A_{0i}}{s^2 + B_{1i}s + B_{0i}} \tag{5.44}$$

where

$$r = \begin{cases} \dfrac{n-1}{2} & \text{for odd } n \\[2mm] \dfrac{n}{2} & \text{for even } n \end{cases}$$

and

$$D_0(s) = \begin{cases} s + \sigma_0 & \text{for odd } n \\ 1 & \text{for even } n \end{cases}$$

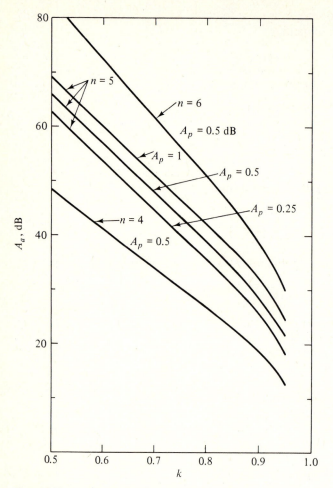

Figure 5.14 A_a versus k for various values of n and A_p.

The transfer-function coefficients and multiplier constant H_0 can be computed by using the following formulas in sequence:

$$k' = \sqrt{1 - k^2} \tag{5.45}$$

$$q_0 = \frac{1}{2}\frac{1 - \sqrt{k'}}{1 + \sqrt{k'}} \tag{5.46}$$

$$q = q_0 + 2q_0^5 + 15q_0^9 + 150q_0^{13} \tag{5.47}$$

$$D = \frac{10^{0.1A_a} - 1}{10^{0.1A_p} - 1} \tag{5.48}$$

$$n \geq \frac{\log 16D}{\log (1/q)} \tag{5.49}$$

$$\Lambda = \frac{1}{2n} \ln \frac{10^{0.05A_p} + 1}{10^{0.05A_p} - 1} \tag{5.50}$$

$$\sigma_0 = \left| \frac{2q^{1/4} \sum\limits_{m=0}^{\infty} (-1)^m q^{m(m+1)} \sinh\left[(2m+1)\Lambda\right]}{1 + 2 \sum\limits_{m=1}^{\infty} (-1)^m q^{m^2} \cosh 2m\Lambda} \right| \tag{5.51}$$

$$W = \sqrt{\left(1 + k\sigma_0^2\right)\left(1 + \frac{\sigma_0^2}{k}\right)} \tag{5.52}$$

$$\Omega_i = \frac{2q^{1/4} \sum\limits_{m=0}^{\infty} (-1)^m q^{m(m+1)} \sin \dfrac{(2m+1)\pi\mu}{n}}{1 + 2 \sum\limits_{m=1}^{\infty} (-1)^m q^{m^2} \cos \dfrac{2m\pi\mu}{n}} \tag{5.53}$$

where
$$\mu = \begin{cases} i & \text{for odd } n \\ i - \frac{1}{2} & \text{for even } n \end{cases} \qquad i = 1, 2, \dots, r$$

$$V_i = \sqrt{\left(1 - k\Omega_i^2\right)\left(1 - \frac{\Omega_i^2}{k}\right)} \tag{5.54}$$

$$A_{0i} = \frac{1}{\Omega_i^2} \tag{5.55}$$

$$B_{0i} = \frac{(\sigma_0 V_i)^2 + (\Omega_i W)^2}{(1 + \sigma_0^2 \Omega_i^2)^2} \tag{5.56}$$

$$B_{1i} = \frac{2\sigma_0 V_i}{1 + \sigma_0^2 \Omega_i^2} \tag{5.57}$$

$$H_0 = \begin{cases} \sigma_0 \prod\limits_{i=1}^{r} \dfrac{B_{0i}}{A_{0i}} & \text{for odd } n \\[2ex] 10^{-0.05A_p} \prod\limits_{i=1}^{r} \dfrac{B_{0i}}{A_{0i}} & \text{for even } n \end{cases} \tag{5.58}$$

The actual minimum stopband loss is given by Eq. (5.43). The series in Eqs. (5.51) and (5.53) converge rapidly, and three or four terms are sufficient for most purposes. $H_N(s)$ can be readily formed by using Program B.7.

Example 5.3 An elliptic filter is required satisfying the following specifications:

$$\omega_p = \sqrt{0.9} \text{ rad/s} \qquad \omega_a = \frac{1}{\sqrt{0.9}} \text{ rad/s} \qquad A_p = 0.1 \text{ dB} \qquad A_a \geq 50.0 \text{ dB}$$

Form $H_N(s)$.

Table 5.1 Coefficients of $H_N(s)$ (Example 5.3)

i	A_{0i}	B_{0i}	B_{1i}
1	1.434825×10	2.914919×10^{-1}	8.711574×10^{-1}
2	2.231643	6.123726×10^{-1}	4.729136×10^{-1}
3	1.320447	8.397386×10^{-1}	1.825141×10^{-1}
4	1.128832	9.264592×10^{-1}	4.471442×10^{-2}

$H_0 = 2.876332 \times 10^{-3}$

SOLUTION From Eqs. (5.45) to (5.49)

$$k = 0.9 \qquad k' = 0.435890 \qquad q_0 = 0.102330$$

$$q = 0.102352 \qquad D = 4{,}293{,}090 \qquad n \geq 7.92 \qquad \text{or} \qquad n = 8$$

From Eqs. (5.50) to (5.58) the transfer-function coefficients in Table 5.1 can be obtained. The corresponding loss characteristic is plotted in Fig. 5.15. The actual value of A_a is

$$A_a = 50.82 \text{ dB}$$

according to Eq. (5.43).

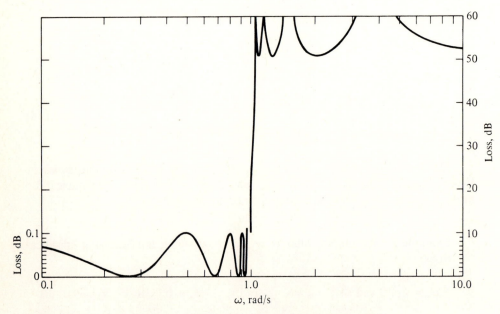

Figure 5.15 Loss characteristic of an eighth-order, elliptic filter (Example 5.3).

5.6 BESSEL APPROXIMATION

Ideally the group delay of a filter should be independent of frequency, or, equivalently, the phase shift should be a linear function of frequency. Since the only objective in the preceding three approximations is to achieve a specific loss characteristic, the phase characteristic turns out to be nonlinear. As a result the delay tends to vary with frequency, in particular in the elliptic approximation.

Consider the transfer function

$$H(s) = \frac{B_0}{\sum\limits_{i=0}^{n} B_i s^i} = \frac{B_0}{s^n B(1/s)} \tag{5.59}$$

where

$$B_i = \frac{(2n-i)!}{2^{n-i} i! (n-i)!} \tag{5.60}$$

$B(s)$ is a Bessel polynomial, and $s^n B(1/s)$ can be shown to have zeros in the left-half s plane. $B(1/j\omega)$ can be expressed in terms of Bessel functions [2, 7] as

$$B\left(\frac{1}{j\omega}\right) = \frac{1}{j^n} \sqrt{\frac{\pi\omega}{2}} [(-1)^n J_{-\nu}(\omega) - j J_\nu(\omega)] e^{j\omega}$$

where $\nu = n + \frac{1}{2}$ and

$$J_\nu(\omega) = \omega^\nu \sum_{i=0}^{\infty} \frac{(-1)^i \omega^{2i}}{2^{2i+\nu} i! \Gamma(\nu+i+1)} \tag{5.61}$$

[$\Gamma(\nu+i+1)$ is the gamma function]. Hence from Eq. (5.59)

$$|H(j\omega)|^2 = \frac{2B_0^2}{\pi\omega^{2n+1}[J_{-\nu}^2(\omega) + J_\nu^2(\omega)]}$$

$$\theta(\omega) = -\omega + \tan^{-1} \frac{(-1)^n J_\nu(\omega)}{J_{-\nu}(\omega)}$$

$$\tau = -\frac{d\theta(\omega)}{d\omega} = 1 - \frac{(-1)^n (J_{-\nu} J_\nu' - J_\nu J_{-\nu}')}{J_{-\nu}^2(\omega) + J_\nu^2(\omega)}$$

Alternatively, from the properties of Bessel functions and Eq. (5.61) [2]

$$|H(j\omega)|^2 = 1 - \frac{\omega^2}{2n-1} + \frac{2(n-1)\omega^4}{(2n-1)^2(2n-3)} + \cdots$$

$$\tau = 1 - \frac{\omega^{2n}}{B_0^2} |H(j\omega)|^2$$

Clearly, as $\omega \to 0$, $|H(j\omega)| \to 1$ and $\tau \to 1$. Furthermore, the first $n-1$ derivatives of τ with respect to ω^2 are zero if $\omega = 0$. This means that there is some frequency range $0 \le \omega < \omega_p$ for which the delay is approximately constant. On the other

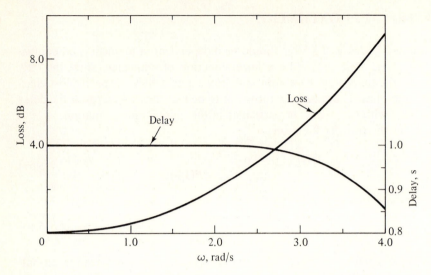

Figure 5.16 Loss and delay characteristics of fifth-order normalized Bessel filter (Example 5.4).

hand, if $\omega \to \infty$, $|H(j\omega)| \to 0$ and therefore $H(s)$ constitutes a lowpass constant-delay approximation. The delay is normalized at 1 s. However, any other delay can be achieved by replacing s by $\tau_0 s$ in Eq. (5.59).

Example 5.4 Form the Bessel transfer function for $n = 5$.

SOLUTION From Eq. (5.60)

$$H(s) = \frac{945}{945 + 945s + 420s^2 + 105s^3 + 15s^4 + s^5}$$

The corresponding loss and delay characteristics are depicted in Fig. 5.16.

5.7 TRANSFORMATIONS

In the preceding sections only normalized lowpass approximations have been considered. The reason is that denormalized lowpass, highpass, bandpass, and bandstop approximations can easily be derived by using transformations of the form

$$s = f(\bar{s})$$

Lowpass-to-Lowpass Transformation

Consider a normalized lowpass transfer function $H_N(s)$ with passband and stopband edges ω_p and ω_a, and let

$$s = \lambda \bar{s} \qquad (5.62)$$

in $H_N(s)$. If $s = j\omega$, $\bar{s} = j\omega/\lambda$ and hence Eq. (5.62) maps the j axis of the s plane onto the j axis of the $\bar{s}$ plane. In particular, ranges 0 to $j\omega_p$ and $j\omega_a$ to $j\infty$ map onto ranges 0 to $j\omega_p/\lambda$ and $j\omega_a/\lambda$ to $j\infty$, respectively, as depicted in Fig. 5.17a.

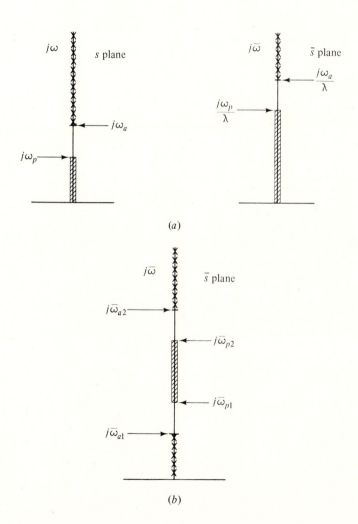

Figure 5.17 Analog-filter transformations: (*a*) lowpass to lowpass. (*b*) lowpass to bandpass.

Therefore

$$H_{LP}(\bar{s}) = H_N(s)\Big|_{s=\lambda\bar{s}}$$

constitutes a denormalized lowpass approximation with passband and stopband edges ω_p/λ and ω_a/λ.

Lowpass-to-Bandpass Transformation

Now let

$$s = \frac{1}{B}\left(\bar{s} + \frac{\omega_0^2}{\bar{s}}\right)$$

in $H_N(s)$, where B and ω_0 are constants. If $s = j\omega$ and $\bar{s} = j\bar{\omega}$, we have

$$j\omega = \frac{j}{B}\left(\bar{\omega} - \frac{\omega_0^2}{\bar{\omega}}\right) \quad \text{or} \quad j\bar{\omega} = j\left[\frac{\omega B}{2} \pm \sqrt{\omega_0^2 + \left(\frac{\omega B}{2}\right)^2}\right]$$

Hence

$$\bar{\omega} = \begin{cases} \omega_0 & \text{if } \omega = 0 \\ \pm\bar{\omega}_{p1}, \pm\bar{\omega}_{p2} & \text{if } \omega = \pm\omega_p \\ \pm\bar{\omega}_{a1}, \pm\bar{\omega}_{a2} & \text{if } \omega = \pm\omega_a \end{cases}$$

where

$$\bar{\omega}_{p1}, \bar{\omega}_{p2} = \mp\frac{\omega_p B}{2} + \sqrt{\omega_0^2 + \left(\frac{\omega_p B}{2}\right)^2}$$

$$\bar{\omega}_{a1}, \bar{\omega}_{a2} = \mp\frac{\omega_a B}{2} + \sqrt{\omega_0^2 + \left(\frac{\omega_a B}{2}\right)^2}$$

Table 5.2 Analog-filter transformations

Type	Transformation
LP to LP	$s = \lambda\bar{s}$
LP to HP	$s = \dfrac{\lambda}{\bar{s}}$
LP to BP	$s = \dfrac{1}{B}\left(\bar{s} + \dfrac{\omega_0^2}{\bar{s}}\right)$
LP to BS	$s = \dfrac{B\bar{s}}{\bar{s}^2 + \omega_0^2}$

The mapping for $s = j\omega$ is thus of the form illustrated in Fig. 5.17b, and consequently

$$H_{\text{BP}}(\bar{s}) = H_N(s)\bigg|_{s = \frac{1}{B}\left(\bar{s} + \frac{\omega_0^2}{\bar{s}}\right)}$$

is a bandpass approximation with passband edges ω_{p1}, ω_{p2} and stopband edges ω_{a1}, ω_{a2}.

Similarly, the transformations in the second and fourth rows of Table 5.2 yield highpass and bandstop approximations. The transformations can be readily applied by using Program B.9.

REFERENCES

1. E. A. Guillemin, "Synthesis of Passive Networks," Wiley, New York, 1957.
2. N. Balabanian, "Network Synthesis," Prentice-Hall, Englewood Cliffs, N.J., 1958.
3. L. Weinberg, "Network Analysis and Synthesis," McGraw-Hill, New York, 1962.
4. J. K. Skwirzynski, "Design Theory and Data for Electrical Filters," Van Nostrand, London, 1965.
5. R. W. Daniels, "Approximation Methods for Electronic Filter Design," McGraw-Hill, New York, 1974.
6. A. J. Grossman, Synthesis of Tchebyscheff Parameter Symmetrical Filters, *Proc. IRE*, vol. 45, pp. 454–473, April 1957.
7. G. N. Watson, "A Treatise on the Theory of Bessel Functions," Cambridge University Press, London, 1948.

PROBLEMS

5.1 A fourth-order normalized lowpass Butterworth filter is required.
    (a) Form $H(s)$.
    (b) Plot the loss characteristic of the filter.

5.2 Filter specifications are often described pictorially as in Fig. P5.2, where $\tilde{\Omega}_p$ and $\tilde{\Omega}_a$ are desired passband and stopband edges, respectively, A_p is the maximum passband loss, and A_a is the minimum stopband loss. Find n, and in turn form $H(s)$, if $\tilde{\Omega}_p = 1$, $\tilde{\Omega}_a = 3$ rad/s, $A_p = 3.0$, $A_a \geq 45$ dB. Use the Butterworth approximation.

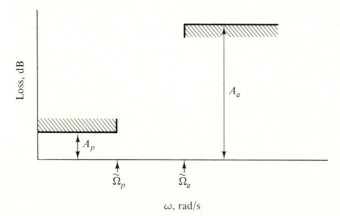

Figure P5.2

5.3 A fifth-order normalized lowpass Tschebyscheff filter is required.

    (*a*) Form $H(s)$ if $A_p = 0.1$ dB.

    (*b*) Plot the loss characteristic of the filter.

5.4 A Tschebyscheff filter satisfying the specifications of Fig. P5.4 is required. Find *n* and in turn form $H(s)$.

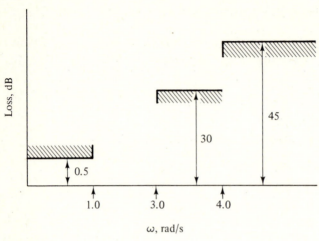

Figure P5.4

5.5 (*a*) Show that

$$T_{n+1}(\omega) = 2\omega T_n(\omega) - T_{n-1}(\omega)$$

    (*b*) Hence demonstrate that the following relation [5] holds:

$$T_n(\omega) = \frac{n}{2} \sum_{r=0}^{K} \frac{(-1)^r(n-r-1)!}{r!(n-2r)!} (2\omega)^{n-2r} \qquad \text{where } K = \text{Int } \frac{n}{2}$$

    (*c*) Obtain $T_{10}(\omega)$.

5.6 (*a*) Find $A(\omega)$ for the normalized lowpass Butterworth and Tschebyscheff approximations if $\omega \gg 1$.

    (*b*) Show that $A(\omega)$ increases at the rate of $20n$ dB/decade in both cases.

5.7 Prove that

    (*a*) $\quad \sqrt{k} \,\text{sn}\,(u + jK', k) = \dfrac{1}{\sqrt{k}\,\text{sn}\,(u, k)}$

    (*b*) $\quad \text{sn}\left(\dfrac{5K_1 u}{K} + jK_1', k_1\right) = \dfrac{1}{k_1 \,\text{sn}\,(5K_1 u/K, k_1)}$

    (*c*) $\quad \text{sn}\left(K_1 + \dfrac{jK_1'}{2}, k_1\right) = \dfrac{1}{\text{dn}\,(K_1'/2, k_1')} = \dfrac{1}{\sqrt{k_1}}$

    Parameter *u* is real in the range $0 \le u \le K$.

5.8 Show that

    (*a*) $\quad \cosh^{-1} x = \pm \ln\left(x + \sqrt{x^2 - 1}\right)$

    (*b*) $\quad \sinh^{-1} x = \ln\left(x + \sqrt{x^2 + 1}\right)$

5.9 Prove that

$$j\sqrt{k}(-1)^i \operatorname{sn}\left(jv_0 \pm \frac{2Ki}{n}, k\right) = \frac{(-1)^i \sigma_0 V_i \pm j\Omega_i W}{1 + \sigma_0^2 \Omega_i^2}$$

where W, V_i and Ω_i are given by Eqs. (5.34) to (5.36).

5.10 Check the derivation of Eqs. (5.37) and (5.39).

5.11 Show that

$$q \approx q_0 + 2q_0^5 + 15q_0^9 + 150q_0^{13} \qquad \text{where } q_0 = \frac{1}{2}\frac{1 - \sqrt{k'}}{1 + \sqrt{k'}}$$

5.12 (a) Sketch the plots of ω and F versus u for values of u in the range 0 to K, if $n = 7$ (Fig. 5.13).
(b) Repeat (a) for $n = 8$.

5.13 Check the derivation of Eqs. (5.56) and (5.57).

5.14 (a) A lowpass elliptic filter is required satisfying the specifications

$$n = 4 \qquad A_p = 1.0 \text{ dB} \qquad k = 0.7$$

Form $H(s)$.
(b) Determine the corresponding minimum stopband loss.
(c) Plot the loss characteristic.

5.15 In a particular application an elliptic lowpass filter is required. The specifications are

$$k = 0.6 \qquad A_p = 0.5 \text{ dB} \qquad A_a \geq 40 \text{ dB}$$

Find n and form $H(s)$.

5.16 (a) Obtain $H(s)$ for the sixth-order normalized Bessel approximation.
(b) Plot the corresponding phase characteristic.

5.17 Show that

$$H(s) = \frac{\sum_{i=0}^{n} B_i(-s)^i}{\sum_{i=0}^{n} B_i s^i}$$

where

$$B_i = \frac{(2n - i)!}{2^{n-i}i!(n - i)!}$$

is a constant-delay, allpass transfer function.

5.18 A normalized elliptic transfer function for which $k = 0.8$ and $A_p = 0.1$ dB is subjected to the lowpass-to-bandpass transformation. Find the resulting passband and stopband edges and also the passband width if $B = 200$, $\omega_0 = 1000$ rad/s.

5.19 A highpass elliptic filter is required satisfying the following specifications:

$$\tilde{\Omega}_p = 3000 \text{ rad/s} \qquad \tilde{\Omega}_a = 1000 \text{ rad/s} \qquad A_p = 0.5 \text{ dB} \qquad A_a \geq 40 \text{ dB}$$

(a) Find the necessary order n and the parameter of transformation λ.
(b) Hence form $H(s)$.

5.20 Figure P5.20 depicts a required bandpass-filter specification. Assuming that the elliptic approximation is to be employed, find ω_0, k, B, and n.

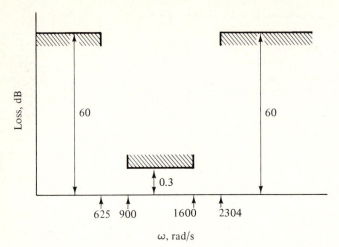

Loss, dB

60

60

0.3

625 900   1600   2304

ω, rad/s

Figure P5.20

5.21 (a) Derive a highpass LC filter from the lowpass filter shown in Fig. P5.21.
(b) Now derive a bandpass LC filter.

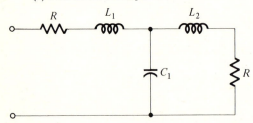

Figure P5.21

5.22 A constant-delay lowpass filter is required with a group delay of 1 ms. Form $H(s)$ using the sixth-order Bessel approximation.

5.23 Not considered in this chapter is the *inverse Tschebyscheff approximation*, in which the loss is given by

$$A(\omega) = 10 \log \left[1 + \frac{1}{\delta^2 T_n^2(\omega)} \right] \qquad \text{where } \delta^2 = \frac{1}{10^{0.1A_a} - 1}$$

(a) Show that $A(\omega)$ represents a normalized highpass filter with an equiripple stopband loss, a monotonic decreasing passband loss, and a stopband edge $\omega_a = 1$ rad/s.

(b) Show that the zeros and poles of $H(s)$ are given by

$$z_k = j\Omega_{\infty k} \qquad p_k = \sigma_k + j\omega_k$$

for $k = 1, 2, \ldots, n$ where

$$\Omega_{\infty k} = \cos \frac{(2k - 1)\pi}{2n}$$

$$\sigma_k = -\sinh \left(\frac{1}{n} \sinh^{-1} \frac{1}{\delta} \right) \sin \frac{(2k - 1)\pi}{2n}$$

$$\omega_k = \cosh \left(\frac{1}{n} \sinh^{-1} \frac{1}{\delta} \right) \cos \frac{(2k - 1)\pi}{2n}$$

5.24 Use the results of Prob. 5.23 to obtain the zeros and poles of a normalized ($\omega_a = 1$ rad/s) lowpass inverse Tschebyscheff transfer function.

CONTINUOUS-TIME, SAMPLED, AND DISCRETE-TIME SIGNALS

6.1 INTRODUCTION

So far the digital filter has been treated as a distinct entity with its own methods of analysis. In this chapter, various interrelations are established between continuous-time, sampled, and discrete-time signals. Through these interrelations many analog techniques can be used in the analysis and design of digital filters. In addition, digital filters can be used for the processing of analog signals.

The chapter begins with a brief review of the Fourier transform, which is the basis of the various interrelations.

6.2 THE FOURIER TRANSFORM

The Fourier transform of a function $f(t)$ is defined by

$$F(j\omega) = \int_{-\infty}^{\infty} f(t)e^{-j\omega t}\, dt \qquad (6.1)$$

In general $F(j\omega)$ is complex and can be written as

$$F(j\omega) = A(\omega)e^{j\phi(\omega)}$$

where $\qquad A(\omega) = |F(j\omega)| \qquad$ and $\qquad \phi(\omega) = \arg F(j\omega)$

The plots of $A(\omega)$ and $\phi(\omega)$ versus ω are called the *amplitude spectrum* and *phase spectrum* of $f(t)$, respectively.

The function $f(t)$ is said to be the *inverse Fourier transform* of $F(j\omega)$ and is given by

$$f(t) = \frac{1}{2\pi} \int_{-\infty}^{\infty} F(j\omega)e^{j\omega t} \, d\omega \qquad (6.2)$$

Equations (6.1) and (6.2) can be represented by

$$F(j\omega) = \mathscr{F}f(t) \qquad \text{and} \qquad f(t) = \mathscr{F}^{-1}F(j\omega)$$

respectively. An alternative notation combining both equations is

$$f(t) \leftrightarrow F(j\omega)$$

The salient properties of the Fourier transform are summarized here through a list of theorems. A thorough treatment of the topic can be found in Papoulis [1] and Lighthill [2].

Theorem 6.1: Convergence If

$$\lim_{T \to \infty} \int_{-T}^{T} |f(t)| \, dt < \infty$$

then the Fourier transform $F(j\omega)$ exists and satisfies Eq. (6.2).

This is a sufficient but not necessary condition for convergence; i.e., there are functions that are not absolutely integrable but have a Fourier transform satisfying Eq. (6.2).

The following theorems hold if

$$f(t) \leftrightarrow F(j\omega) \qquad \text{and} \qquad g(t) \leftrightarrow G(j\omega)$$

The parameters a, b, t_0, and ω_0 are arbitrary real constants.

Theorem 6.2: Linearity

$$af(t) + bg(t) \leftrightarrow aF(j\omega) + bG(j\omega)$$

Theorem 6.3: Symmetry If

$$f(t) \leftrightarrow F(j\omega)$$

then another transform pair can be generated as

$$F(jt) \leftrightarrow 2\pi f(-\omega)$$

Theorem 6.4: Time scaling

$$f(at) \leftrightarrow \frac{1}{|a|} F\left(\frac{j\omega}{a}\right)$$

Theorem 6.5: Time shifting

$$f(t - t_0) \leftrightarrow F(j\omega)e^{-j\omega t_0}$$

Theorem 6.6: Frequency shifting

$$e^{j\omega_0 t} f(t) \leftrightarrow F(j\omega - j\omega_0)$$

Theorem 6.7a: Time convolution If

$$f(t) \leftrightarrow F(j\omega) \qquad \text{and} \qquad g(t) \leftrightarrow G(j\omega)$$

then

$$\int_{-\infty}^{\infty} f(\tau)g(t - \tau) \, d\tau \leftrightarrow F(j\omega)G(j\omega)$$

If the input, output, and impulse response of a filter are denoted by $x(t)$, $y(t)$, and $h(t)$, respectively, it can be shown that

$$y(t) = \int_{-\infty}^{\infty} x(\tau)h(t - \tau) \, d\tau$$

Hence the above theorem gives

$$Y(j\omega) = H(j\omega)X(j\omega)$$

where $H(j\omega)$ is the transfer function of the filter.

Theorem 6.7b: Frequency convolution

$$f(t)g(t) \leftrightarrow \frac{1}{2\pi} \int_{-\infty}^{\infty} F(jv)G(j\omega - jv) \, dv$$

Theorem 6.8: Parseval's formula

$$\int_{-\infty}^{\infty} |f(t)|^2 \, dt = \frac{1}{2\pi} \int_{-\infty}^{\infty} |F(j\omega)|^2 \, d\omega$$

If $f(t)$ represents a voltage or current waveform, the left-hand integral represents the total energy that would be delivered to a 1-Ω resistor. Thus the quantity $|F(j\omega)|^2$ represents the energy density per unit bandwidth (in hertz), and is said to be the *energy spectral density*. A plot of $|F(j\omega)|^2$ versus ω is called the *energy spectrum* of $x(t)$.

Theorem 6.9: Poisson's summation formula (*a*) If $f(t)$ is an arbitrary function and

$$f(t) \leftrightarrow F(j\omega)$$

then
$$\sum_{n=-\infty}^{\infty} f(nT) = \frac{1}{T} \sum_{n=-\infty}^{\infty} F(jn\omega_s) \qquad \text{where } \omega_s = \frac{2\pi}{T}$$

(b) If $f(t) = 0$ for $t \leq 0-$, then

$$\sum_{n=0}^{\infty} f(nT) = \frac{f(0+)}{2} + \frac{1}{T} \sum_{n=-\infty}^{\infty} F(jn\omega_s) \qquad \text{where } f(0) \equiv f(0+)$$

Theorem 6.9b is actually a corollary of Theorem 6.9a.

Example 6.1 Find $F(j\omega)$ if

$$f(t) = p_{t_0}(t) = \begin{cases} 1 & \text{for } |t| \leq t_0 \\ 0 & \text{for } |t| > t_0 \end{cases}$$

SOLUTION
$$F(j\omega) = \int_{-\infty}^{\infty} p_{t_0}(t)e^{-j\omega t}\, dt = \int_{-t_0}^{t_0} e^{-j\omega t}\, dt$$

$$= \frac{2 \sin \omega t_0}{\omega}$$

or
$$p_{t_0}(t) \longleftrightarrow \frac{2 \sin \omega t_0}{\omega}$$

Function $p_{t_0}(t)$ and its amplitude spectrum are illustrated in Fig. 6.1.

Example 6.2 Find $F(j\omega)$ if

$$f(t) = \begin{cases} e^{-at} & \text{for } t \geq 0+ \\ 0 & \text{otherwise} \end{cases}$$

where a is a positive constant.

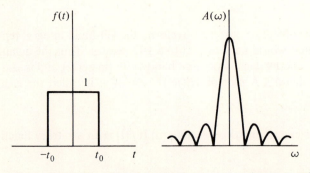

Figure 6.1 Function $f(t)$ and its amplitude spectrum (Example 6.1).

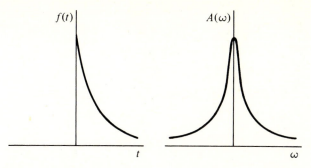

Figure 6.2 Function $f(t)$ and its amplitude spectrum (Example 6.2).

SOLUTION We can write

$$f(t) = u(t)e^{-at}$$

where $u(t)$ is the unit-step function defined by

$$u(t) = \begin{cases} 1 & \text{for } t \geq 0+ \\ 0 & \text{otherwise} \end{cases}$$

Hence

$$F(j\omega) = \int_{-\infty}^{\infty} [u(t)e^{-at}]e^{-j\omega t} \, dt = \int_{0}^{\infty} e^{-(a+j\omega)t} \, dt$$

$$= \frac{1}{a + j\omega}$$

or

$$u(t)e^{-at} \leftrightarrow \frac{1}{a + j\omega}$$

(see Fig. 6.2).

Example 6.3 Find $F(j\omega)$ if

$$f(t) = \begin{cases} e^{-at} \sin \omega_0 t & \text{for } t \geq 0+ \\ 0 & \text{otherwise} \end{cases}$$

and $a > 0$.

SOLUTION

$$f(t) = \frac{u(t)}{2j} (e^{j\omega o t} - e^{-j\omega o t})e^{-at}$$

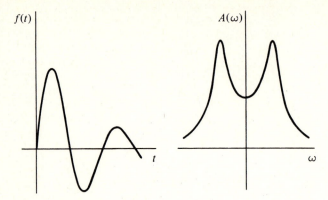

Figure 6.3 Function $f(t)$ and its amplitude spectrum (Example 6.3).

Hence from Theorem 6.6 and Example 6.2

$$F(j\omega) = \frac{1}{2j}\left[\frac{1}{a + j(\omega - \omega_0)} - \frac{1}{a + j(\omega + \omega_0)}\right] = \frac{\omega_0}{(a + j\omega)^2 + \omega_0^2}$$

or

$$u(t)e^{-at}\sin\omega_0 t \leftrightarrow \frac{\omega_0}{(a + j\omega)^2 + \omega_0^2}$$

(see Fig. 6.3).

6.3 GENERALIZED FUNCTIONS

Some of the most important functions in signal analysis have no Fourier transform in ordinary function theory; i.e., either $F(j\omega)$ does not exist or it does not satisfy Eq. (6.2), e.g., the function

$$p_0(t) = \lim_{\varepsilon \to 0} \begin{cases} \dfrac{1}{2\varepsilon} & \text{for } |t| \leq \varepsilon \\ 0 & \text{otherwise} \end{cases}$$

illustrated in Fig. 6.4a or any infinite-energy signal such as a periodic signal. This difficulty can be overcome in a somewhat subtle mathematical way by redefining $f(t)$ and $F(j\omega)$ as sequences of well-behaved functions called *generalized functions*. A brief outline of this theory based on Lighthill's approach [2] follows.

Definition 1 A function $g(t)$ is said to be a *good function* if it is everywhere differentiable any number of times such that

$$\lim_{t \to \infty} |t|^N \left| \frac{d^k g(t)}{dt^k} \right| < \infty$$

for all N. For example, e^{-t^2} is a good function.

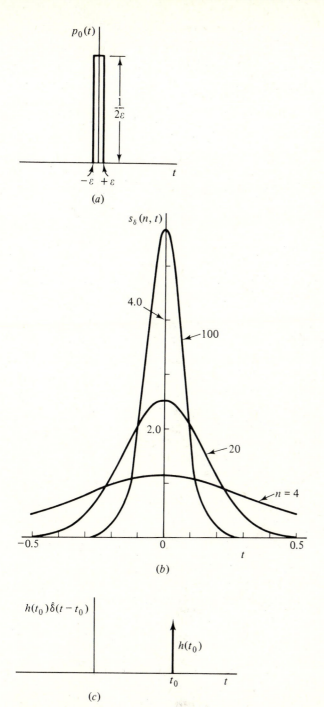

Figure 6.4 (a) Function $p_0(t)$. (b) A possible sequence of good functions defining $\hat{\delta}(t)$. (c) Simplified graph for an impulse.

Theorem 6.10 (*a*) The derivative of a good function is a good function.

(*b*) The sum and product of two good functions are good functions.

(*c*) The Fourier transform $F(j\omega)$ of a good function $f(t)$ exists, is a good function, and satisfies Eq. (6.2).

Definition 2 A sequence of good functions $s(n, t)$ is said to be *regular* if the limit

$$L = \lim_{n \to \infty} \int_{-\infty}^{\infty} s(n, t)\phi(t) \, dt \tag{6.3}$$

exists for any good function $\phi(t)$. For example, $s(n, t) = e^{-t^2/n^2}$ is regular since

$$\lim_{n \to \infty} \int_{-\infty}^{\infty} e^{-t^2/n^2}\phi(t) \, dt = \int_{-\infty}^{\infty} \phi(t) \, dt$$

Definition 3 Two or more sequences of good functions are called *equivalent* if for any good function $\phi(t)$ limit L in Eq. (6.3) exists and is the same for each sequence. For example, e^{-t^4/n^4} is equivalent to e^{-t^2/n^2}.

Definition 4 (*a*) A regular sequence $s_f(n, t)$ and each of its equivalent sequences define a unique generalized function $\hat{f}(t)$.

(*b*) The integral of $\hat{f}(t)\phi(t)$ between $-\infty$ and ∞, where $\phi(t)$ is any good function, is defined by

$$\int_{-\infty}^{\infty} \hat{f}(t)\phi(t) \, dt = \lim_{n \to \infty} \int_{-\infty}^{\infty} s_f(n, t)\phi(t) \, dt$$

This can serve as a means of testing two or more generalized functions for equality. Hence $\phi(t)$ is said to be a *test function*.

Impulse Function

A generalized impulse function $\hat{\delta}(t)$ is defined by

$$\int_{-\infty}^{\infty} \hat{\delta}(t)\phi(t) \, dt = \lim_{n \to \infty} \int_{-\infty}^{\infty} s_\delta(n, t)\phi(t) \, dt = \phi(0) \tag{6.4}$$

A possible sequence of good functions defining $\hat{\delta}(t)$ is

$$s_\delta(n, t) = \sqrt{\frac{n}{\pi}} e^{-nt^2} \tag{6.5}$$

and is illustrated in Fig. 6.4*b*. Some fundamental properties of $\hat{\delta}(t)$ follow readily from Eq. (6.4), for example,

$$\int_{-\infty}^{\infty} \hat{\delta}(t - t_0)\phi(t) \, dt = \int_{-\infty}^{\infty} \hat{\delta}(t)\phi(t + t_0) \, dt = \phi(t_0)$$

Also, for a good function $h(t)$

$$\int_{-\infty}^{\infty} \hat{\delta}(t - t_0)h(t)\phi(t) \, dt = h(t_0)\phi(t_0)$$

and

$$\int_{-\infty}^{\infty} \hat{\delta}(t - t_0)h(t_0)\phi(t) \, dt = h(t_0)\phi(t_0)$$

Therefore

$$h(t)\hat{\delta}(t - t_0) = h(t_0)\hat{\delta}(t - t_0) \tag{6.6}$$

The quantity on the right-hand side is an impulse of strength $h(t_0)$ located at $t = t_0$. It can be represented by the simplified graph of Fig. 6.4c.

Unity Function

The generalized function $\hat{i}(t)$, defined by

$$\int_{-\infty}^{\infty} \hat{i}(t)\phi(t) \, dt = \lim_{n \to \infty} \int_{-\infty}^{\infty} s_i(n, t)\phi(t) \, dt = \int_{-\infty}^{\infty} \phi(t) \, dt$$

has the effect of multiplying $\phi(t)$ by 1 and is the generalized function for unity. A possible sequence of good functions defining $\hat{i}(t)$ is

$$s_i(n, t) = e^{-t^2/4n} \tag{6.7}$$

This is illustrated in Fig. 6.5a. The generalized function $\hat{i}(t)$ can be represented by the simplified graph of Fig. 6.5b.

Definition 5 If $\hat{f}(t)$ and $\hat{g}(t)$ are defined by sequences $s_f(n, t)$ and $s_g(n, t)$, respectively, then $\hat{f}(t) + \hat{g}(t)$, $\hat{f}'(t)$, $\hat{f}(at + b)$, and $\hat{f}(t)\phi(t)$ are defined by sequences $s_f(n, t) + s_g(n, t)$, $s_f'(n, t)$, $s_f(n, at + b)$, and $s_f(n, t)\phi(t)$, respectively. Also the Fourier transform of $\hat{f}(t)$, namely

$$\mathscr{F}\hat{f}(t) = \hat{F}(j\omega)$$

is defined by

$$\mathscr{F}s_f(n, t) = S_f(n, j\omega)$$

Theorem 6.11 From Theorem 6.10c

$$s_f(n, t) \leftrightarrow S_f(n, j\omega)$$

Therefore, from Definition 5,

$$\hat{f}(t) \leftrightarrow \hat{F}(j\omega)$$

Example 6.4 Find $F(j\omega)$ for (a) $f(t) = \hat{\delta}(t)$ and (b) $f(t) = \hat{\delta}(t - t_0)$.

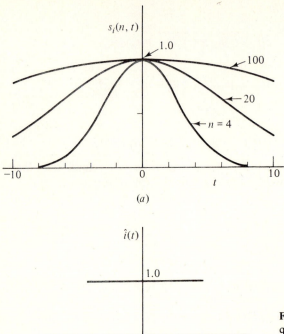

(a)

(b)

Figure 6.5 (a) A possible sequence of good functions defining $\hat{i}(t)$. (b) Simplified graph for the unity function.

SOLUTION (a) From Eq. (6.5)

$$S_\delta(n, j\omega) = \int_{-\infty}^{\infty} s_\delta(n, t) e^{-j\omega t}\, dt = \sqrt{\frac{n}{\pi}} \int_{-\infty}^{\infty} e^{-nt^2 - j\omega t} dt$$

and from standard tables

$$S_\delta(n, j\omega) = e^{-\omega^2/4n}$$

Hence from Eq. (6.7)

$$S_\delta(n, t) \leftrightarrow s_i(n, \omega)$$

and therefore

$$\hat{\delta}(t) \leftrightarrow \hat{i}(\omega)$$

or in a simplified notation

$$\delta(t) \leftrightarrow 1$$

(b) By Theorem 6.5

$$\delta(t - t_0) \leftrightarrow e^{-j\omega t_0}$$

(see Fig. 6.6).

Example 6.5 Find $F(j\omega)$ for (a) $f(t) = e^{j\omega_0 t}$ and (b) $f(t) = \cos \omega_0 t$.

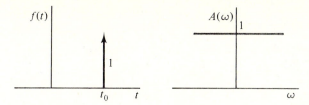

Figure 6.6 Function $f(t)$ and its amplitude spectrum (part (b) Example 6.4).

SOLUTION (a) From Theorem 6.3 and Example 6.4

$$e^{-jtt_0} \leftrightarrow 2\pi\delta(-\omega - t_0)$$

and if $-t_0 = \omega_0$,

$$e^{j\omega_0 t} \leftrightarrow 2\pi\delta(\omega - \omega_0)$$

(b) $$\mathscr{F} \cos \omega_0 t = \mathscr{F}\tfrac{1}{2}(e^{j\omega_0 t} + e^{-j\omega_0 t})$$

$$= \pi[\delta(\omega - \omega_0) + \delta(\omega + \omega_0)]$$

or $$\cos \omega_0 t \leftrightarrow \pi[\delta(\omega - \omega_0) + \delta(\omega + \omega_0)]$$

(see Fig. 6.7a and b).

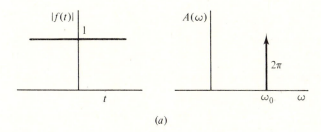

(a)

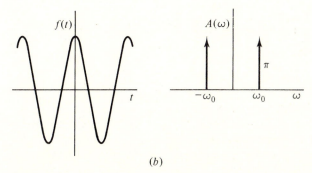

(b)

Figure 6.7 Function $f(t)$ and its amplitude spectrum: (a) part (a) and (b) part (b) Example 6.5.

Table 6.1 Fourier transforms

$f(t)$	$F(j\omega)$
$\delta(t)$	1
1	$2\pi\delta(\omega)$
$\delta(t - t_0)$	$e^{-j\omega t_0}$
$e^{j\omega_0 t}$	$2\pi\delta(\omega - \omega_0)$
$\cos \omega_0 t$	$\pi[\delta(\omega + \omega_0) + \delta(\omega - \omega_0)]$
$\sin \omega_0 t$	$j\pi[\delta(\omega + \omega_0) - \delta(\omega - \omega_0)]$
$u(t)e^{-at}$	$\dfrac{1}{a + j\omega}$
$u(t)e^{-at} \sin \omega_0 t$	$\dfrac{\omega_0}{(a + j\omega)^2 + \omega_0^2}$
$p_{t_0}(t) = \begin{cases} 1 & \lvert t\rvert \le t_0 \\ 0 & \lvert t\rvert > t_0 \end{cases}$	$\dfrac{2 \sin \omega t_0}{\omega}$
$\dfrac{\sin \omega_0 t}{\pi t}$	$p_{\omega_0}(\omega) = \begin{cases} 1 & \lvert\omega\rvert \le \omega_0 \\ 0 & \lvert\omega\rvert > \omega_0 \end{cases}$
$\sqrt{\dfrac{n}{\pi}}e^{-nt^2}$	$e^{-\omega^2/4n}$

The theory of generalized functions can be extended to any ordinary function $f(t)$ for which

$$\lim_{T \to \infty} \int_{-T}^{T} \left| \frac{f(t)}{(1 + t^2)^N} \right| < \infty$$

for some N [2].

A number of useful Fourier transform pairs can be found in Table 6.1.

6.4 SAMPLED SIGNALS

A sampled signal, denoted as $x^*(t)$, can be generated by sampling a continuous-time signal $x(t)$ using an ideal impulse sampler like that depicted in Fig. 6.8a. An impulse sampler is essentially a modulator characterized by the equation

$$x^*(t) = c(t)x(t) \tag{6.8}$$

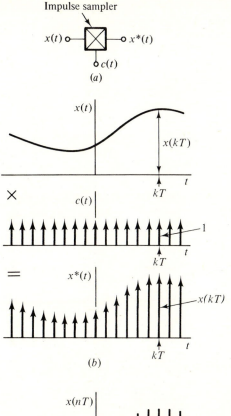

Figure 6.8 Generation of a sampled signal: (a) ideal impulse modulator, (b) sampled signal $x^*(t)$, (c) discrete-time signal $x(nT)$.

where $c(t)$ is a carrier given by

$$c(t) = \sum_{n=-\infty}^{\infty} \delta(t - nT)$$

Hence from Eqs. (6.6) and (6.8)

$$x^*(t) = \sum_{n=-\infty}^{\infty} x(nT)\delta(t - nT) \qquad (6.9)$$

In effect, a sampled signal is a sequence of impulses, like that illustrated in Fig. 6.8b. Note that a sampled signal can be converted into a discrete-time signal by replacing each impulse $x(nT)$ by a number $x(nT)$, as shown in Fig. 6.8c.

The Fourier transform of $x^*(t)$ is

$$X^*(j\omega) = \mathscr{F} \sum_{n=-\infty}^{\infty} x(nT)\delta(t-nT) = \sum_{n=-\infty}^{\infty} x(nT)\mathscr{F}\delta(t-nT)$$

$$= \sum_{n=-\infty}^{\infty} x(nT)e^{-j\omega nT} \tag{6.10}$$

Clearly

$$X^*(j\omega) = X_D(e^{j\omega T}) \tag{6.11}$$

where

$$X_D(z) = \mathscr{Z}x(nT)$$

i.e., the Fourier transform of sampled signal $x^*(t)$ is numerically equal to the z transform of the corresponding discrete-time signal $x(nT)$ evaluated on the unit circle $|z| = 1$.

Now consider the transform pair

$$x(t)e^{-j\Omega t} \leftrightarrow X(j\Omega + j\omega) \tag{6.12}$$

where

$$x(t) \leftrightarrow X(j\omega)$$

according to Theorem 6.6. On using Theorem 6.9a and then letting $\Omega = \omega$ we have

$$\sum_{n=-\infty}^{\infty} x(nT)e^{-j\omega nT} = \frac{1}{T} \sum_{n=-\infty}^{\infty} X(j\omega + jn\omega_s) \tag{6.13}$$

where $\omega_s = 2\pi/T$. Therefore, from Eqs. (6.11) and (6.13)

$$X^*(j\omega) = X_D(e^{j\omega T}) = \frac{1}{T} \sum_{n=-\infty}^{\infty} X(j\omega + jn\omega_s) \tag{6.14}$$

i.e., if the Fourier transform of $x(t)$ is known, that of $x^*(t)$ is uniquely determined. As is to be expected, $X^*(j\omega)$ is a periodic function of ω with period ω_s. Indeed

$$X^*(j\omega + jm\omega_s) = \frac{1}{T} \sum_{n=-\infty}^{\infty} X[j\omega + j(m+n)\omega_s] = \frac{1}{T} \sum_{n'=-\infty}^{\infty} X[j\omega + jn'\omega_s]$$

$$= X^*(j\omega)$$

Often $x(t) = 0$ for $t \leq 0-$. If such is the case, Eq. (6.9) becomes

$$x^*(t) = \sum_{n=0}^{\infty} x(nT)\delta(t-nT) \tag{6.15}$$

where $x(0) \equiv x(0+)$, and from Eqs. (6.10) and (6.11)

$$X^*(j\omega) = \sum_{n=0}^{\infty} x(nT)e^{-j\omega nT} = X_D(e^{j\omega T})$$

Now from Eq. (6.12) and Theorem 6.9b we deduce

$$X^*(j\omega) = X_D(e^{j\omega T}) = \frac{x(0+)}{2} + \frac{1}{T} \sum_{n=-\infty}^{\infty} X(j\omega + jn\omega_s) \tag{6.16}$$

With $j\omega = s$ and $e^{sT} = z$, Eqs. (6.14) and (6.16) become

$$X^*(s) = X_D(z) = \frac{1}{T} \sum_{n=-\infty}^{\infty} X(s + jn\omega_s) \tag{6.17}$$

$$X^*(s) = X_D(z) = \frac{x(0+)}{2} + \frac{1}{T} \sum_{n=-\infty}^{\infty} X(s + jn\omega_s) \tag{6.18}$$

where the transforms in the second equation are one-sided since $x(t) = 0$ for $t \le 0-$. The value of $x(0+)$ can be deduced from $X(s)$ as

$$x(0+) = \lim_{s \to \infty} [sX(s)]$$

by using the initial-value theorem of the Laplace transform.

Example 6.6 (a) Find $X^*(j\omega)$ if $x(t) = \cos \omega_0 t$. (b) Repeat part (a) for $x(t) = u(t)e^{-t}$.

SOLUTION (a) From Eq. (6.14) and Table 6.1

$$X^*(j\omega) = \frac{\pi}{T} \sum_{n=-\infty}^{\infty} [\delta(\omega + n\omega_s + \omega_0) + \delta(\omega + n\omega_s - \omega_0)]$$

(b) From Eq. (6.16) and Table 6.1

$$X^*(j\omega) = \frac{1}{2} + \frac{1}{T} \sum_{n=-\infty}^{\infty} \frac{1}{1 + j(\omega + n\omega_s)}$$

(see Fig. 6.9a and b).

6.5 THE SAMPLING THEOREM

The application of digital filters for the processing of continuous-time signals is made possible by the sampling theorem. This states that a bandlimited signal $x(t)$ for which

$$X(j\omega) = 0 \qquad \text{for } |\omega| \ge \frac{\omega_s}{2} \tag{6.19}$$

where $\omega_s = 2\pi/T$, can be uniquely determined from its values $x(nT)$.

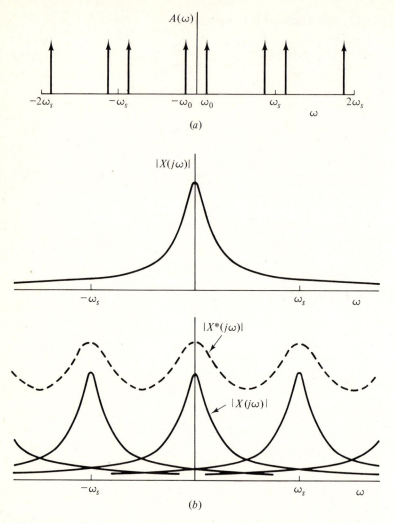

Figure 6.9 Amplitude spectrum of $x^*(t)$: (a) part (a) and (b) part (b) Example 6.6.

The validity of this theorem can easily be demonstrated. With Eq. (6.19) satisfied, $TX^*(j\omega)$ given by Eq. (6.14) as

$$TX^*(j\omega) = \sum_{n=-\infty}^{\infty} X(j\omega + jn\omega_s)$$

is a periodic continuation of $X(j\omega)$. Hence

$$X(j\omega) = H(j\omega)TX^*(j\omega) \tag{6.20}$$

where
$$H(j\omega) = \begin{cases} 1 & \text{for } |\omega| < \dfrac{\omega_s}{2} \\ 0 & \text{for } |\omega| \geq \dfrac{\omega_s}{2} \end{cases}$$

as depicted in Fig. 6.10. Thus from Eqs. (6.10) and (6.20) we can write

$$X(j\omega) = H(j\omega)T \sum_{n=-\infty}^{\infty} x(nT)e^{-j\omega nT}$$

and consequently

$$x(t) = \mathcal{F}^{-1}\left[H(j\omega)T \sum_{n=-\infty}^{\infty} x(nT)e^{-j\omega nT} \right]$$

$$= T \sum_{n=-\infty}^{\infty} x(nT)\mathcal{F}^{-1}[H(j\omega)e^{-j\omega nT}] \tag{6.21}$$

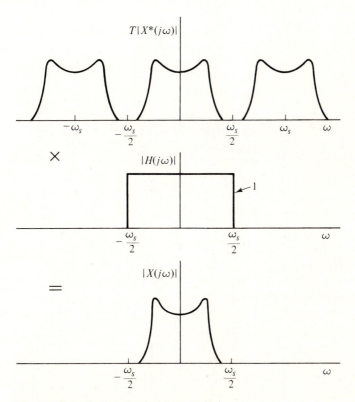

Figure 6.10 Derivation of $X(j\omega)$ from $X^*(j\omega)$.

Now from Table 6.1

$$\frac{\sin (\omega_s t/2)}{\pi t} \leftrightarrow H(j\omega)$$

and by Theorem 6.5

$$\frac{\sin [\omega_s(t - nT)/2]}{\pi(t - nT)} \leftrightarrow H(j\omega)e^{-j\omega nT}$$

Therefore, from Eq. (6.21)

$$x(t) = \sum_{n=-\infty}^{\infty} x(nT) \frac{\sin [\omega_s(t - nT)/2]}{\omega_s(t - nT)/2} \tag{6.22}$$

According to Eq. (6.20), signals $x^*(t)$ and $x(t)$ can be regarded as the input and output of a lowpass filter characterized by $TH(j\omega)$. In effect, the sampling theorem provides a means by which a sampled signal can be converted back into the original continuous-time signal.

Aliasing

If

$$X(j\omega) \neq 0 \qquad \text{for } |\omega| \geq \frac{\omega_s}{2}$$

as in the example of Fig. 6.9b, the tails of $X(j\omega)$ extend outside the baseband. As a result, $X^*(j\omega)$ (dotted curve in Fig. 6.9b) is no longer a periodic continuation of $X(j\omega)/T$, and the use of an ideal lowpass filter will at best yield a distorted version of $x(t)$. This effect is known as *aliasing* or *frequency folding*.

6.6 INTERRELATIONS

Various important interrelations have been established in the preceding sections between continuous-time, sampled, and discrete-time signals. They are illustrated pictorially in Fig. 6.11. The two-directional paths between $x^*(t)$ and $x(nT)$ and between $X^*(j\omega)$ and $X_D(z)$ render the Fourier transform applicable to digital filters. On the other hand, the two-directional paths between $x(t)$ and $x(nT)$ and between $X(j\omega)$ and $X_D(z)$ will allow us to use digital filters for the processing of continuous-time signals. Finally, the path between $X(s)$ and $X_D(z)$ will allow us to derive digital fiters from analog filters.

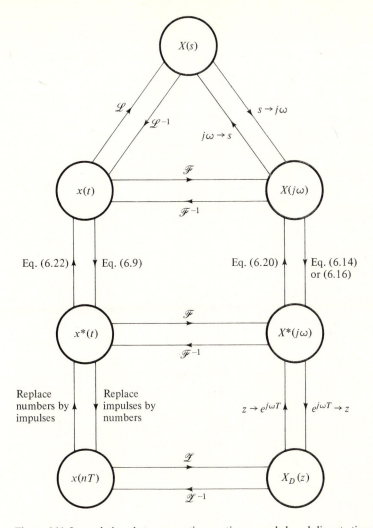

Figure 6.11 Interrelations between continuous-time, sampled, and discrete-time signals.

6.7 THE PROCESSING OF CONTINUOUS-TIME SIGNALS

Consider the filtering scheme of Fig. 6.12a, where S_1 and S_2 are impulse samplers, F_A is an analog filter characterized by $H_A(s)$, and F_{LP} is a lowpass filter for which

$$H_{LP}(s) = \begin{cases} T^2 & \text{for } |\omega| < \dfrac{\omega_s}{2} \\ 0 & \text{otherwise} \end{cases} \qquad (6.23)$$

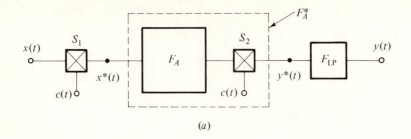

(a)

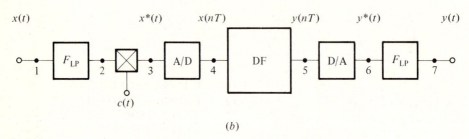

(b)

Figure 6.12 The processing of continuous-time signals: analog-filter configuration using (a) a sampled-data filter and (b) a digital filter.

F_A and S_2 constitute a sampled-data filter F_A^*. By analogy with Eqs. (6.15) and (6.18) the impulse response and transfer function of F_A^* can be expressed as

$$h_A^*(t) = \sum_{n=0}^{\infty} h_A(nT)\delta(t - nT)$$

and

$$H_A^*(s) = H_D(z) = \frac{h_A(0+)}{2} + \frac{1}{T} \sum_{n=-\infty}^{\infty} H_A(s + jn\omega_s) \qquad (6.24)$$

respectively, where

$$h_A(t) = \mathcal{L}^{-1} H_A(s) \qquad h_A(0+) = \lim_{s \to \infty} [sH_A(s)]$$

$$H_D(z) = \mathscr{Z} h_A(nT) \qquad z = e^{sT}$$

The Fourier transform of $y(t)$ in Fig. 6.12a is

$$Y(j\omega) = H_A^*(j\omega)H_{LP}(j\omega)X^*(j\omega) \qquad (6.25)$$

and if

$$x(0+) = h_A(0+) = 0$$

and

$$X(j\omega) = H_A(j\omega) = 0 \qquad \text{for } |\omega| \geq \frac{\omega_s}{2}$$

then $TX^*(j\omega)$ and $TH_A^*(j\omega)$ will be periodic continuations of $X(j\omega)$ and $H_A(j\omega)$, respectively, in which case

$$X^*(j\omega) = \frac{1}{T}X(j\omega) \left. \right\}$$

$$H_A^*(j\omega) = \frac{1}{T}H_A(j\omega) \left. \right\} \quad \text{for } |\omega| < \frac{\omega_s}{2}$$

Therefore, from Eqs. (6.23) and (6.25)

$$Y(j\omega) = H_A(j\omega)X(j\omega)$$

In effect, the response of the configuration will be identical with the response of filter F_A to excitation $x(t)$. We thus conclude that sampled-data filters can be used for the processing of continuous-time signals.

A sampled-data filter, like a digital filter, can be characterized by a discrete-time transfer function, according to Eq. (6.24). Hence, given a sampled-data filter, an equivalent digital filter can be derived and vice versa. Consequently, by implication, digital filters can be used for the processing of continuous-time signals. In addition, analog filters can be used to derive digital filters (see Chap. 7).

A digital-filter implementation of Fig. 6.12a can be obtained by replacing the sampled-data filter by a digital filter together with suitable interfacing devices, as shown in Fig. 6.12b. The analog-to-digital and digital-to-analog converters are required to convert impulses into numbers and numbers into impulses. The input lowpass filter is used to bandlimit $x(t)$ (if it is not already bandlimited) to prevent aliasing errors.

Example 6.7 The configuration of Fig. 6.12b is used to filter the periodic signal given by

$$x(t) = \begin{cases} \sin \omega_0 t & \text{for } 0 \leq \omega_0 t < \pi \\ 0 & \text{for } \pi \leq \omega_0 t < 2\pi \end{cases}$$

The lowpass filters are characterized by

$$H_{LP}(j\omega) = \begin{cases} 1 & \text{for } 0 \leq |\omega| < 6\omega_0 \\ 0 & \text{otherwise} \end{cases}$$

and the digital filter has a baseband response

$$H_D(e^{j\omega T}) = \begin{cases} 1 & \text{for } 0.95\omega_0 < |\omega| < 1.05\omega_0 \\ 0 & \text{otherwise} \end{cases}$$

Assuming that $\omega_s = 12\omega_0$, find the time- and frequency-domain representations of the signals at nodes 1, 2, ..., 7.

SOLUTION **Node 1** The Fourier series gives

$$x_1(t) = \sum_{k=-\infty}^{\infty} A_k e^{jk\omega_0 t}$$

where
$$A_k = \frac{1}{T_0} \int_0^{T_0} x_1(t) e^{-jk\omega_0 t} \, dt \qquad T_0 = \frac{2\pi}{\omega_0}$$

and hence $\quad A_0 = \frac{1}{\pi} \qquad A_1 = -A_{-1} = \frac{-j}{4}$

$$A_2 = A_{-2} = \frac{-1}{3\pi} \qquad A_3 = A_{-3} = 0$$

$$A_4 = A_{-4} = \frac{-1}{15\pi} \qquad A_5 = A_{-5} = 0 \qquad A_6 = A_{-6} = \frac{-1}{35\pi}$$

and so forth, or

$$x_1(t) = \frac{1}{\pi} + \frac{1}{2} \sin \omega_0 t - \frac{2}{3\pi} \cos 2\omega_0 t - \frac{2}{15\pi} \cos 4\omega_0 t$$

$$- \frac{2}{35\pi} \cos 6\omega_0 t - \cdots$$

and from Table 6.1

$$X_1(j\omega) = 2\pi \sum_{k=-\infty}^{\infty} A_k \delta(\omega - k\omega_0)$$

Node 2 The output of the bandlimiting filter is

$$X_2(j\omega) = 2\pi \sum_{k=-4}^{4} A_k \delta(\omega - k\omega_0)$$

in which case

$$x_2(t) = \frac{1}{\pi} + \frac{1}{2} \sin \omega_0 t - \frac{2}{3\pi} \cos 2\omega_0 t - \frac{2}{15\pi} \cos 4\omega_0 t$$

Nodes 3 and 4 The output of the sampler is obtained from Eq. (6.9) as

$$x_3^*(t) = \sum_{n=-\infty}^{\infty} \left(\frac{1}{\pi} + \frac{1}{2} \sin \omega_0 nT - \frac{2}{3\pi} \cos 2\omega_0 nT - \frac{2}{15\pi} \cos 4\omega_0 nT \right) \delta(t - nT)$$

Thus

$$x_4(nT) = \frac{1}{\pi} + \frac{1}{2} \sin \omega_0 nT - \frac{2}{3\pi} \cos 2\omega_0 nT - \frac{2}{15\pi} \cos 4\omega_0 nT$$

From Eq. (6.14) and Fig. 6.11

$$X_3^*(j\omega) = X_4(e^{j\omega T}) = \frac{2\pi}{T} \sum_{n=-\infty}^{\infty} \sum_{k=-4}^{4} A_k \delta(\omega + n\omega_s - k\omega_0)$$

Figure 6.13 Example 6.7: (a) time-domain representations of signals at nodes 1, 2, ..., 7; (b) amplitude spectra of signals at nodes 1, 2, ..., 7.

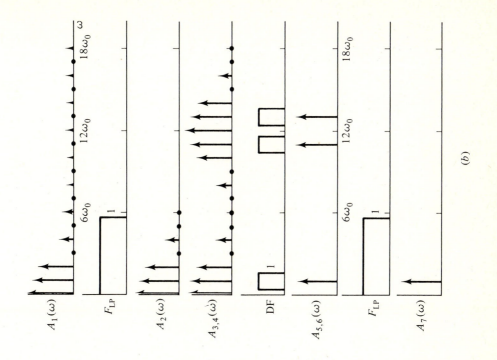

(b)

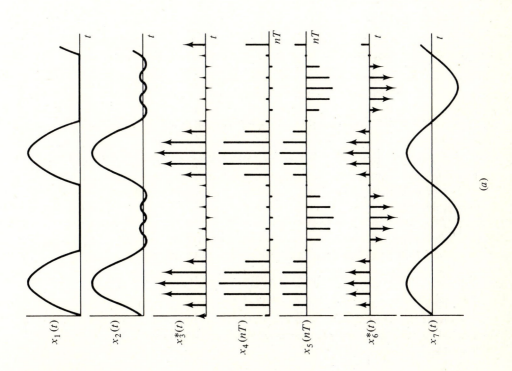

(a)

Nodes 5 and 6 The digital filter will reject all components except those with frequencies $\pm \omega_0 \pm n\omega_s$, and so

$$X_5(e^{j\omega T}) = X_6^*(j\omega) = \frac{2\pi}{T} \sum_{n=-\infty}^{\infty} A_1 \delta(\omega + n\omega_s - \omega_0) + A_{-1}\delta(\omega + n\omega_s + \omega_0)$$

$$= \frac{j\pi}{2T} \sum_{n=-\infty}^{\infty} [\delta(\omega + n\omega_s + \omega_0) - \delta(\omega + n\omega_s - \omega_0)]$$

Thus
$$x_6^*(t) = \frac{1}{2} \sum_{n=-\infty}^{\infty} \sin(\omega_0 nT)\delta(t - nT)$$

$$x_5(nT) = \tfrac{1}{2} \sin \omega_0 nT$$

Node 7 Finally, the output lowpass filter will reject all components with frequencies outside the baseband, and as a result

$$X_7(j\omega) = \frac{j\pi}{2T}[\delta(\omega + \omega_0) - \delta(\omega - \omega_0)]$$

and
$$x_7(t) = \frac{1}{2T} \sin \omega_0 t$$

The various signal waveforms and amplitude spectra are illustrated in Fig. 6.13a and b, respectively.

Practical Considerations

A practical implementation of the analog-to-digital interface is shown in Fig. 6.14a. The function of the sample-and-hold device is to generate a signal

$$\tilde{x}(t) = \sum_{n=-\infty}^{\infty} x(nT)p_{T/2}\left(t - nT - \frac{T}{2}\right)$$

like that illustrated in Fig. 6.14b. The function of the encoder, on the other hand, is to convert each signal level $x(nT)$ into a corresponding binary number. Since the number of bits in the binary representation must be finite, the response of the encoder denoted by $x_q(nT)$ can assume only a finite number of discrete levels; that is, $x_q(nT)$ will be a quantized signal. Assuming that the encoder is designed so that each value of $x(nT)$ is rounded to the nearest discrete level, the response of the encoder will be of the form depicted in Fig. 6.14b. We can write

$$x_q(nT) = x(nT) - e(nT)$$

where $e(nT)$ is the quantization error. Hence a practical A/D converter can be represented by the model of Fig. 6.14c, where $-e(nT)$ can be regarded as a noise source. The effect of this noise source on the filter response will be considered in Chap. 11.

Practical A/D converter

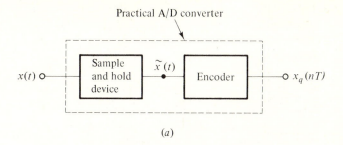

(a)

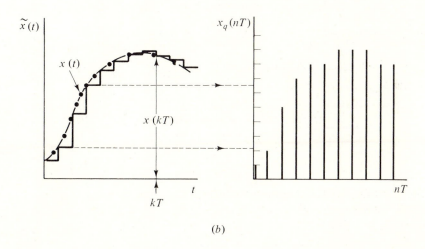

(b)

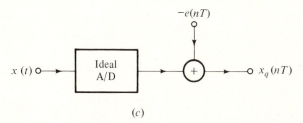

(c)

Figure 6.14 Analog-to-digital interface: (a) practical A/D converter, (b) response of a practical A/D converter, (c) model for a practical A/D converter.

The function of the D/A converter in Fig. 6.12b is to generate a sampled signal $y^*(t)$ like that in Fig. 6.15a. In a practical D/A converter, however, the response is of the form illustrated in Fig. 6.15b, where

$$\tilde{y}(t) = \sum_{n=-\infty}^{\infty} y(nT)p_{\tau/2}\left(t - nT - \frac{\tau}{2}\right)$$

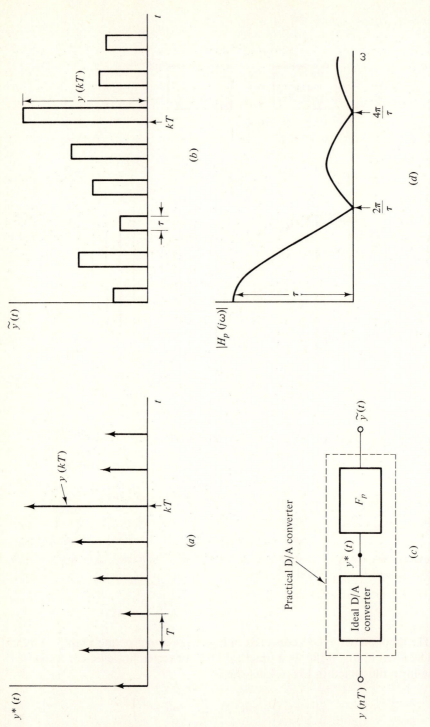

Figure 6.15 Digital-to-analog interface: (*a*) response of an ideal D/A converter, (*b*) response of a practical D/A converter, (*c*) model for a practical D/A converter, (*d*) amplitude response of fictitious filter F_p.

The Fourier transform of $\tilde{y}(t)$ is

$$\tilde{Y}(j\omega) = \sum_{n=-\infty}^{\infty} y(nT)\mathscr{F}p_{\tau/2}\left(t - nT - \frac{\tau}{2}\right)$$

$$= \frac{2\sin(\omega\tau/2)e^{-j\omega\tau/2}}{\omega}\sum_{n=-\infty}^{\infty} y(nT)e^{-j\omega nT}$$

and from Eq. (6.10)

$$\tilde{Y}(j\omega) = H_p(j\omega)Y^*(j\omega)$$

where

$$H_p(j\omega) = \frac{\tau\sin(\omega\tau/2)e^{-j\omega\tau/2}}{\omega\tau/2} \qquad Y^*(j\omega) = \mathscr{F}y^*(t)$$

Consequently, a practical D/A converter can be represented by an ideal D/A converter followed by a fictitious filter F_p characterized by $H_p(j\omega)$, as depicted in Fig. 6.15c. The amplitude response of this filter is given by

$$|H_p(j\omega)| = \tau\left|\frac{\sin(\omega\tau/2)}{\omega\tau/2}\right|$$

and is sketched in Fig. 6.15d. Clearly, a practical D/A converter will introduce distortion in the overall amplitude response. Fortunately, however, this effect can to a large extent be eliminated by designing the output lowpass filter so that

$$|H_p(j\omega)H_{\text{LP}}(j\omega)| \approx \begin{cases} 1 & \text{for } |\omega| \leq \dfrac{\omega_s}{2} \\ 0 & \text{otherwise} \end{cases}$$

The above effects and some others that may arise in the digital filtering of continuous-time signals are discussed by Stockham [3].

REFERENCES

1. A. Papoulis, "The Fourier Integral and Its Applications," McGraw-Hill, New York, 1962.
2. M. J. Lighthill, "Introduction to Fourier Analysis and Generalized Functions," Cambridge University Press, London, 1958.
3. T. G. Stockham Jr., A-D and D-A Converters: Their Effect on Digital Audio Fidelity, *1971 Proc. 41st Conv. Audio Eng. Soc., New York.*

PROBLEMS

6.1 Prove Theorems 6.3, 6.4, and 6.7a.

6.2 Show that for a real function $f(t)$

$$|F(j\omega)| = |F(-j\omega)| \qquad \text{and} \qquad \arg F(j\omega) = -\arg F(-j\omega)$$

6.3 (a) Show that for an even $f(t)$

$$\operatorname{Re} F(j\omega) = 2 \int_0^{\infty} f(t) \cos \omega t \, dt \qquad \operatorname{Im} F(j\omega) = 0$$

(b) Show that for an odd $f(t)$

$$\operatorname{Re} F(j\omega) = 0 \qquad \operatorname{Im} F(j\omega) = 2 \int_0^{\infty} f(t) \sin \omega t \, dt$$

6.4 Find the Fourier transforms of
   (a) $f(t) = u(t)e^{-at} \cos \omega_0 t$
   (b) $f(t) = p_\tau(t - nT - \tau)$

6.5 (a) Assuming that the Fourier transform of $d^n f(t)/dt^n$ exists, prove that

$$\frac{d^n f(t)}{dt^n} \leftrightarrow (j\omega)^n F(j\omega)$$

(b) Hence show that

$$\mathscr{F} \int_{-\infty}^{t} f(\tau) \, d\tau = \frac{F(j\omega)}{j\omega}$$

(c) Use this relation to obtain the Fourier transform of

$$\phi(t) = \begin{cases} 1 - \dfrac{|t|}{T} & \text{for } |t| < T \\ 0 & \text{otherwise} \end{cases}$$

6.6 An analog filter is characterized by the transfer function

$$H(j\omega) = \sum_{i=1}^{N} \frac{A_i}{j\omega - p_i}$$

Find its impulse response.

6.7 Signal $x(t)$ is periodic with period T_0. Show that

$$x(t) \leftrightarrow 2\pi \sum_{k=-\infty}^{\infty} A_k \delta(\omega - k\omega_0)$$

where $\omega_0 = 2\pi/T_0$ and

$$A_k = \frac{1}{T_0} \int_0^{T_0} x(t)e^{-jk\omega_0 t} \, dt$$

6.8 Find the Fourier transforms of the periodic signals shown in Fig. P6.8a to d. Sketch the amplitude spectra for the first two cases.

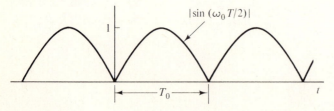

Figure P6.8(a)

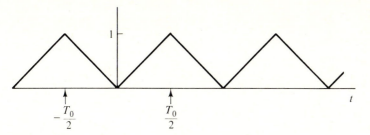

Figure P6.8(*b*)

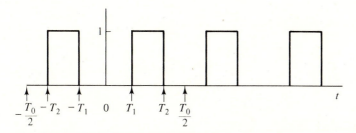

Figure P6.8(*c*)

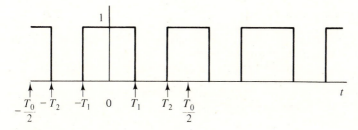

Figure P6.8(*d*)

6.9 Obtain the Fourier transforms of the following:

(*a*) $f(t) = \cos^2 \omega_0 t$

(*b*) $f(t) = \cos at^2$

(*c*) $f(t) = \begin{cases} \frac{1}{2}(1 + \cos \omega_0 t) & \text{for } |t| \le \dfrac{\pi}{\omega_0} \\ 0 & \text{otherwise} \end{cases}$

6.10 (*a*) Find the Fourier transform of $x*(t)$ in closed form if

$$x(t) = p_\tau(t)$$

where $\tau = (N - 1)T/2$. The sampling frequency is $\omega_s = 2\pi/T$.

(*b*) Repeat part (*a*) if

$$x(t) = \begin{cases} \alpha + (1 - \alpha) \cos \dfrac{\pi t}{\tau} & \text{for } |t| \le \tau \\ 0 & \text{otherwise} \end{cases}$$

6.11 Signal $x(t)$ is given by

$$x(t) = \begin{cases} 1 - \dfrac{|t|}{\tau} & \text{for } |t| \leq \tau \\ 0 & \text{otherwise} \end{cases}$$

where $\tau = (N-1)T/2$ and $T = 2\pi/\omega_s$. Show that

$$X^*(j\omega) \simeq \frac{8}{\omega^2(N-1)T^2}\sin^2\frac{\omega(N-1)T}{4} \qquad \text{for } |\omega| < \frac{\omega_s}{2}$$

if $N \gg \omega_s$.

6.12 The signal

$$x(t) = u(t)e^{-t}\cos 2t$$

is sampled at a rate of 2π rad/s.

(a) Show that

$$X^*(j\omega) = X_D(e^{j\omega T}) = \frac{1}{2} + \sum_{k=-\infty}^{\infty}\frac{1 + j(\omega + 2\pi k)}{[1 + j(\omega + 2\pi k)]^2 + 4}$$

(b) Demonstrate the validity of this relation by computation [first evaluate the right-hand summation and then evaluate the z transform of $x(nT)$ on the unit circle $|z| = 1$].

6.13 The filter of Example 6.7 is used to process the signal of Fig. P6.8b.

(a) Assuming that $\omega_0 = 2\pi/T_0$, find the time- and frequency-domain representations of the signals at nodes 1, 2, ..., 7.

(b) Sketch the various waveforms and amplitude spectra.

6.14 The configuration of Fig. 6.12b employs a digital filter with an amplitude response like that depicted in Fig. P6.14. The signal generated by the D/A converter is of the form shown in Fig. 6.15b, where $\tau = 3.0$ ms and the sampling frequency is 1000 rad/s.

(a) Assuming ideal lowpass filters and A/D converter, sketch the overall amplitude response of the configuration. Indicate relevant quantities.

(b) The gain at $\omega = 300$ rad/s is required to be equal to or greater than 0.99 times the gain at $\omega = 200$ rad/s. Find the maximum permissible value of τ.

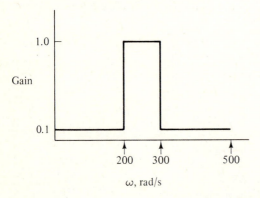

Figure P6.14

SEVEN

APPROXIMATIONS FOR RECURSIVE FILTERS

7.1 INTRODUCTION

As in analog filters, the approximation step in the design of digital filters is the process whereby a realizable transfer function satisfying prescribed specifications is obtained. Approximations methods differ radically between recursive and non-recursive filters.

Recursive-filter approximations can be obtained from analog-filter approximations using the following methods [1–9]:

1. Invariant-impulse-response method
2. Modified version of method 1
3. Matched-z transformation
4. Bilinear transformation

Nonrecursive-filter approximations, on the other hand, can be obtained by using Fourier series and numerical-analysis formulas.

This chapter considers the approximation problem for recursive filters. The realizability constraints imposed on the transfer function are first outlined. Then the details of the above approximation methods are discussed. Subsequently, a set of z-domain transformations is described which can be used to design filters with prescribed passband edges.

Nonrecursive-filter approximation techniques will be considered later, in Chap. 9.

7.2 REALIZABILITY CONSTRAINTS

In order to be realizable as a recursive filter, a transfer function must satisfy the following constraints:

1. It must be a rational function of z with real coefficients.
2. Its poles must lie within the unit circle of the z plane.
3. The degree of the numerator polynomial must be equal to or less than that of the denominator polynomial.

The coefficients must be real, so that inputs and outputs of unit delays, adders, and multipliers can be real numbers. The second and third constraints will ensure a stable and causal filter (see Chap. 3).

7.3 INVARIANT-IMPULSE-RESPONSE METHOD

Consider the sampled-data filter F_A^* of Fig. 7.1, where S is an ideal impulse sampler and F_A is an analog filter characterized by $H_A(s)$. F_A^* can be represented by a continuous-time transfer function $H_A^*(s)$ or, equivalently, by a discrete-time transfer function $H_D(z)$, as shown in Sec. 6.7. From Eq. (6.24)

$$H_A^*(j\omega) = H_D(e^{j\omega T}) = \frac{h_A(0+)}{2} + \frac{1}{T}\sum_{k=-\infty}^{\infty} H_A(j\omega + jk\omega_s) \tag{7.1}$$

where $\omega_s = 2\pi/T$ is the sampling frequency and

$$h_A(t) = \mathscr{L}^{-1}H_A(s)$$

$$h_A(0+) = \lim_{s\to\infty}[sH_A(s)] \tag{7.2}$$

$$H_D(z) = \mathscr{Z}h_A(nT)$$

Therefore, given an analog filter F_A, a corresponding digital filter, represented by $H_D(z)$, can be derived by using the following procedure:

1. Deduce $h_A(t)$, the impulse response of the analog filter.
2. Replace t by nT in $h_A(t)$.
3. Form the z transform of $h_A(nT)$.

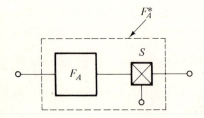

Figure 7.1 Sampled-data filter.

If

$$H_A(j\omega) \approx 0 \qquad \text{for } |\omega| \geq \frac{\omega_s}{2}$$

then

$$\sum_{\substack{k=-\infty \\ k \neq 0}}^{\infty} H_A(j\omega + jk\omega_s) \approx 0 \qquad \text{for } |\omega| \leq \frac{\omega_s}{2} \tag{7.3}$$

If, in addition,

$$h_A(0+) = 0 \tag{7.4}$$

Eqs. (7.1), (7.3), and (7.4) yield

$$H_A^*(j\omega) = H_D(e^{j\omega T}) \approx \frac{1}{T} H_A(j\omega) \qquad \text{for } |\omega| \leq \frac{\omega_s}{2} \tag{7.5}$$

i.e., if $H_A(j\omega)$ is bandlimited, the baseband frequency response of the derived digital filter is approximately the same as that of the analog filter (to within a multiplier constant $1/T$).

If the denominator degree in $H_A(s)$ exceeds the numerator degree by at least 2, the basic assumptions in Eqs. (7.3) and (7.4) hold for some value of ω_s. If, in addition, the poles of $H_A(s)$ are simple, we can write

$$H_A(s) = \sum_{i=1}^{N} \frac{A_i}{s - p_i} \tag{7.6}$$

Hence from steps 1 and 2 above

$$h_A(t) = \mathscr{L}^{-1} H_A(s) = \sum_{i=1}^{N} A_i e^{p_i t} \qquad \text{and} \qquad h_A(nT) = \sum_{i=1}^{N} A_i e^{p_i nT}$$

Subsequently, from step 3

$$H_D(z) = \mathscr{Z} h_A(nT) = \sum_{i=1}^{N} \frac{A_i z}{z - e^{T p_i}} \tag{7.7}$$

Since complex-conjugate pairs of poles in $H_A(s)$ yield complex-conjugate values of A_i and $e^{T p_i}$, the coefficients in $H_D(z)$ are real. Pole $p_i = \sigma_i + j\omega_i$ gives rise to a pole z_i in $H_D(z)$, where

$$z_i = e^{T p_i} = e^{T(\sigma_i + j\omega_i)}$$

and for $\sigma_i < 0$, $|z_i| < 1$. Hence a stable analog filter yields a stable digital filter. Also the denominator degree in $H_D(z)$ cannot exceed the numerator degree, and $H_D(z)$ is therefore realizable.

The invariant-impulse-response method yields good results for Butterworth, Bessel, or Tschebyscheff lowpass and bandpass filters for which the basic assumptions of Eqs. (7.3) and (7.4) hold. An advantage of the method is that it preserves the phase as well as the loss characteristic of the analog filter.

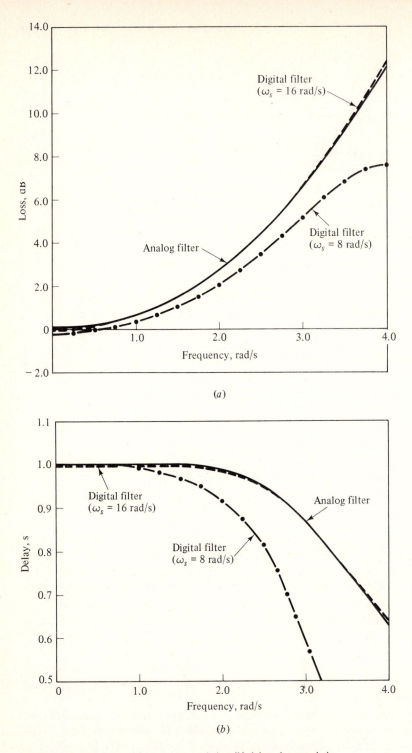

Figure 7.2 Example 7.1: (*a*) loss characteristics, (*b*) delay characteristics.

Table 7.1 Coefficients of $TH_D(z)$ (Example 7.1)

ω_s	j	A_{1j}	A_{2j}	B_{0j}	B_{1j}
8	1	6.452333×10^{-1}	2.612851	1.057399×10^{-2}	-1.597700×10^{-1}
	2	-8.345233×10^{-1}	-2.612851	3.671301×10^{-2}	1.891907×10^{-1}
16	1	3.114550×10^{-1}	1.306425	1.028299×10^{-1}	-6.045080×10^{-1}
	2	-3.790011×10^{-1}	-1.306425	1.916064×10^{-1}	-4.404794×10^{-1}

Example 7.1 Design a digital filter using the Bessel transfer function

$$H_A(s) = \frac{105}{105 + 105s + 45s^2 + 10s^3 + s^4}$$

(see Sec. 5.6). Employ a sampling frequency $\omega_s = 8$ rad/s; repeat with $\omega_s = 16$ rad/s.

SOLUTION The poles of $H_A(s)$ and the residues in Eq. (7.6) are

$$p_1, p_1^* = -2.896211 \pm j0.8672341$$
$$p_2, p_2^* = -2.103789 \pm j2.657418$$
$$A_1, A_1^* = 1.663392 \mp j8.396299$$
$$A_2, A_2^* = -1.663392 \pm j2.244076$$

Hence from Eq. (7.7)

$$TH_D(z) = \sum_{j=1}^{2} \frac{A_{1j}z + A_{2j}z^2}{B_{0j} + B_{1j}z + z^2}$$

where coefficients A_{ij} and B_{ij} are given in Table 7.1. The transfer function is multiplied by T to eliminate constant $1/T$ in Eq. (7.5). The loss and delay characteristics obtained are plotted in Fig. 7.2a and b. The higher sampling frequency gives better results because aliasing errors are less pronounced.

The invariant-impulse-response method can be readily applied by using Programs B.12 and B.13.

7.4 MODIFIED INVARIANT-IMPULSE-RESPONSE METHOD

Aliasing errors tend to restrict the invariant-impulse-response method to the design of allpole filters. A modified version of the method however, can be applied to filters with finite transmission zeros.

Consider the transfer function

$$H_A(s) = \frac{HN(s)}{D(s)} = \frac{H\prod_{i=1}^{M}(s - s_i)}{\prod_{i=1}^{N}(s - p_i)} \tag{7.8}$$

where M can be as high as N. We can write

$$H_A(s) = \frac{HH_{A1}(s)}{H_{A2}(s)}$$

where

$$H_{A1}(s) = \frac{1}{D(s)} \tag{7.9}$$

$$H_{A2}(s) = \frac{1}{N(s)} \tag{7.10}$$

Clearly, with $M, N \geq 2$ Eq. (7.2) yields

$$h_{A1}(0+) = 0 \qquad h_{A2}(0+) = 0$$

and furthermore

$$\left. \begin{array}{l} H_{A1}(j\omega) \approx 0 \\ H_{A2}(j\omega) \approx 0 \end{array} \right\} \quad \text{for } |\omega| \geq \frac{\omega_s}{2}$$

for some value of ω_s. Consequently, from Eq. (7.1) we can write

$$\left. \begin{array}{l} H_{A1}^*(j\omega) = H_{D1}(e^{j\omega T}) \approx \dfrac{1}{T} H_{A1}(j\omega) \\[2mm] H_{A2}^*(j\omega) = H_{D2}(e^{j\omega T}) \approx \dfrac{1}{T} H_{A2}(j\omega) \end{array} \right\} \quad \text{for } |\omega| \leq \frac{\omega_s}{2}$$

Therefore, we can form

$$H_D(z) = \frac{HH_{D1}(z)}{H_{D2}(z)} \tag{7.11}$$

such that

$$H_D(e^{j\omega T}) = \frac{HH_{D1}(e^{j\omega T})}{H_{D2}(e^{j\omega T})} \approx H_A(j\omega) \qquad \text{for } |\omega| \leq \frac{\omega_s}{2}$$

If the zeros and poles of $H_A(s)$ are simple, Eq. (7.7) gives

$$H_{D1}(z) = \sum_{i=1}^{N} \frac{A_i z}{z - e^{T p_i}} = \frac{N_1(z)}{D_1(z)} \tag{7.12}$$

$$H_{D2}(z) = \sum_{i=1}^{M} \frac{B_i z}{z - e^{T s_i}} = \frac{N_2(z)}{D_2(z)} \tag{7.13}$$

Thus from Eqs. (7.11) to (7.13)

$$H_D(z) = \frac{HN_1(z)D_2(z)}{N_2(z)D_1(z)} \qquad (7.14)$$

The derived filter can be unstable, as some of the zeros of $N_2(z)$ may be located on or outside the unit circle of the z plane, but the problem can easily be overcome. For a real pole of $H_D(z)$, say z_i

$$|e^{j\omega T} - z_i| = |z_i| \cdot \left| e^{-j\omega T} - \frac{1}{z_i} \right| = |z_i| \cdot \left| \left(e^{j\omega T} - \frac{1}{z_i} \right)^* \right|$$

$$= |z_i| \cdot \left| e^{j\omega T} - \frac{1}{z_i} \right|$$

and, similarly, for a pair of complex conjugate poles, say z_i and z_i^*

$$|(e^{j\omega T} - z_i)(e^{j\omega T} - z_i^*)| = |z_i|^2 \left| \left(e^{j\omega T} - \frac{1}{z_i} \right) \left(e^{j\omega T} - \frac{1}{z_i^*} \right) \right|$$

Hence poles of $H_D(z)$ outside the unit circle can be replaced by their reciprocals without changing the shape of the loss characteristic. Poles on the unit circle, on the other hand, can be adjusted by decreasing their magnitude slightly.

The method yields good results for lowpass and bandpass elliptic filters. Its disadvantage is that polynomials $N_1(z)$ and $N_2(z)$ increase the order of $H_D(z)$.

Example 7.2 The transfer function

$$H_A(s) = H \prod_{j=1}^{3} \frac{A_{0j} + s^2}{B_{0j} + B_{1j}s + s^2}$$

where H, A_{0j}, and B_{ij} are given in Table 7.2, represents a lowpass elliptic filter satisfying the following specifications:

Passband ripple: 0.1 dB
Minimum stopband loss: 43.46 dB
Passband edge: $\sqrt{0.8}$ rad/s
Stopband edge: $1/\sqrt{0.8}$ rad/s

Table 7.2 Coefficients of $H_A(s)$ (Example 7.2)

j	A_{0j}	B_{0j}	B_{1j}
1	1.199341×10	3.581929×10^{-1}	9.508335×10^{-1}
2	2.000130	6.860742×10^{-1}	4.423164×10^{-1}
3	1.302358	8.633304×10^{-1}	1.088749×10^{-1}

$H = 6.713267 \times 10^{-3}$

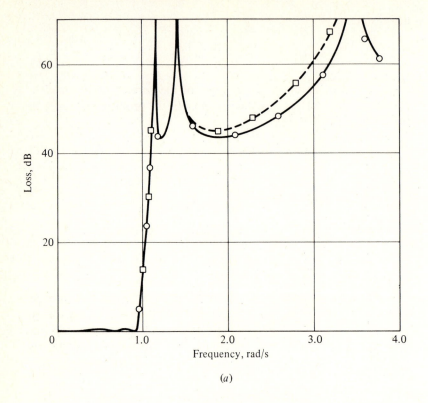

(a)

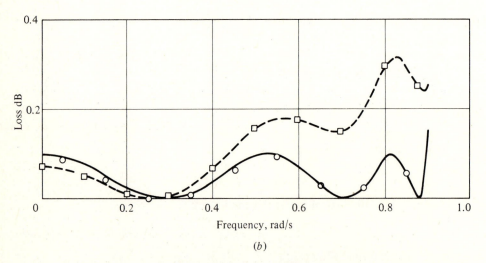

(b)

Figure 7.3 Examples 7.2 and 7.4: (a) loss characteristics, (b) passband characteristics. ————-analog filter, ○ ○ ○-modified impulse-invariant-response method, ---□---□---□----matched-z-transformation method.

Table 7.3 Coefficients of $H_D(z)$ (Example 7.2)

j	A_{0j}	A_{1j}	B_{0j}	B_{1j}
1	1.0	1.942528	4.508735×10^{-1}	-1.281134
2	1.0	-7.529504×10^{-1}	6.903517×10^{-1}	-1.303834
3	1.0	-1.153491	9.128252×10^{-1}	-1.362371
4	6.603146×10^{-1}	2.193514	4.821261×10^{-1}	1.388706
5	6.552540×10^{-1}	1.775846×10	6.530851×10^{-3}	1.616274×10^{-1}

$H = 3.846783 \times 10^{-4}$

Employing the modified invariant-impulse-response method, design a corresponding digital filter. Use $\omega_s = 7.5$ rad/s.

SOLUTION From Eqs. (7.9) and (7.10)

$$H_{A1}(s) = \prod_{j=1}^{3} \frac{1}{B_{0j} + B_{1j}s + s^2}$$

$$H_{A2}(s) = \prod_{j=1}^{3} \frac{1}{A_{0j} + s^2}$$

The design can be accomplished by using the following procedure:

1. Find the poles and residues of $H_{A1}(s)$ and $H_{A2}(s)$.
2. Form $H_{D1}(z)$ and $H_{D2}(z)$ using Eqs. (7.12) and (7.13).
3. Replace zeros of $N_2(z)$ outside the unit circle by their reciprocals.
4. Adjust constant H to achieve zero minimum passband loss.

With this procedure $H_D(z)$ can be deduced as

$$H_D(z) = H \prod_{j=1}^{5} \frac{A_{0j} + A_{1j}z + z^2}{B_{0j} + B_{1j}z + z^2}$$

where H, A_{ij}, and B_{ij} are given in Table 7.3.

The loss characteristic achieved, plotted in Fig. 7.3a and b, is seen to be a faithful reproduction of the analog loss characteristic. For this filter, the conventional invariant-impulse-response method gives unsatisfactory results because the assumptions of Eqs. (7.3) and (7.4) are violated.

7.5 MATCHED-z-TRANSFORMATION METHOD

An alternative approximation method for the design of recursive filters is the so-called *matched-z-transformation method* [5, 8]. In this method, given a continuous-time transfer function like that in Eq. (7.8), a corresponding discrete-time transfer function can be formed as

$$H_D(z) = (z + 1)^L \frac{H \prod_{i=1}^{M}(z - e^{s_i T})}{\prod_{i=1}^{N}(z - e^{p_i T})} \tag{7.15}$$

where L is an integer. The method is closely related to the modified invariant-impulse-response method. The difference between the two is that $N_1(z)/N_2(z)$ in Eq. (7.14) is replaced by $(z + 1)^L$. The value of L is equal to the number of zeros at $s = \infty$ in $H_A(s)$. Typical values for L are given in Table 7.4.

The method gives reasonable results for highpass and bandstop filters, although it tends to distort the passband ripple in Tschebyscheff and elliptic filters. For lowpass and bandpass filters, better approximations can be obtained by using the modified invariant-impulse-response method.

Example 7.3 The transfer function

$$H_A(s) = \frac{Hs^4}{\prod_{j=1}^{2}(s - p_j)(s - p_j^*)}$$

where $\qquad H = 0.9885531 \qquad p_1, p_1^* = -2.047535 \pm j1.492958$

$$p_2, p_2^* = -0.3972182 \pm j1.688095$$

represents a highpass Tschebyscheff filter with a passband edge of 2 rad/s and a passband ripple of 0.1 dB. Obtain a corresponding discrete-time transfer function employing the matched-z-transformation method. Use a sampling frequency of 10 rad/s.

SOLUTION The value of L in Eq. (7.15) is generally zero for highpass filters, according to Table 7.4. Hence $H_D(z)$ can be readily formed as

$$H_D(z) = H \frac{(1 - 2z + z^2)^2}{\prod_{j=1}^{2}(B_{0j} + B_{1j}z + z^2)}$$

Table 7.4 Typical values of L in Eq. (7.15)

Type of filter	Lowpass	Highpass	Bandpass	Bandstop
All-pole	N	0	$N/2$	0
Elliptic,				
$\quad N$ odd	1	0		
$\quad N$ even	0	0	1 for $N/2$ odd	0
			0 for $N/2$ even	

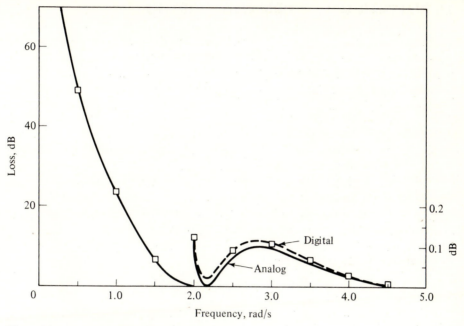

Figure 7.4 Loss characteristic (Example 7.3).

where $B_{01} = 7.630567 \times 10^{-2}$ $B_{11} = -3.267079 \times 10^{-1}$

$B_{02} = 6.070409 \times 10^{-1}$ $B_{12} = -7.608887 \times 10^{-1}$

$H = 2.076398 \times 10^{-1}$

(see Program B.14).

The loss characteristic of the derived filter is compared with that of the analog filter in Fig. 7.4.

Example 7.4 Redesign the lowpass filter of Example 7.2 employing the matched-z-transformation method.

SOLUTION From Eq. (7.15) $H_D(z)$ can be formed as

$$H_D(z) = H \prod_{j=1}^{3} \frac{A_{0j} + A_{1j}z + z^2}{B_{0j} + B_{1j}z + z^2}$$

where A_{ij} and B_{ij} are given by the first three rows in Table 7.3. For zero minimum passband loss H is given by

$$H = 8.604492 \times 10^{-3}$$

The loss characteristic achieved is shown in Fig. 7.3 (dotted curve). As can be seen, this is inferior to the loss characteristic obtained by using the modified invariant-impulse-response method

7.6 BILINEAR-TRANSFORMATION METHOD

In the approximation method of Sec. 7.3 the derived digital filter has exactly the same impulse response as the original analog filter for $t = nT$. An approximation method will now be described whereby a digital filter is derived which has approximately the same time-domain response as the original analog filter for any excitation.

Derivation

Consider an analog integrator characterized by the transfer function

$$H_I(s) = \frac{1}{s}$$

The impulse response of the integrator is

$$\mathscr{L}^{-1}H_I(s) = h_I(t) = \begin{cases} 1 & \text{for } t \geq 0+ \\ 0 & \text{for } t \leq 0- \end{cases}$$

and its response to an arbitrary excitation $x(t)$ is given by the convolution integral (see Theorem 6.7a)

$$y(t) = \int_0^t x(\tau)h_I(t - \tau)\, d\tau$$

If $0+ < t_1 < t_2$, we can write

$$y(t_2) - y(t_1) = \int_0^{t_2} x(\tau)h_I(t_2 - \tau)\, d\tau - \int_0^{t_1} x(\tau)h_I(t_1 - \tau)\, d\tau \tag{7.16}$$

For $0+ < \tau \leq t_1, t_2$

$$h_I(t_2 - \tau) = h_I(t_1 - \tau) = 1$$

and thus Eq. (7.16) simplifies to

$$y(t_2) - y(t_1) = \int_{t_1}^{t_2} x(\tau)\, d\tau$$

As $t_1 \to t_2$, from Fig. 7.5

$$y(t_2) - y(t_1) \approx \frac{t_2 - t_1}{2}[x(t_1) + x(t_2)]$$

and on letting $t_1 = nT - T$ and $t_2 = nT$ the difference equation

$$y(nT) - y(nT - T) = \frac{T}{2}[x(nT - T) + x(nT)] \tag{7.17}$$

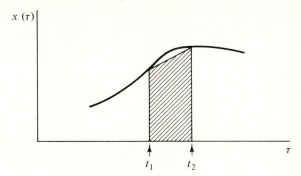

Figure 7.5 Response of analog integrator.

can be formed. This equation represents a digital integrator which has approximately the same time-domain response as the analog integrator for any excitation. From Eq. (7.17)

$$Y(z) - z^{-1}Y(z) = \frac{T}{2}[z^{-1}X(z) + X(z)]$$

and hence the transfer function of the digital integrator can be derived as

$$H_I(z) = \frac{Y(z)}{X(z)} = \frac{T}{2}\frac{z+1}{z-1}$$

An analog filter characterized by

$$H_A(s) = \frac{\displaystyle\sum_{i=0}^{N} a_i s^{N-i}}{s^N + \displaystyle\sum_{i=1}^{N} b_i s^{N-i}} \tag{7.18}$$

can be realized in terms of analog adders, multipliers, and integrators by using the configuration of Fig. 7.6. If each analog element in Fig. 7.6 is replaced by a corresponding digital element, a digital filter will result. If the digital integrator derived above is used, the transfer function of the resulting digital filter can be formed by applying the bilinear transformation

$$s = \frac{2}{T}\frac{z-1}{z+1} \tag{7.19}$$

to $H_A(s)$, that is,

$$H_D(z) = H_A(s)\Big|_{s = \frac{2}{T}\frac{z-1}{z+1}} \tag{7.20}$$

The derived digital filter will of course have approximately the same time-domain response as the original analog filter for any excitation.

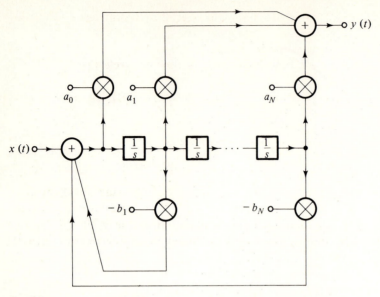

Figure 7.6 Realization of $H_A(s)$.

Mapping Properties of Bilinear Transformation

The relation between the frequency response of the derived digital filter and that of the original analog filter can be established by examining the mapping properties of the bilinear transformation.

Equation (7.19) can be put in the form

$$z = \frac{2/T + s}{2/T - s}$$

and with $s = \sigma + j\omega$ we have

$$z = re^{j\theta}$$

where

$$r = \left[\frac{\left(\dfrac{2}{T} + \sigma\right)^2 + \omega^2}{\left(\dfrac{2}{T} - \sigma\right)^2 + \omega^2} \right]^{1/2}$$

and

$$\theta = \tan^{-1} \frac{\omega}{2/T + \sigma} + \tan^{-1} \frac{\omega}{2/T - \sigma} \tag{7.21}$$

Clearly

$$r > 1 \qquad \text{for } \sigma > 0$$

$$r = 1 \qquad \text{for } \sigma = 0$$

$$r < 1 \qquad \text{for } \sigma < 0$$

i.e., the bilinear transformation maps

1. The open right-half s plane onto the region exterior to the unit circle $|z| = 1$ of the z plane
2. The j axis of the s plane onto the unit circle $|z| = 1$
3. The open left-half s plane onto the interior of the unit circle $|z| = 1$

For $\sigma = 0$, $r = 1$, and from Eq. (7.21) $\theta = 2 \tan^{-1}(\omega T/2)$. Hence

$$\theta = 0 \qquad \text{for } \omega = 0$$

$$\theta \to \pi \qquad \text{as } \omega \to +\infty$$

$$\theta \to -\pi \qquad \text{as } \omega \to -\infty$$

i.e., the origin of the s plane maps onto point $(1, 0)$ of the z plane and the positive and negative j axes of the s plane map onto the upper and lower semicircles $|z| = 1$, respectively. The transformation is illustrated in Fig. 7.7.

From property 2 above it follows that the maxima and minima of $|H_A(j\omega)|$ will be preserved in $|H_D(e^{j\Omega T})|$. Also if

$$M_1 \le |H_A(j\omega)| \le M_2$$

for some frequency range $\omega_1 \le \omega \le \omega_2$, then

$$M_1 \le |H_D(e^{j\Omega T})| \le M_2$$

for a corresponding frequency range $\Omega_1 \le \Omega \le \Omega_2$. Consequently, passbands or stopbands in the analog filter translate into passbands or stopbands in the digital filter.

From property 3 it follows that a stable analog filter will yield a stable digital filter, and since the transformation has real coefficients, $H_D(z)$ will have real coefficients. Finally, the numerator degree in $H_D(z)$ cannot exceed the denominator degree, and therefore $H_D(z)$ is a realizable transfer function.

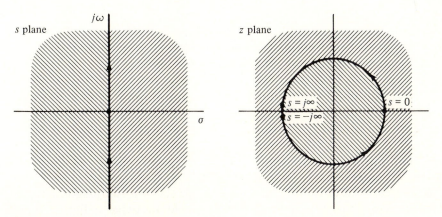

Figure 7.7 Bilinear transformation.

The Warping Effect

Let ω and Ω represent the frequency variable in the analog filter and the derived digital filter, respectively. From Eq. (7.20)

$$H_D(e^{j\Omega T}) = H_A(j\omega)$$

provided that

$$\omega = \frac{2}{T} \tan \frac{\Omega T}{2} \tag{7.22}$$

For $\Omega < 0.3/T$

$$\omega \approx \Omega$$

As a result, the digital filter will have the same amplitude response as the analog filter at approximately the same frequency. For higher frequencies, however, the relation between ω and Ω becomes nonlinear, as illustrated in Fig. 7.8, and distortion is introduced in the frequency scale of the digital filter relative to that of the analog filter. This is known as the *warping effect* [2, 5].

The influence of the warping effect on the amplitude response can be demonstrated by considering an analog filter with a number of passbands centered at regular intervals, as in Fig. 7.8. The derived digital filter will have the same number of passbands, but the center frequencies and bandwidths of higher-frequency passbands will tend to be reduced disproportionately, as shown in Fig. 7.8.

If only the amplitude response is of concern, the warping effect can for all practical purposes be eliminated by *prewarping* the analog filter [2, 5]. Let $\omega_1, \omega_2, \ldots, \omega_i, \ldots$ be the passband and stopband edges in the analog filter. The corresponding passband and stopband edges in the digital filter are given by Eq. (7.22) as

$$\Omega_i = \frac{2}{T} \tan^{-1} \frac{\omega_i T}{2} \qquad i = 1, 2, \ldots \tag{7.23}$$

Consequently, if prescribed passband and stopband edges $\tilde{\Omega}_1, \tilde{\Omega}_2, \ldots, \tilde{\Omega}_i, \ldots$ are to be achieved in the digital filter, the analog filter must be prewarped before application of the bilinear transformation to ensure that

$$\omega_i = \frac{2}{T} \tan \frac{\tilde{\Omega}_i T}{2} \tag{7.24}$$

Under these circumstances

$$\Omega_i = \tilde{\Omega}_i$$

according to Eqs. (7.23) and (7.24), as required.

The bilinear transformation together with the above prewarping technique are used in Chap. 8 to develop a detailed procedure for the design of Butterworth, Tschebyscheff, and elliptic filters satisfying prescribed loss specifications.

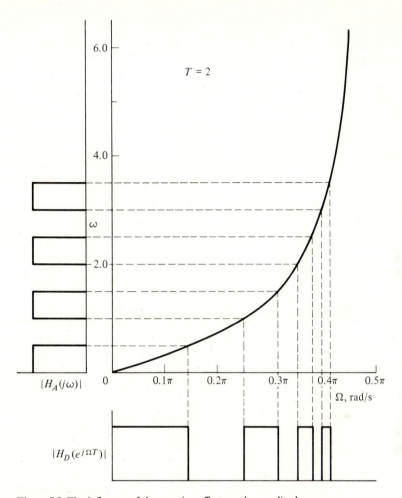

Figure 7.8 The influence of the warping effect on the amplitude response.

The influence of the warping effect on the phase response can be demonstrated by considering an analog filter with linear phase response. As illustrated in Fig. 7.9, the phase response of the derived digital filter will be nonlinear. Furthermore, little can be done to linearize it except by employing delay equalization (see Sec. 8.5). Consequently, if it is mandatory to preserve a linear phase response, the alternative methods of Sec. 7.3 to 7.5 should be considered.

Example 7.5 The transfer function

$$H_A(s) = \prod_{j=1}^{3} \frac{A_{0j} + s^2}{B_{0j} + B_{1j}s + s^2}$$

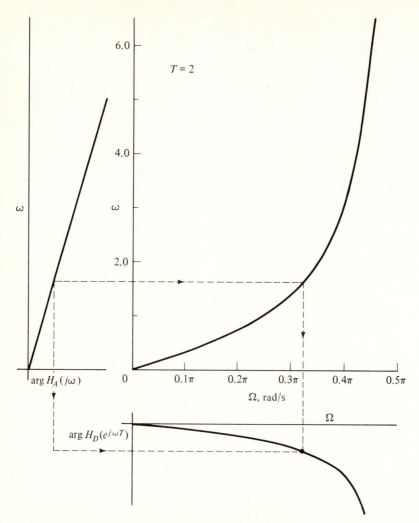

Figure 7.9 The influence of the warping effect on the phase response.

where A_{0j} are B_{ij} are given in Table 7.5, represents an elliptic bandstop filter with a passband ripple of 1 dB and a minimum stopband loss of 34.45 dB. Use the bilinear transformation to obtain a corresponding digital filter. Assume a sampling frequency of 10 rad/s.

Table 7.5 Coefficients of $H_A(s)$ (Example 7.5)

j	A_{0j}	B_{0j}	B_{1j}
1	6.250	6.250	2.618910
2	8.013554	1.076433×10	3.843113×10^{-1}
3	4.874554	3.628885	2.231394×10^{-1}

Table 7.6 Coefficients of $H_D(z)$ (Example 7.5)

j	A'_{0j}	A'_{1j}	B'_{0j}	B'_{1j}
1	6.627508×10^{-1}	-3.141080×10^{-1}	3.255016×10^{-1}	-3.141080×10^{-1}
2	8.203382×10^{-1}	-1.915542×10^{-1}	8.893929×10^{-1}	5.716237×10^{-2}
3	1.036997	-7.266206×10^{-1}	9.018366×10^{-1}	-8.987781×10^{-1}

SOLUTION From Eq. (7.20)

$$H_D(z) = \prod_{j=1}^{3} \frac{A'_{0j} + A'_{1j}z + A'_{0j}z^2}{B'_{0j} + B'_{1j}z + z^2}$$

where
$$A'_{0j} = \frac{A_{0j} + 4/T^2}{C_j} \qquad A'_{1j} = \frac{2(A_{0j} - 4/T^2)}{C_j}$$

$$B'_{0j} = \frac{B_{0j} - 2B_{1j}/T + 4/T^2}{C_j} \qquad B'_{1j} = \frac{2(B_{0j} - 4/T^2)}{C_j}$$

$$C_j = B_{0j} + \frac{2B_{1j}}{T} + \frac{4}{T^2}$$

The numerical values of A'_{ij} and B'_{ij} are given in Table 7.6 (see Program B.11). The loss characteristic of the derived digital filter is compared with that of the analog filter in Fig. 7.10. The expected lateral displacement in the characteristic of the digital filter is evident.

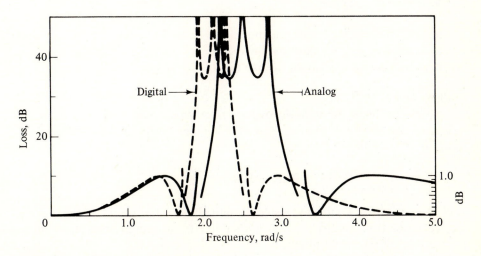

Figure 7.10 Loss characteristic (Example 7.5).

7.7 DIGITAL-FILTER TRANSFORMATIONS

A normalized lowpass analog filter can be transformed into a denormalized lowpass, highpass, bandpass, or bandstop filter by employing the transformations described in Sec. 5.7. Analogous transformations can be derived for digital filters as we shall now show. These were first proposed by Constantinides [10].

General Transformation

Consider the transformation

$$z = f(\bar{z}) = e^{j\zeta\pi} \prod_{i=1}^{m} \frac{\bar{z} - a_i^*}{1 - a_i \bar{z}} \tag{7.25}$$

where ζ and m are integers and a_i^* is the complex conjugate of a_i. With $z = Re^{j\Omega T}$, $\bar{z} = re^{j\omega T}$, and $a_i = c_i e^{j\psi_i}$ Eq. (7.25) becomes

$$Re^{j\Omega T} = e^{j\zeta\pi} \prod_{i=1}^{m} \frac{re^{j\omega T} - c_i e^{-j\psi_i}}{1 - rc_i e^{j(\omega T + \psi_i)}}$$

and hence
$$R^2 = \prod_{i=1}^{m} \frac{r^2 + c_i^2 - 2rc_i \cos(\omega T + \psi_i)}{1 + (rc_i)^2 - 2rc_i \cos(\omega T + \psi_i)} \tag{7.26}$$

Equation (7.26) yields

$$r^2 + c_i^2 > 1 + (rc_i)^2 \qquad \text{or} \qquad r > 1 \qquad \text{if } R > 1$$

$$r^2 + c_i^2 = 1 + (rc_i)^2 \qquad \text{or} \qquad r = 1 \qquad \text{if } R = 1$$

$$r^2 + c_i^2 < 1 + (rc_i)^2 \qquad \text{or} \qquad r < 1 \qquad \text{if } R < 1$$

In effect, Eq. (7.25) maps

1. The unit circle $|z| = 1$ onto the unit circle $|\bar{z}| = 1$
2. The interior of $|z| = 1$ onto the interior of $|\bar{z}| = 1$
3. The exterior of $|z| = 1$ onto the exterior of $|\bar{z}| = 1$

as illustrated in Fig. 7.11.

Now consider a normalized lowpass filter characterized by $H_N(z)$ with a passband extending from 0 to Ω_p. On applying the above transformation we can form

$$H(\bar{z}) = H_N(z)\Big|_{z=f(\bar{z})} \tag{7.27}$$

With the poles of $H_N(z)$ located inside the unit circle $|z| = 1$ those of $H(\bar{z})$ will be located inside the unit circle $|\bar{z}| = 1$; that is, $H(\bar{z})$ will represent a stable filter. Furthermore, from item 1 above, if

$$M_1 \leq |H_N(e^{j\Omega T})| \leq M_2$$

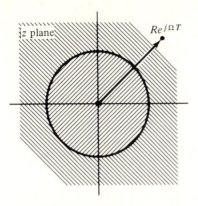
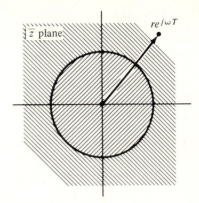

Figure 7.11 General z-domain transformation.

for some frequency range $\Omega_1 \leq \Omega \leq \Omega_2$, then

$$M_1 \leq |H(e^{j\omega T})| \leq M_2$$

for one or more corresponding ranges of ω; that is, the passband (stopband) in $H_N(z)$ will translate into one or more passbands (stopbands) in $H(\bar{z})$. Therefore, the above transformation can form the basis of a set of transformations which can be used to derive denormalized lowpass, highpass, bandpass, and bandstop filters from a given normalized lowpass filter.

Lowpass-to-Lowpass Transformation

The appropriate values for ζ, m, and a_i in Eq. (7.25) can be determined by examining the details of the necessary mapping. If $H(\bar{z})$ is to represent a lowpass filter with a passband edge ω_p, the mapping must be of the form shown in Fig. 7.12a, where solid lines denote passbands. As $e^{j\Omega T}$ traces the unit circle in the z plane once, $e^{j\omega T}$ must trace the unit circle in the $\bar{z}$ plane once in the same sense. The transformation must thus be bilinear ($m = 1$) of the form

$$z = e^{j\zeta\pi} \frac{\bar{z} - a^*}{1 - a\bar{z}} \tag{7.28}$$

At points A and A', $z = \bar{z} = 1$, and at C and C', $z = \bar{z} = -1$. Hence Eq. (7.28) gives

$$1 = e^{j\zeta\pi} \frac{1 - a^*}{1 - a} \quad \text{and} \quad 1 = e^{j\zeta\pi} \frac{1 + a^*}{1 + a}$$

By solving these equations we obtain

$$a = a^* \equiv \alpha \qquad \zeta = 0$$

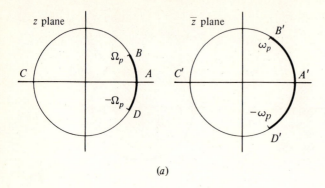

(a)

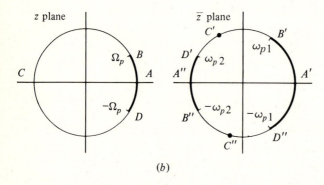

(b)

Figure 7.12 (a) Lowpass-to-lowpass transformation, (b) lowpass-to-bandstop transformation.

where α is a real constant. Thus Eq. (7.28) becomes

$$z = \frac{\bar{z} - \alpha}{1 - \alpha \bar{z}}$$

The necessary value for α can be determined by noting that at points B and B' $\Omega = \Omega_p$ and $\omega = \omega_p$, in which case

$$e^{j\Omega_p T} = \frac{e^{j\omega_p T} - \alpha}{1 - \alpha e^{j\omega_p T}}$$

or

$$\alpha = \frac{\sin\left[(\Omega_p - \omega_p)T/2\right]}{\sin\left[(\Omega_p + \omega_p)T/2\right]}$$

Lowpass-to-Bandstop Transformation

If a bandstop filter is required with passband edges ω_{p1} and ω_{p2}, the mapping must have the form shown in Fig. 7.12b. In order to introduce an upper passband in $H(\bar{z})$, $e^{j\Omega T}$ must trace the unit circle of the z plane twice for each revolution of

$e^{j\omega T}$ in the $\bar{z}$ plane. Consequently, in this case, the transformation must be biquadratic ($m = 2$) of the form

$$z = e^{j\zeta\pi} \frac{\bar{z}^2 + \beta\bar{z} + \gamma}{1 + \beta\bar{z} + \gamma\bar{z}^2}$$

where β and γ are real constants. At points A and A', $z = \bar{z} = 1$ and

$$e^{j\zeta\pi} = 1$$

so that

$$z = \frac{\bar{z}^2 + \beta\bar{z} + \gamma}{1 + \beta\bar{z} + \gamma\bar{z}^2}$$

With $z = e^{j\Omega T}$ and $\bar{z} = e^{j\omega T}$

$$e^{j\Omega T} = \frac{e^{j2\omega T} + \beta e^{j\omega T} + \gamma}{e^{j2\omega T}(e^{-j2\omega T} + \beta e^{-j\omega T} + \gamma)}$$

Hence

$$\frac{\Omega T}{2} = \tan^{-1} \frac{\sin 2\omega T + \beta \sin \omega T}{\cos 2\omega T + \beta \cos \omega T + \gamma} - \omega T$$

and after some manipulation

$$\tan \frac{\Omega T}{2} = \frac{(1 - \gamma) \sin \omega T}{(1 + \gamma) \cos \omega T + \beta}$$

At points B and B', $\Omega = \Omega_p$ and $\omega = \omega_{p1}$, respectively, and as a result

$$\tan \frac{\Omega_p T}{2} = \frac{(1 - \gamma) \sin \omega_{p1} T}{(1 + \gamma) \cos \omega_{p1} T + \beta} \qquad (7.29)$$

Likewise, at points D and D', $\Omega = -\Omega_p$ and $\omega = \omega_{p2}$ so that

$$\tan \frac{-\Omega_p T}{2} = \frac{(1 - \gamma) \sin \omega_{p2} T}{(1 + \gamma) \cos \omega_{p2} T + \beta} \qquad (7.30)$$

Now by solving Eqs. (7.29) and (7.30), β and γ can be deduced as

$$\beta = \frac{-2\alpha}{1 + k} \qquad \gamma = \frac{1 - k}{1 + k}$$

where $\quad \alpha = \dfrac{\cos\left[(\omega_{p2} + \omega_{p1})T/2\right]}{\cos\left[(\omega_{p2} - \omega_{p1})T/2\right]} \quad$ and $\quad k = \tan \dfrac{\Omega_p T}{2} \tan \dfrac{(\omega_{p2} - \omega_{p1})T}{2}$

Lowpass-to-highpass and lowpass-to-bandpass transformations can similarly be derived. The complete set of transformations is summarized in Table 7.7.

Table 7.7 Constantinides transformations

Type	Transformation	α, k
LP to LP	$z = \dfrac{\bar{z} - \alpha}{1 - \alpha\bar{z}}$	$\alpha = \dfrac{\sin\,[(\Omega_p - \omega_p)T/2]}{\sin\,[(\Omega_p + \omega_p)T/2]}$
LP to HP	$z = -\dfrac{\bar{z} - \alpha}{1 - \alpha\bar{z}}$	$\alpha = \dfrac{\cos\,[(\Omega_p - \omega_p)T/2]}{\cos\,[(\Omega_p + \omega_p)T/2]}$
LP to BP	$z = -\dfrac{\bar{z}^2 - \dfrac{2\alpha k}{k+1}\bar{z} + \dfrac{k-1}{k+1}}{1 - \dfrac{2\alpha k}{k+1}\bar{z} + \dfrac{k-1}{k+1}\bar{z}^2}$	$\alpha = \dfrac{\cos\,[(\omega_{p2} + \omega_{p1})T/2]}{\cos\,[(\omega_{p2} - \omega_{p1})T/2]}$ $k = \tan\dfrac{\Omega_p T}{2}\cot\dfrac{(\omega_{p2} - \omega_{p1})T}{2}$
LP to BS	$z = \dfrac{\bar{z}^2 - \dfrac{2\alpha}{1+k}\bar{z} + \dfrac{1-k}{1+k}}{1 - \dfrac{2\alpha}{1+k}\bar{z} + \dfrac{1-k}{1+k}\bar{z}^2}$	$\alpha = \dfrac{\cos\,[(\omega_{p2} + \omega_{p1})T/2]}{\cos\,[(\omega_{p2} - \omega_{p1})T/2]}$ $k = \tan\dfrac{\Omega_p T}{2}\tan\dfrac{(\omega_{p2} - \omega_{p1})T}{2}$

Application

The Constantinides transformations can readily be applied to design filters with prescribed passband edges. The following procedure can be employed:

1. Obtain a lowpass transfer function $H_N(z)$ using any approximation method.
2. Determine the passband edge Ω_p in $H_N(z)$.
3. Form $H(\bar{z})$ according to Eq. (7.27) using the appropriate transformation.

An important feature of filters designed by using the above procedure is that the passband edge in lowpass or highpass filters can be varied by varying a single parameter, namely α. Similarly, both the lower and upper passband edges in bandpass or bandstop filters can be varied by varying only a pair of parameters, namely α and k [14, 15].

An alternative design procedure by which prescribed passband as well as stopband edges can be achieved is described in Chap. 8.

REFERENCES

1. J. F. Kaiser, Design Methods for Sampled Data Filters, *Proc. 1st Allerton Conf. Circuit Syst. Theory*, pp. 221–236, November 1963.
2. R. M. Golden and J. F. Kaiser, Design of Wideband Sampled-Data Filters, *Bell Syst. Tech. J.*, vol. 43, pp. 1533–1546, July 1964.

3. C. M. Rader and B. Gold, Digital Filter Design Techniques in the Frequency Domain, *Proc. IEEE*, vol. 55, pp. 149–171, February 1967.
4. D. J. Nowak and P. E. Schmid, Introduction to Digital Filters, *IEEE Trans. Electromagn. Compat.*, vol. EMC-10, pp. 210–220, June 1968.
5. R. M. Golden, Digital Filter Synthesis by Sampled-Data Transformation, *IEEE Trans. Audio Electroacoust.*, vol. AU-16, pp. 321–329, September 1968.
6. A. J. Gibbs, An Introduction to Digital Filters, *Aust. Telecommun. Res.*, vol. 3, pp. 3–14, November 1969.
7. A. J. Gibbs, The Design of Digital Filters, *Aust. Telecommun. Res.*, vol. 4, pp. 29–34, March 1970.
8. R. L. Rabiner and B. Gold, "Theory and Application of Digital Signal Processing," Prentice-Hall, Englewood Cliffs, N.J., 1975.
9. A. V. Oppenheim and R. W. Schafer, "Digital Signal Processing," Prentice-Hall, Englewood Cliffs, N.J., 1975.
10. A. G. Constantinides, Spectral Transformations for Digital Filters, *Proc. IEE*, vol. 117, pp. 1585–1590, August 1970.
11. S. S. Haykin and R. Carnegie, New Method of Synthetising Linear Digital Filters Based on Convolution Integral, *Proc. IEE*, vol. 117, pp. 1063–1072, June 1970.
12. S. A. White, New Method of Synthetising Linear Digital Filters Based on Convolution Integral, *Proc. IEE (Corr.)*, vol. 118, p. 348, February 1971.
13. A. Antoniou, Invariant-sinusoid Approximation Method for Recursive Digital Filters, *Electron. Lett.*, vol. 9, pp. 498–500, October 1973.
14. R. E. Crochiere and P. Penfield, Jr., On the Efficient Design of Bandpass Digital Filter Structures, *IEEE Trans. Acoust., Speech, Signal Process.*, vol. ASSP-23, pp. 380–381, August 1975.
15. M. N. S. Swamy and K. S. Thyagarajan, Digital Bandpass and Bandstop Filters with Variable Center Frequency and Bandwidth, *Proc. IEEE*, vol. 64, pp. 1632–1634, November 1976.

ADDITIONAL REFERENCES

Blinchikoff, H., and M. Savetman: Least-Squares Approximation to Wide-Band Constant Delay, *IEEE Trans. Circuit Theory*, vol. CT-19, pp. 387–389, July 1972.

Broome, P.: A Frequency Transformation for Numerical Filters, *Proc. IEEE*, vol. 54, pp. 326–327, February 1966.

Brophy, F., and A. C. Salazar: Considerations of the Padé Approximant Technique in the Synthesis of Recursive Digital Filters, *IEEE Trans. Audio Electroacoust.*, vol. AU-21, pp. 500–505, December 1973.

Constantinides, A. G.: Family of Equiripple Lowpass Digital Filters, *Electron. Lett.*, vol. 6, pp. 351–353, May 28, 1970.

Deczky, A. G.: Synthesis of Recursive Digital Filters Using the Minimum *p*-Error Criterion, *IEEE Trans. Audio Electroacoust.*, vol. AU-20, pp. 257–263, October 1972.

Fettweis, A.: A Simple Design of Maximally Flat Delay Digital Filters, *IEEE Trans. Audio Electroacoust.*, vol. AU-20, pp. 112–114, June 1972.

Haykin, S. S., and R. Carnegie: Improved Analogue-Digital Filter Transformation, *Proc. IEE*, vol. 118, pp. 759–761, June 1971.

Maenhout, G. C., and W. Steenaart: A Direct Approximation Technique for Digital Filters and Equalizers, *IEEE Trans. Circuit Theory*, vol. CT-20, pp. 548–555, September 1973.

Rabiner, L. R., N. Y. Graham, and H. D. Helms: Linear Programming Design of IIR Digital Filters with Arbitrary Magnitude Function, *IEEE Trans. Acoust., Speech, Signal Process.*, vol. ASSP-22, pp. 117–123, April 1974.

Swamy, M. N. S., and K. S. Thyagarajan: Frequency Transformations for Digital Filters, *Proc. IEEE*, vol. 65, pp. 165–166, January 1977.

Thiran, J. P.: Recursive Digital Filters with Maximally Flat Group Delay, *IEEE Trans. Circuit Theory*, vol. CT-18, pp. 659–664, November 1971.

————: Equal-Ripple Delay Recursive Digital Filters, *IEEE Trans. Circuit Theory*, vol. CT-18, pp. 664–669, November 1971.

PROBLEMS

7.1 By using the invariant-impulse-response method, derive a discrete-time transfer function from the continuous-time transfer function

$$H_A(s) = \frac{1}{(s + 1)(s^2 + s + 1)}$$

The sampling frequency is 10 rad/s.

7.2 The sixth-order normalized Bessel transfer function can be expressed as

$$H_A(s) = \sum_{i=1}^{3} \left(\frac{A_i}{s - p_i} + \frac{A_i^*}{s - p_i^*} \right)$$

where A_i and p_i are given in Table P7.2.

(a) Design a digital filter by using the invariant-impulse-response method, assuming a sampling frequency of 10 rad/s.

(b) Plot the phase characteristic of the resulting filter.

Table P7.2

i	p_i	A_i
1	$-4.248359 + j8.675097 \times 10^{-1}$	$1.095923 \times 10 - j3.942517 \times 10$
2	$-3.735708 + j2.626272$	$-1.412677 \times 10 + j1.270117 \times 10$
3	$-2.515932 + j4.492673$	$3.167539 - j2.024596 \times 10^{-1}$

7.3 The transfer function

$$H_A(s) = \frac{s^2 - 3s + 3}{s^2 + 3s + 3}$$

is a constant-delay approximation (see Prob. 5.17). Can one design a corresponding digital filter by using the invariant-impulse-response method? If so, carry out the design employing a sampling frequency of 10 rad/s. Otherwise, explain the reasons for the failure of the method.

7.4 A bandpass filter is required with passband edges of 900 and 1600 rad/s and a maximum passband loss of 1.0 dB. Obtain a design by employing the invariant-impulse-response method. Start with a second-order normalized lowpass Tschebyscheff approximation and neglect the effects of aliasing. A suitable sampling frequency is 10,000 rad/s.

7.5 Given an analog filter characterized by

$$H_A(s) = \sum_{i=1}^{N} \frac{A_i}{s - p_i}$$

a corresponding digital filter characterized by $H_D(z)$ can be derived such that

$$\mathcal{R}_A u(t) \bigg|_{t=nT} = \mathcal{R}_D u(nT)$$

This is called the *invariant-unit-step-response approximation method* [11, 12].

(a) Show that

$$H_D(e^{j\omega T}) \approx H_A(j\omega) \qquad \text{for } |\omega| < \frac{\omega_s}{2}$$

if $\omega \ll 1/T$ and

$$\frac{H_A(j\omega)}{j\omega} \approx 0 \qquad \text{for } |\omega| \geq \frac{\omega_s}{2}$$

(b) Show that

$$H_D(z) = \sum_{i=1}^{N} \frac{A_i'}{z - e^{p_i T}} \qquad \text{where } A_i' = \frac{(e^{p_i T} - 1)A_i}{p_i}$$

7.6 Given an analog filter characterized by

$$H_A(s) = H_0 + \sum_{i=1}^{N} \frac{A_i}{s - p_i}$$

a corresponding digital filter characterized by $H_D(z)$ can be derived such that

$$\mathcal{R}_A u(t) \sin \omega_0 t \bigg|_{t=nT} = \mathcal{R}_D u(nT) \sin \omega_0 nT$$

This is the so-called *invariant-sinusoid-response approximation method* [13].

(a) Show that

$$H_D(e^{j\omega T}) \approx \frac{2\omega_0(1 - \cos \omega_0 T)}{(\omega_0^2 - \omega^2)T \sin \omega_0 T} H_A(j\omega) \qquad \text{for } |\omega| < \frac{\omega_s}{2}$$

if $\omega \ll 1/T$ and

$$\frac{\omega_0 H_A(j\omega)}{\omega_0^2 - \omega^2} \approx 0 \qquad \text{for } |\omega| \geq \frac{\omega_s}{2}$$

(b) Show that

$$H_D(z) = H_0 + \sum_{i=1}^{N} \frac{U_i z + V_i}{z - e^{p_i T}}$$

where

$$U_i = (\omega_0 e^{p_i T} - p_i \sin \omega_0 T - \omega_0 \cos \omega_0 T)A_i'$$

$$V_i = [e^{p_i T}(p_i \sin \omega_0 T - \omega_0 \cos \omega_0 T) + \omega_0]A_i'$$

$$A_i' = \frac{A_i}{(p_i^2 + \omega_0^2) \sin \omega_0 T}$$

7.7 (a) Design a third-order digital filter by applying the invariant-unit-step-response method (see Prob. 7.5) to the transfer function in Prob. 7.1. Assume that $\omega_s = 10$ rad/s.

(b) Compare the design with that obtained in Prob. 7.1.

7.8 (a) Redesign the filter of Prob. 7.2 by employing the invariant-sinusoid-response method (see Prob. 7.6). The value of ω_0 may be assumed to be 1 rad/s.

(b) Plot the resulting phase characteristic.

7.9 A lowpass filter is required satisfying the specifications of Fig. P7.9. Obtain a design by applying the modified invariant-impulse-response method to an elliptic approximation. The sampling frequency is 20,000 rad/s.

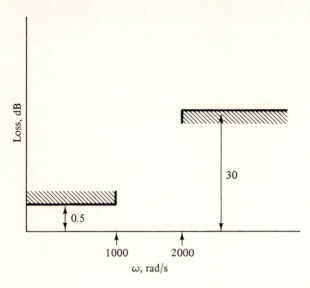

Figure P7.9

7.10 Redesign the filter of Prob. 7.9 by using the matched-*z*-transformation method.

7.11 Design a highpass filter satisfying the specifications of Fig. P7.11. Use the matched-*z*-transformation method along with an elliptic approximation. Assume that $\omega_s = 6000$ rad/s.

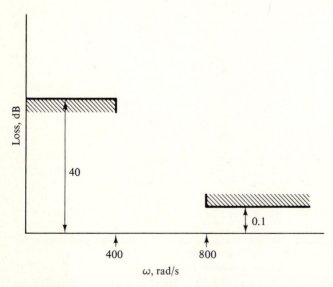

Figure P7.11

7.12 Design a digital filter by applying the bilinear transformation to the Tschebyscheff transfer function of Example 5.2. The sampling frequency is 10 rad/s. Adjust the multiplier constant to achieve a minimum passband loss of zero.

7.13 The lowpass transfer function of Example 5.3 is subjected to the bilinear transformation.

(a) Assuming that $\omega_s = 10$ rad/s, find the resulting passband and stopband edges and also the infinite-loss frequencies.

(b) Determine the effective selectivity factor for the digital filter.

(c) Find the minimum value of ω_s if the passband and stopband edges in the digital filter are to be within ± 1 percent of the corresponding values in the analog filter.

7.14 $H_D(z)$ represents a lowpass filter with a passband edge Ω_p. Show that $H_D(-z)$ represents a highpass filter with a passband edge $\Omega_p - \omega_s/2$.

7.15 The transfer function

$$H_D(z) = H \prod_{j=1}^{2} \frac{A_{0j} + A_{1j}z + A_{0j}z^2}{B_{0j} + B_{1j}z + z^2}$$

where A_{ij} and B_{ij} are given in Table P7.15, represents a lowpass filter with a passband edge of 1 rad/s if $\omega_s = 2\pi$ rad/s. By using the lowpass-to-highpass transformation, design a highpass filter with a passband edge of 2 rad/s if $\omega_s = 2\pi$ rad/s.

Table P7.15

j	A_{0j}	A_{1j}	B_{0j}	B_{1j}
1	1.722415×10^{-1}	3.444829×10^{-1}	4.928309×10^{-1}	-1.263032
2	1.727860×10^{-1}	3.455720×10^{-1}	7.892595×10^{-1}	-9.753164×10^{-1}

$H = 3.500865 \times 10^{-1}$

7.16 By using the transfer function of Prob. 7.15 obtain a lowpass cascade canonic structure whose passband edge can be varied by varying a single parameter.

7.17 Derive the lowpass-to-highpass transformation of Table 7.7.

7.18 Derive the lowpass-to-bandpass transformation of Table 7.7.

EIGHT

RECURSIVE FILTERS SATISFYING PRESCRIBED SPECIFICATIONS

8.1 INTRODUCTION

The most important of the design methods described in Chap. 7 is the bilinear-transformation method. As was demonstrated, the main problem with the method is the warping effect which introduces frequency-scale distortion. If $\omega_1, \omega_2, \ldots, \omega_i, \ldots$ are the passband and stopband edges in the analog filter, the corresponding passband and stopband edges in the derived digital filter are given by

$$\Omega_i = \frac{2}{T} \tan^{-1} \frac{\omega_i T}{2} \qquad i = 1, 2, \ldots$$

according to Eq. (7.23). Consequently, if prescribed passband and stopband edges $\tilde{\Omega}_1, \tilde{\Omega}_2, \ldots, \tilde{\Omega}_i, \ldots$ are to be achieved, the analog filter must be prewarped before application of the bilinear transformation to ensure that

$$\omega_i = \frac{2}{T} \tan \frac{\tilde{\Omega}_i T}{2}$$

in which case

$$\Omega_i = \tilde{\Omega}_i$$

The design of lowpass, highpass, etc., filters is usually accomplished in two steps. First a normalized lowpass transfer function is transformed into a denormalized lowpass, highpass, etc., transfer function employing the standard analog-filter transformations described in Sec. 5.7. Then the bilinear transformation is applied. Prewarping can be effected by choosing the parameters in the analog-filter transformations appropriately.

This chapter considers the details of the above design procedure. Formulas are derived for the parameters of the analog-filter transformations for Butterworth, Tschebyscheff, and elliptic filters, which simplify the design of filters satisfying prescribed specifications [1].

8.2 DESIGN PROCEDURE

Consider a normalized analog filter characterized by $H_N(s)$, with a loss

$$A_N(\omega) = 20 \log \frac{1}{|H_N(j\omega)|}$$

and assume that

$$0 \leq A_N(\omega) \leq A_p \qquad \text{for } 0 \leq |\omega| \leq \omega_p$$
$$A_N(\omega) \geq A_a \qquad \text{for } \omega_a \leq |\omega| \leq \infty$$

as illustrated in Fig. 8.1. A corresponding lowpass, highpass, bandpass, or bandstop filter with the same passband ripple and the same minimum stopband loss can be derived by using the following steps:

1. Form

$$H_X(\bar{s}) = H_N(s) \Big|_{s = f_X(\bar{s})} \tag{8.1}$$

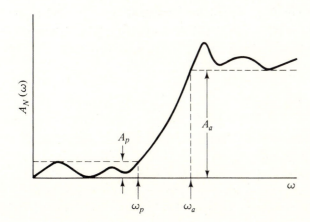

Figure 8.1 Loss characteristic of normalized lowpass filter.

Table 8.1 Standard forms of $f_X(\bar{s})$

X	$f_X(\bar{s})$
LP	$\lambda \bar{s}$
HP	$\dfrac{\lambda}{\bar{s}}$
BP	$\dfrac{1}{B}\left(\bar{s} + \dfrac{\omega_0^2}{\bar{s}}\right)$
BS	$\dfrac{B\bar{s}}{\bar{s}^2 + \omega_0^2}$

where $f_X(\bar{s})$ is given in Table 8.1 (see Sec. 5.7).

2. Form

$$H_D(z) = H_X(\bar{s})\bigg|_{\bar{s} = \frac{2}{T}\frac{z-1}{z+1}} \tag{8.2}$$

If the derived filter is to have prescribed passband and stopband edges, the parameters λ, ω_0, and B in Table 8.1 and the order of $H_N(s)$ must be chosen appropriately. Formulas for these parameters for Butterworth, Tschebyscheff, and elliptic filters are derived in the following section.

8.3 DESIGN FORMULAS

Lowpass and Highpass Filters

Consider the lowpass-filter specification of Fig. 8.2, where $\tilde{\Omega}_p$ and $\tilde{\Omega}_a$ are the desired passband and stopband edges, and assume that the above design procedure yields a transfer function $H_D(z)$ such that

$$A_D(\Omega) = 20 \log \frac{1}{|H_D(e^{j\Omega T})|}$$

where

$$0 \le A_D(\Omega) \le A_p \qquad \text{for } 0 \le |\Omega| \le \Omega_p$$

$$A_D(\Omega) \ge A_a \qquad \text{for } \Omega_a \le |\Omega| \le \frac{\omega_s}{2}$$

(see Fig. 8.2).

From Eq. (8.1) and Table 8.1

$$|H_{\text{LP}}(j\bar{\omega})| = |H_N(j\omega)|$$

if

$$\omega = \lambda\bar{\omega}$$

and hence

$$\omega_p = \lambda\bar{\omega}_p \tag{8.3}$$

$$\omega_a = \lambda\bar{\omega}_a \tag{8.4}$$

where $\bar{\omega}_p$ and $\bar{\omega}_a$ denote the passband and stopband edges, respectively, in $H_{LP}(\bar{s})$. From Eq. (8.2)

$$|H_D(e^{j\Omega T})| = |H_{LP}(j\bar{\omega})|$$

if

$$\bar{\omega} = \frac{2}{T} \tan \frac{\Omega T}{2}$$

and thus

$$\bar{\omega}_p = \frac{2}{T} \tan \frac{\Omega_p T}{2} \tag{8.5}$$

$$\bar{\omega}_a = \frac{2}{T} \tan \frac{\Omega_a T}{2} \tag{8.6}$$

Hence Eqs. (8.3) to (8.6) yield

$$\omega_p = \frac{2}{T} \lambda \tan \frac{\Omega_p T}{2} \tag{8.7}$$

$$\omega_a = \frac{2}{T} \lambda \tan \frac{\Omega_a T}{2} \tag{8.8}$$

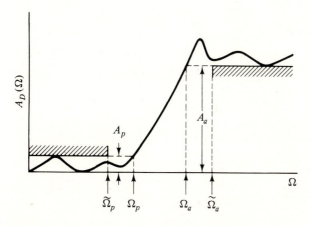

Figure 8.2 Loss characteristic of lowpass digital filter.

Now on assigning

$$\Omega_p = \tilde{\Omega}_p \qquad \Omega_a \leq \tilde{\Omega}_a$$

the desired specifications will be met. The appropriate value for λ is obtained from Eq. (8.7) as

$$\lambda = \frac{T\omega_p}{2 \tan (\tilde{\Omega}_p T/2)}$$

Since $\tan (\tilde{\Omega}_a T/2) \geq \tan (\Omega_a T/2)$ for $0 < \Omega_a, \tilde{\Omega}_a < \omega_s/2$, Eqs. (8.7) and (8.8) give

$$\omega_a \leq \frac{\omega_p}{K_0}$$

where

$$K_0 = \frac{\tan (\tilde{\Omega}_p T/2)}{\tan (\tilde{\Omega}_a T/2)} \qquad (8.9)$$

This inequality imposes a lower limit on the order of $H_N(s)$, as will be shown later.

The preceding approach can readily be extended to highpass filters. The resulting formulas for λ and ω_a are given in Table 8.2.

Bandpass and Bandstop Filters

Now consider the bandpass-filter specification of Fig. 8.3, where $\tilde{\Omega}_{p1}, \tilde{\Omega}_{p2}$ and $\tilde{\Omega}_{a1}, \tilde{\Omega}_{a2}$ represent the desired passband and stopband edges, respectively, and assume that the derived filter has

$$0 \leq A_D(\Omega) \leq A_p \qquad \text{for } \Omega_{p1} \leq |\Omega| \leq \Omega_{p2}$$

$$A_D(\Omega) \geq A_a \quad \begin{cases} \text{for } 0 \leq |\Omega| \leq \Omega_{a1} \\ \\ \text{and } \Omega_{a2} \leq |\Omega| \leq \dfrac{\omega_s}{2} \end{cases}$$

as shown in Fig. 8.3.

Table 8.2 Lowpass and highpass filters

LP	$\omega_a \leq \dfrac{\omega_p}{K_0}$
	$\lambda = \dfrac{\omega_p T}{2 \tan (\tilde{\Omega}_p T/2)}$
HP	$\omega_a \leq \omega_p K_0$
	$\lambda = \dfrac{2\omega_p \tan (\tilde{\Omega}_p T/2)}{T}$
where	$K_0 = \dfrac{\tan (\tilde{\Omega}_p T/2)}{\tan (\tilde{\Omega}_a T/2)}$

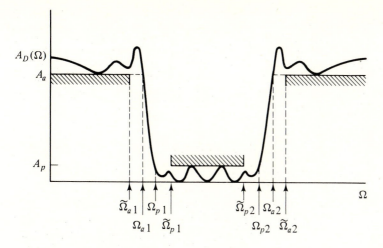

Figure 8.3 Loss characteristic of bandpass digital filter.

From Eq. (8.1) and Table 8.1

$$|H_{BP}(j\bar\omega)| = |H_N(j\omega)|$$

provided that

$$\omega = \frac{1}{B}\left(\bar\omega - \frac{\omega_0^2}{\bar\omega}\right)$$

or by solving for $\bar\omega$ if

$$\bar\omega = \frac{\omega B}{2} \pm \sqrt{\omega_0^2 + \left(\frac{\omega B}{2}\right)^2} \tag{8.10}$$

With

$$\omega = \pm\omega_p \text{ or } \pm\omega_a$$

the positive passband and stopband edges in $H_{BP}(\bar s)$ can be deduced from Eq. (8.10) as

$$\bar\omega_{p1}, \bar\omega_{p2} = \mp\frac{\omega_p B}{2} + \sqrt{\omega_0^2 + \omega_p^2\left(\frac{B}{2}\right)^2}$$

and

$$\bar\omega_{a1}, \bar\omega_{a2} = \mp\frac{\omega_a B}{2} + \sqrt{\omega_0^2 + \omega_a^2\left(\frac{B}{2}\right)^2}$$

respectively. Evidently

$$\bar\omega_{p1}\bar\omega_{p2} = \omega_0^2 \tag{8.11}$$

$$\bar\omega_{a1}\bar\omega_{a2} = \omega_0^2 \tag{8.12}$$

$$\bar\omega_{p2} - \bar\omega_{p1} = \omega_p B \tag{8.13}$$

$$\bar\omega_{a2} - \bar\omega_{a1} = \omega_a B \tag{8.14}$$

From Eq. (8.2)

$$|H_D(e^{j\Omega T})| = |H_{BP}(j\bar{\omega})|$$

if

$$\bar{\omega} = \frac{2}{T} \tan \frac{\Omega T}{2}$$

and hence

$$\bar{\omega}_{p1} = \frac{2}{T} \tan \frac{\Omega_{p1} T}{2} \tag{8.15}$$

$$\bar{\omega}_{p2} = \frac{2}{T} \tan \frac{\Omega_{p2} T}{2} \tag{8.16}$$

$$\bar{\omega}_{a1} = \frac{2}{T} \tan \frac{\Omega_{a1} T}{2} \tag{8.17}$$

$$\bar{\omega}_{a2} = \frac{2}{T} \tan \frac{\Omega_{a2} T}{2} \tag{8.18}$$

We can now assign

$$\Omega_{p1} = \tilde{\Omega}_{p1} \qquad \text{and} \qquad \Omega_{p2} = \tilde{\Omega}_{p2}$$

From Eqs. (8.13), (8.15), and (8.16)

$$B = \frac{2K_A}{T\omega_p} \tag{8.19}$$

where

$$K_A = \tan \frac{\tilde{\Omega}_{p2} T}{2} - \tan \frac{\tilde{\Omega}_{p1} T}{2} \tag{8.20}$$

Also from Eqs. (8.11), (8.15), and (8.16)

$$\omega_0 = \frac{2\sqrt{K_B}}{T} \tag{8.21}$$

where

$$K_B = \tan \frac{\tilde{\Omega}_{p1} T}{2} \tan \frac{\tilde{\Omega}_{p2} T}{2} \tag{8.22}$$

From Eqs. (8.11) and (8.12)

$$\bar{\omega}_{a1} \bar{\omega}_{a2} = \bar{\omega}_{p1} \bar{\omega}_{p2}$$

but since $\bar{\omega}_{p1}$ and $\bar{\omega}_{p2}$ have already been fixed by assigning values to Ω_{p1} and Ω_{p2}, it is not in general possible to achieve the prescribed stopband edges exactly. The alternative is to assign

$$\Omega_{a1} \geq \tilde{\Omega}_{a1} \qquad \text{or} \qquad \Omega_{a2} \leq \tilde{\Omega}_{a2}$$

without violating Eqs. (8.12) and (8.14). In this way the loss at $\Omega = \tilde{\Omega}_{a1}$ or $\tilde{\Omega}_{a2}$ will equal or exceed the minimum specified.

With

$$\Omega_{a1} \geq \tilde{\Omega}_{a1}$$

and Ω_{a2} chosen such that Eqs. (8.12) and (8.14) hold, Eq. (8.17) gives

$$\bar{\omega}_{a1} = \frac{2}{T} \tan \frac{\Omega_{a1} T}{2} \geq \frac{2}{T} \tan \frac{\tilde{\Omega}_{a1} T}{2} \tag{8.23}$$

Subsequently, from Eqs. (8.12), (8.14), and (8.17)

$$\omega_a = \frac{\omega_0^2 - (2/T)^2 \tan^2 (\Omega_{a1} T/2)}{(2B/T) \tan (\Omega_{a1} T/2)} \tag{8.24}$$

Therefore, from Eqs. (8.23) and (8.24)

$$\omega_a \leq \frac{\omega_0^2 - (2/T)^2 \tan^2 (\tilde{\Omega}_{a1} T/2)}{(2B/T) \tan (\tilde{\Omega}_{a1} T/2)}$$

and on eliminating ω_0 and B using Eqs. (8.19) and (8.21) we deduce

$$\omega_a \leq \frac{\omega_p}{K_1}$$

where

$$K_1 = \frac{K_A \tan (\tilde{\Omega}_{a1} T/2)}{K_B - \tan^2 (\tilde{\Omega}_{a1} T/2)} \tag{8.25}$$

The other possibility is to let

$$\Omega_{a2} \leq \tilde{\Omega}_{a2}$$

and choose Ω_{a1} such that Eqs. (8.12) and (8.14) hold. From Eq. (8.18)

$$\bar{\omega}_{a2} = \frac{2}{T} \tan \frac{\Omega_{a2} T}{2} \leq \frac{2}{T} \tan \frac{\tilde{\Omega}_{a2} T}{2} \tag{8.26}$$

and from Eqs. (8.12), (8.14), and (8.18)

$$\omega_a = \frac{(2/T)^2 \tan^2 (\Omega_{a2} T/2) - \omega_0^2}{(2B/T) \tan (\Omega_{a2} T/2)} \tag{8.27}$$

Therefore, from Eqs. (8.26) and (8.27)

$$\omega_a \leq \frac{(2/T)^2 \tan^2 (\tilde{\Omega}_{a2} T/2) - \omega_0^2}{(2B/T) \tan (\tilde{\Omega}_{a2} T/2)}$$

and on eliminating ω_0 and B we have

$$\omega_a \leq \frac{\omega_p}{K_2}$$

where

$$K_2 = \frac{K_A \tan (\tilde{\Omega}_{a2} T/2)}{\tan^2 (\tilde{\Omega}_{a2} T/2) - K_B} \tag{8.28}$$

Table 8.3 Bandpass and bandstop filters

	$\omega_0 = \dfrac{2\sqrt{K_B}}{T}$
BP	$\omega_a \leq \begin{cases} \dfrac{\omega_p}{K_1} & \text{if } K_C \geq K_B \\[2mm] \dfrac{\omega_p}{K_2} & \text{if } K_C < K_B \end{cases}$
	$B = \dfrac{2K_A}{T\omega_p}$
	$\omega_0 = \dfrac{2\sqrt{K_B}}{T}$
BS	$\omega_a \leq \begin{cases} \omega_p K_2 & \text{if } K_C \geq K_B \\ \omega_p K_1 & \text{if } K_C < K_B \end{cases}$
	$B = \dfrac{2K_A\omega_p}{T}$

where $K_A = \tan \dfrac{\tilde{\Omega}_{p2} T}{2} - \tan \dfrac{\tilde{\Omega}_{p1} T}{2}$ $K_B = \tan \dfrac{\tilde{\Omega}_{p1} T}{2} \tan \dfrac{\tilde{\Omega}_{p2} T}{2}$

$K_C = \tan \dfrac{\tilde{\Omega}_{a1} T}{2} \tan \dfrac{\tilde{\Omega}_{a2} T}{2}$ $K_1 = \dfrac{K_A \tan (\tilde{\Omega}_{a1} T/2)}{K_B - \tan^2 (\tilde{\Omega}_{a1} T/2)}$

$K_2 = \dfrac{K_A \tan (\tilde{\Omega}_{a2} T/2)}{\tan^2 (\tilde{\Omega}_{a2} T/2) - K_B}$

Summarizing, if $\Omega_{a1} \geq \tilde{\Omega}_{a1}$ then $\omega_a \leq \omega_p/K_1$; and if $\Omega_{a2} \leq \tilde{\Omega}_{a2}$, then $\omega_a \leq \omega_p/K_2$. Therefore, if we ensure that

$$\omega_a \leq \min\left(\frac{\omega_p}{K_1}, \frac{\omega_p}{K_2}\right) \quad \text{or} \quad \omega_a \leq \frac{\omega_p}{K} \quad \text{where } K = \max\,(K_1, K_2)$$

then $$\Omega_{a1} \geq \tilde{\Omega}_{a1} \quad \text{and} \quad \Omega_{a2} \leq \tilde{\Omega}_{a2}$$

as required. The appropriate value for K is easily deduced from Eqs. (8.25) and (8.28) as

$$K = \begin{cases} K_1 & \text{if } K_C \geq K_B \\ K_2 & \text{if } K_C < K_B \end{cases}$$

where $$K_C = \tan \frac{\tilde{\Omega}_{a1} T}{2} \tan \frac{\tilde{\Omega}_{a2} T}{2} \tag{8.29}$$

The same approach can also be applied to bandstop filters. The relevant design formulas are summarized in Table 8.3.

The formulas derived so far are general in the sense that they apply equally well to Butterworth, Tschebyscheff, and elliptic filters. Let us now consider the detailed implications for each of the three types of approximation.

Butterworth Filters

The loss in a normalized Butterworth filter is given by

$$A_N(\omega) = 10 \log (1 + \omega^{2n})$$

(see Sec. 5.3), where n is the order of the transfer function. For $\omega = \omega_p$ or ω_a

$$A_N(\omega_p) = A_p = 10 \log (1 + \omega_p^{2n}) \qquad A_N(\omega_a) = A_a = 10 \log (1 + \omega_a^{2n})$$

or

$$\omega_p = (10^{0.1A_p} - 1)^{1/2n} \qquad \omega_a = (10^{0.1A_a} - 1)^{1/2n}$$

Thus from Tables 8.2 and 8.3

$$\frac{\omega_a}{\omega_p} = \left(\frac{10^{0.1A_a} - 1}{10^{0.1A_p} - 1} \right)^{1/2n} \leq \frac{1}{K}$$

where K is given in Table 8.4. Therefore, the necessary value for n in order to meet the prescribed specifications is given by

$$n \geq \frac{\log D}{2 \log (1/K)}$$

where

$$D = \frac{10^{0.1A_a} - 1}{10^{0.1A_p} - 1} \qquad\qquad (8.30)$$

Table 8.4 Butterworth filters

$$n \geq \frac{\log D}{2 \log (1/K)}$$

$$\omega_p = (10^{0.1A_p} - 1)^{1/2n}$$

LP	$K = K_0$
HP	$K = \dfrac{1}{K_0}$
BP	$K = \begin{cases} K_1 & \text{if } K_C \geq K_B \\ K_2 & \text{if } K_C < K_B \end{cases}$
BS	$K = \begin{cases} \dfrac{1}{K_2} & \text{if } K_C \geq K_B \\ \dfrac{1}{K_1} & \text{if } K_C < K_B \end{cases}$
where	$D = \dfrac{10^{0.1A_a} - 1}{10^{0.1A_p} - 1}$

Tschebyscheff Filters

In normalized Tschebyscheff filters

$$A_N(\omega) = 10 \log [1 + \varepsilon^2 T_n^2(\omega)]$$

where

$$T_n(\omega) = \cosh (n \cosh^{-1} \omega) \qquad \text{for } \omega_p \leq \omega < \infty$$

$$\varepsilon^2 = 10^{0.1A_p} - 1 \qquad \omega_p = 1$$

(see Sec. 5.4). For $\omega = \omega_a$

$$A_N(\omega_a) = A_a = 10 \log \{1 + (10^{0.1A_p} - 1)[\cosh (n \cosh^{-1} \omega_a)]^2\}$$

or

$$\omega_a = \cosh \left(\frac{1}{n} \cosh^{-1} \sqrt{D} \right)$$

Thus from Tables 8.2 and 8.3

$$\frac{\omega_a}{\omega_p} = \cosh \left(\frac{1}{n} \cosh^{-1} \sqrt{D} \right) \leq \frac{1}{K}$$

where K is given in Table 8.5. Therefore

$$n \geq \frac{\cosh^{-1} \sqrt{D}}{\cosh^{-1} (1/K)}$$

where $\cosh^{-1} x$ can be evaluated using the identity

$$\cosh^{-1} x = \ln (x + \sqrt{x^2 - 1})$$

Table 8.5 Tschebyscheff filters

	$n \geq \dfrac{\cosh^{-1} \sqrt{D}}{\cosh^{-1} (1/K)}$
	$\omega_p = 1$
LP	$K = K_0$
HP	$K = \dfrac{1}{K_0}$
BP	$K = \begin{cases} K_1 & \text{if } K_C \geq K_B \\ K_2 & \text{if } K_C < K_B \end{cases}$
BS	$K = \begin{cases} \dfrac{1}{K_2} & \text{if } K_C \geq K_B \\ \dfrac{1}{K_1} & \text{if } K_C < K_B \end{cases}$

Table 8.6 Elliptic filters

	k	ω_p
LP	K_0	$\sqrt{K_0}$
HP	$\dfrac{1}{K_0}$	$\dfrac{1}{\sqrt{K_0}}$
BP	K_1 if $K_C \geq K_B$ K_2 if $K_C < K_B$	$\sqrt{K_1}$ $\sqrt{K_2}$
BS	$\dfrac{1}{K_2}$ if $K_C \geq K_B$ $\dfrac{1}{K_1}$ if $K_C < K_B$	$\dfrac{1}{\sqrt{K_2}}$ $\dfrac{1}{\sqrt{K_1}}$

$$n \geq \frac{\log 16D}{\log (1/q)}$$

Elliptic Filters

The selectivity factor in elliptic filters is defined as

$$k = \frac{\omega_p}{\omega_a}$$

(see Sec. 5.5). Thus from Tables 8.2 and 8.3

$$k \geq K$$

where $K = K_0, 1/K_0, \ldots$. Since any value in the range 0 to 1 (except for unity) is a permissible value for k we can assign

$$k = K$$

as in Table 8.6. With k chosen, the value of ω_p is fixed, i.e.,

$$\omega_p = \sqrt{k}$$

Finally, with k, A_p, and A_a known the necessary value for n can be computed by using the formula in Table 8.6 (see Sec. 5.5).

8.4 DESIGN USING THE FORMULAS AND TABLES

The formulas and tables developed in the preceding section lead to the following simple design procedure:

1. Using the prescribed specifications, determine k (for elliptic filters only), n, and ω_p from Tables 8.4 to 8.6.

2. Determine λ for lowpass and highpass filters or B and ω_0 for bandpass and bandstop filters using Table 8.2 or 8.3.
3. Form the normalized transfer function (see Chap. 5).
4. Apply the transformation in Eq. (8.1).
5. Apply the transformation in Eq. (8.2).

The design can be readily accomplished by using Programs B.6, B.7, B.9 and B.11 in sequence.

Example 8.1 Design a highpass filter satisfying the following specifications:

$$A_p = 1 \text{ dB} \qquad A_a = 45 \text{ dB} \qquad \tilde{\Omega}_p = 3.5 \text{ rad/s}$$
$$\tilde{\Omega}_a = 1.5 \text{ rad/s} \qquad \omega_s = 10 \text{ rad/s}$$

Use a Butterworth, a Tschebyscheff, and then an elliptic approximation.

SOLUTION **Butterworth filter** From Eqs. (8.9) and (8.30)

$$K_0 = \frac{\tan (3.5\pi/10)}{\tan (1.5\pi/10)} = 3.85184 \qquad D = \frac{10^{4.5} - 1}{10^{0.1} - 1} = 1.22127 \times 10^5$$

Hence from Table 8.4

$$n \geq \frac{\log D}{2 \log K_0} \approx 4.34$$
$$= 5$$
$$\omega_p = (10^{0.1} - 1)^{0.1} = 0.8736097$$

Now from Table 8.2

$$\lambda = \frac{2}{T} \omega_p \tan \frac{\tilde{\Omega}_p T}{2} = 5.4576$$

Tschebyscheff filter From Table 8.5

$$n \geq \frac{\cosh^{-1} \sqrt{D}}{\cosh^{-1} K_0} = \frac{\ln (\sqrt{D} + \sqrt{D-1})}{\ln (K_0 + \sqrt{K_0^2 - 1})} \approx 3.24$$
$$= 4$$
$$\omega_p = 1$$

Hence from Table 8.2

$$\lambda = 6.247183$$

Elliptic filter From Table 8.6

$$k = \frac{1}{K_0} = 0.2596162$$

From Eqs. (5.45) to (5.47)

$$k' = \sqrt{1 - k^2} = 0.9657119$$

$$q_0 = \frac{1}{2} \frac{1 - \sqrt{k'}}{1 + \sqrt{k'}} = 4.361108 \times 10^{-3}$$

$$q = q_0 + 2q_0^5 + \cdots \approx q_0$$

Hence

$$n \geq \frac{\log 16D}{\log (1/q)} \approx 2.67$$

$$= 3$$

$$\omega_p = \sqrt{k} = 0.5095255$$

and from Table 8.2

$$\lambda = 3.183099$$

The first two steps of the design can be carried out using Program B.6. The transfer functions $H_N(s)$, $H_{HP}(\bar{s})$, and $H_D(z)$ can be put in the form

$$H \prod_{j=1}^{t} \frac{A_{0j} + A_{1j}w + A_{2j}w^2}{B_{0j} + B_{1j}w + B_{2j}w^2}$$

where H is a multiplier constant and $w = s$, $\bar{s}$, or z. The coefficients of $H_N(s)$ can be computed as in Table 8.7 using Program B.7, those of $H_{HP}(\bar{s})$ as in

Table 8.7 Coefficients of $H_N(s)$ (Example 8.1)

	j	A_{0j}	A_{1j}	A_{2j}	B_{0j}	B_{1j}	B_{2j}
Butterworth	1	1	0	0	1	1	0
	2	1	0	0	1	1.618034	1
	3	1	0	0	1	0.618034	1
	$H = 1.0$						
Tschebyscheff	1	1	0	0	0.279398	0.673739	1
	2	1	0	0	0.986505	0.279072	1
	$H = 0.245653$						
Elliptic	1	1	0	0	0.257305	1	0
	2	5.091668	0	1	0.259234	0.244205	1
	$H = 0.0131003$						

Table 8.8 Coefficients of $H_{HP}(\bar{s})$ (Example 8.1)

	j	A_{0j}	A_{1j}	A_{2j}	B_{0j}	B_{1j}	B_{2j}
Butterworth	1	0	1	0	5.45760	1	0
	2	0	0	1	29.7854	8.83058	1
	3	0	0	1	29.7854	3.37298	1
	$H = 1.0$						
Tschebyscheff	1	0	0	1	139.684	15.0644	1
	2	0	0	1	39.5612	1.76726	1
	$H = 0.891251$						
Elliptic	1	0	1	0	12.3709	1	0
	2	1.98994	0	1	39.0848	2.99855	1
	$H = 1.0$						

Table 8.8 using Program B.9, and those of $H_D(z)$ as in Table 8.9 using Program B.11.

The loss characteristics of the three filters are plotted in Fig. 8.4.

Table 8.9 Coefficients of $H_D(z)$ (Example 8.1)

	j	A_{0j}	A_{1j}	A_{2j}	B_{0j}	B_{1j}	B_{2j}
Butterworth	1	−0.368384	−A_{01}	0	0.263231	1	0
	2	0.148945	−0.297889	A_{02}	0.173594	0.577816	1
	3	0.200026	−0.400052	A_{03}	0.576084	0.775981	1
	$H = 1.0$						
Tschebyscheff	1	0.0512326	−0.102465	A_{01}	0.515070	1.31014	1
	2	0.183159	−0.366318	A_{02}	0.796619	1.06398	1
	$H = 0.891251$						
Elliptic	1	−0.204648	−A_{01}	0	0.590704	1	0
	2	0.206292	−0.277126	A_{02}	0.675139	0.985428	1
	$H = 1.0$						

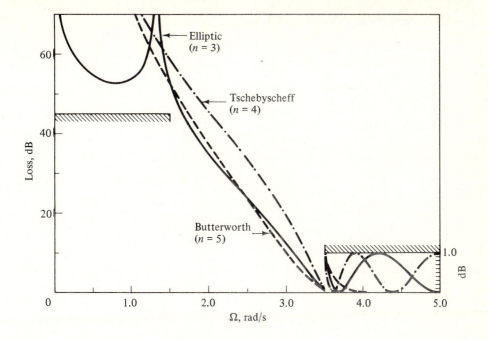

Figure 8.4 Loss characteristics of highpass filters (Example 8.1).

Example 8.2 Design an elliptic bandpass filter satisfying the following specifications:

$$A_p = 1 \text{ dB} \qquad A_a = 45 \text{ dB} \qquad \tilde{\Omega}_{p1} = 900 \text{ rad/s} \qquad \tilde{\Omega}_{p2} = 1100 \text{ rad/s}$$

$$\tilde{\Omega}_{a1} = 800 \text{ rad/s} \qquad \tilde{\Omega}_{a2} = 1200 \text{ rad/s} \qquad \omega_s = 6000 \text{ rad/s}$$

SOLUTION From Eqs. (8.20), (8.22), and (8.29)

$$K_A = \tan \frac{1100\pi}{6000} - \tan \frac{900\pi}{6000} = 0.1398821$$

$$K_B = \tan \frac{900\pi}{6000} \tan \frac{1100\pi}{6000} = 0.3308897$$

$$K_C = \tan \frac{800\pi}{6000} \tan \frac{1200\pi}{6000} = 0.3234776$$

Hence $K_B > K_C$ and from Table 8.6

$$k = K_2 = \frac{K_A \tan (\tilde{\Omega}_{a2} T/2)}{\tan^2 (\tilde{\Omega}_{a2} T/2) - K_B} = 0.5159572 \qquad \omega_p = \sqrt{K_2} = 0.7183016$$

D is the same as in Example 8.1, and

$$k' = \sqrt{1 - k^2} = 0.8566144$$

$$q_0 = \frac{1}{2}\frac{1 - \sqrt{k'}}{1 + \sqrt{k'}} = 0.01933628$$

$$q = q_0 + 2q_0^5 + \cdots \approx 0.01933629$$

Hence

$$n \geq \frac{\log 16D}{\log (1/q)} \approx 3.67$$

$$= 4$$

Now from Table 8.3

$$\omega_0 = \frac{2\sqrt{K_B}}{T} = 1098.609 \qquad B = \frac{2K_A}{T\omega_p} = 371.9263$$

On using Programs B.7, B.9, and B.11 in sequence, $H_D(z)$ can be formed as

$$H_D(z) = H \prod_{j=1}^{4} \frac{A_{0j} + A_{1j}z + A_{0j}z^2}{B_{0j} + B_{1j}z + z^2}$$

The numerical values of the coefficients are given in Table 8.10. The loss characteristic of the filter is plotted in Fig. 8.5.

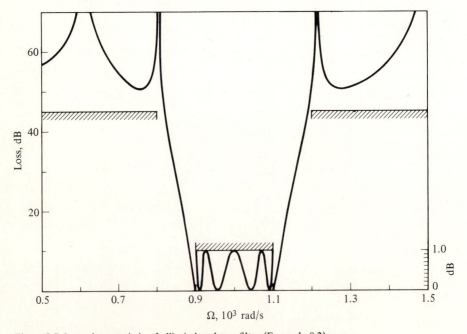

Figure 8.5 Loss characteristic of elliptic bandpass filter (Example 8.2).

Table 8.10 Coefficients of $H_D(z)$ (Example 8.2)

j	A_{0j}	A_{1j}	B_{0j}	B_{1j}
1	1.40266	−0.0102157	0.926867	−0.888660
2	0.826391	−1.32443	0.930606	−1.04661
3	1.07347	−0.631800	0.973854	−0.804891
4	0.941754	−1.25358	0.976782	−1.16031

$H = 0.00293912$

Example 8.3 Design a Tschebyscheff bandstop filter satisfying the following specifications:

$$A_p = 0.5 \text{ dB} \qquad A_a = 40 \text{ dB} \qquad \tilde{\Omega}_{p1} = 350 \text{ rad/s} \qquad \tilde{\Omega}_{p2} = 700 \text{ rad/s}$$

$$\tilde{\Omega}_{a1} = 430 \text{ rad/s} \qquad \tilde{\Omega}_{a2} = 600 \text{ rad/s} \qquad \omega_s = 3000 \text{ rad/s}$$

SOLUTION From Tables 8.3 and 8.5 (using Program B.6)

$$n = 5 \qquad \omega_0 = 561.4083 \qquad B = 493.2594$$

Hence Programs B.7, B.9, and B.11 yield

$$H_D(z) = \prod_{j=1}^{5} \frac{A_{0j} + A_{1j}z + A_{0j}z^2}{B_{0j} + B_{1j}z + z^2}$$

where A_{ij} and B_{ij} are given in Table 8.11. The loss characteristic of the filter is plotted in Fig. 8.6.

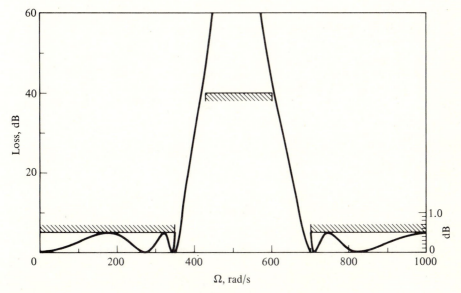

Figure 8.6 Loss characteristic of Tschebyscheff bandstop filter (Example 8.3).

Table 8.11 Coefficients of $H_D(z)$ (Example 8.3)

j	A_{0j}	A_{1j}	B_{0j}	B_{1j}
1	0.485564	−0.472249	−0.0288728	−0.472249
2	0.529076	−0.514569	0.623010	0.0502889
3	1.06121	−1.03211	0.754357	−1.40016
4	0.718033	−0.698344	0.916899	−0.217511
5	1.13666	−1.10549	0.942893	−1.43593

8.5 DELAY EQUALIZATION

The phase response in filters designed by using the preceding method is in general nonlinear because of the warping effect and because the Butterworth, Tschebyscheff, and elliptic approximations are inherently nonlinear-phase approximations. As a consequence, the group delay in these filters given by

$$\tau_D(\omega) = -\frac{d\theta_D(\omega)}{d\omega} \qquad \text{where } \theta_D(\omega) = \arg H_D(e^{j\omega T})$$

tends to vary with frequency.

Constant group-delay filters can sometimes be designed by using constant-delay approximations such as the Bessel approximation with design methods that preserve the linearity in the phase response, e.g., the invariant-impulse-response method. However, a constant delay and prescribed loss specifications are usually difficult to achieve simultaneously, particularly if bandpass or bandstop high-selectivity filters are desired.

The design of constant-delay analog filters satisfying prescribed loss specifications is almost invariably accomplished in two steps. First a filter is designed satisfying the loss specifications ignoring the group delay. Then a delay equalizer is designed which can be used in cascade with the filter to compensate for variations in the group delay of the filter. The same technique can also be used in digital filters.

Let $H_E(z)$ be the transfer function of the equalizer. The group delay of the equalizer is given by

$$\tau_E(\omega) = -\frac{d\theta_E(\omega)}{d\omega} \qquad \text{where } \theta_E(\omega) = \arg H_E(e^{j\omega T})$$

The overall transfer function of the filter-equalizer combination is

$$H_T(z) = H_D(z)H_E(z)$$

Hence $\left|H_T(e^{j\omega T})\right| = \left|H_D(e^{j\omega T})\right| \cdot \left|H_E(e^{j\omega T})\right|$ and $\theta_T(\omega) = \theta_D(\omega) + \theta_E(\omega)$

or $$\tau_T(\omega) = \tau_D(\omega) + \tau_E(\omega)$$

Therefore, a constant-delay digital filter satisfying prescribed loss specifications can be designed by using the following steps:

1. Design a filter satisfying the loss specifications using the procedure in Sec. 8.4.
2. Design an equalizer with

$$\left|H_E(e^{j\omega T})\right| = 1 \qquad \text{for } 0 \le \omega \le \frac{\omega_s}{2}$$

and

$$\tau_E(\omega) = \tau - \tau_D(\omega) \qquad \text{for } \omega_{p1} \le \omega \le \omega_{p2} \tag{8.31}$$

where τ is a constant and ω_{p1}, ω_{p2} are the passband edges.

$H_E(z)$ must thus be an allpass transfer function of the form

$$H_E(z) = \prod_{i=1}^{m} \frac{z^2 + a_{1i}z + a_{2i}}{1 + a_{1i}z + a_{2i}z^2}$$

Values for a_{1i}, a_{2i}, m, and τ such that Eq. (8.31) holds to within a prescribed error can be determined by using optimization techniques. To ensure stability in the equalizer the coefficients a_{1i} and a_{2i} must satisfy the inequality

$$a_{2i} > \max\left[1, (a_{1i} - 1), -(a_{1i} + 1)\right]$$

(see Sec. 3.3).

An alternative approach to the design of constant-delay filters is to use non-recursive approximations. This possibility will be considered in the following chapter.

REFERENCE

1. A. Antoniou, Design of Elliptic Digital Filters: Prescribed Specifications, *Proc. IEE*, vol. 124, pp. 341–344, April 1977.

PROBLEMS

8.1 Design a lowpass digital filter satisfying the specifications of Fig. P8.1. Use a Butterworth approximation.

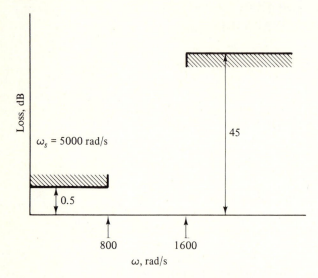

Figure P8.1

8.2 Redesign the filter of Prob. 8.1 using a Tschebyscheff approximation.

8.3 Redesign the filter of Prob. 8.1 using an elliptic approximation.

8.4 Design a highpass digital filter satisfying the specifications of Fig. P8.4. Use a Butterworth approximation.

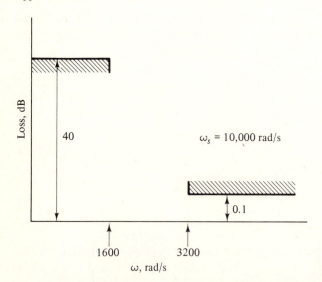

Figure P8.4

8.5 Redesign the filter of Prob. 8.4 using a Tschebyscheff approximation.

8.6 Redesign the filter of Prob. 8.4 using an elliptic approximation.

8.7 Design a bandpass digital filter satisfying the specifications of Fig. P8.7. Use a Butterworth approximation.

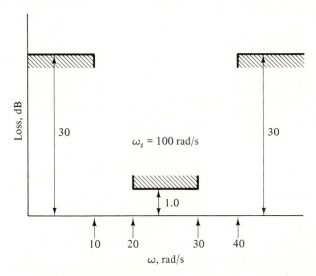

Figure P8.7

8.8 Redesign the filter of Prob. 8.7 using a Tschebyscheff approximation.

8.9 Redesign the filter of Prob. 8.7 using an elliptic approximation.

8.10 Design a bandstop digital filter satisfying the specifications of Fig. P8.10. Use a Butterworth approximation.

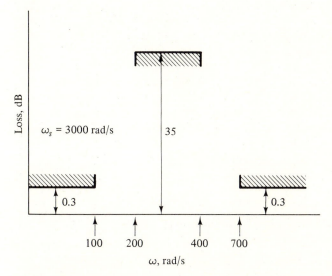

Figure P8.10

8.11 Redesign the filter of Prob. 8.10 using a Tschebyscheff approximation.

8.12 Redesign the filter of Prob. 8.10 using an elliptic approximation.

8.13 Derive the formulas of Table 8.2 for highpass filters.

8.14 Derive the formulas of Table 8.3 for bandstop filters.

# NINE

DESIGN OF NONRECURSIVE FILTERS

9.1 INTRODUCTION

The preceding two chapters have demonstrated that highly selective filters can readily be designed as recursive filters. Although constant-delay filters can also be designed by using the invariant-impulse-response method, constant delay and prescribed loss specifications are difficult to achieve simultaneously. In contrast, nonrecursive filters can readily be designed to have constant delay as well as prescribed loss specifications.

The approximation problem in nonrecursive filters is usually solved by using Fourier series or numerical-analysis formulas. The details of these methods and their application will be examined in the following pages. An alternative possibility is to use the discrete Fourier transform [1, 2]. The details of this approach will be described later, in Sec. 13.8.

9.2 PROPERTIES OF NONRECURSIVE FILTERS

Constant-Delay Filters

A nonrecursive causal filter can be characterized by the transfer function

$$H(z) = \sum_{n=0}^{N-1} h(nT)z^{-n} \tag{9.1}$$

Its frequency response is given by

$$H(e^{j\omega T}) = M(\omega)e^{j\theta(\omega)} = \sum_{n=0}^{N-1} h(nT)e^{-j\omega nT} \qquad (9.2)$$

where

$$M(\omega) = |H(e^{j\omega T})|$$

and

$$\theta(\omega) = \arg H(e^{j\omega T}) \qquad (9.3)$$

The phase and group delays of a filter are given by

$$\tau_p = -\frac{\theta(\omega)}{\omega} \quad \text{and} \quad \tau_g = -\frac{d\theta(\omega)}{d\omega}$$

respectively.

For constant phase delay as well as group delay the phase response must be linear, i.e.,

$$\theta(\omega) = -\tau\omega$$

and thus from Eqs. (9.2) and (9.3)

$$\theta(\omega) = -\tau\omega = \tan^{-1} \frac{-\sum_{n=0}^{N-1} h(nT)\sin \omega nT}{\sum_{n=0}^{N-1} h(nT)\cos \omega nT}$$

Consequently

$$\tan \omega\tau = \frac{\sum_{n=0}^{N-1} h(nT)\sin \omega nT}{\sum_{n=0}^{N-1} h(nT)\cos \omega nT}$$

and accordingly

$$\sum_{n=0}^{N-1} h(nT)(\cos \omega nT \sin \omega\tau - \sin \omega nT \cos \omega\tau) = 0$$

or

$$\sum_{n=0}^{N-1} h(nT)\sin (\omega\tau - \omega nT) = 0$$

The solution of this equation can be shown to be

$$\tau = \frac{(N-1)T}{2} \qquad (9.4)$$

$$h(nT) = h[(N-1-n)T] \qquad \text{for } 0 \le n \le N-1 \qquad (9.5)$$

Therefore, a nonrecursive filter, unlike a recursive filter, can have constant phase and group delays over the entire baseband. It is only necessary for the impulse response to be symmetrical about the midpoint between samples $(N-2)/2$ and

$N/2$ for even N or about sample $(N-1)/2$ for odd N. The required symmetry is illustrated in Fig. 9.1 for $N = 10$ and 11.

In many applications only the group delay need be constant, in which case the phase response can have the form

$$\theta(\omega) = \theta_0 - \tau\omega$$

where θ_0 is a constant. On using the above procedure a second class of constant-delay nonrecursive filters can be obtained. With $\theta_0 = \pm\pi/2$, the solution is

$$\tau = \frac{(N-1)T}{2} \tag{9.6}$$

$$h(nT) = -h[(N-1-n)T] \tag{9.7}$$

In this case the impulse response is antisymmetrical about the midpoint between samples $(N-2)/2$ and $N/2$ for even N or about sample $(N-1)/2$ for odd N, as illustrated in Fig. 9.2.

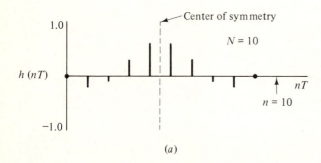

(a)

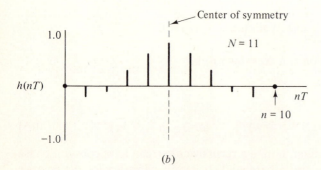

(b)

Figure 9.1 Impulse response for constant phase and group delays: (a) even N, (b) odd N.

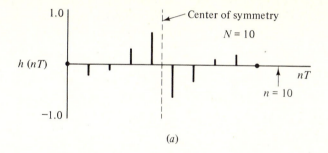

(a)

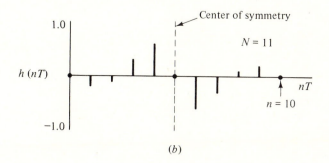

(b)

Figure 9.2 Alternative impulse response for constant group delay: (a) even N, (b) odd N.

Frequency Response

Equations (9.5) and (9.7) lead to some simple expressions for the frequency response. For a symmetrical impulse response with N odd, Eq. (9.2) can be expressed as

$$H(e^{j\omega T}) = \sum_{n=0}^{(N-3)/2} h(nT)e^{-j\omega nT} + h\left[\frac{(N-1)T}{2}\right]e^{-j\omega(N-1)T/2}$$

$$+ \sum_{n=(N+1)/2}^{N-1} h(nT)e^{-j\omega nT} \qquad (9.8)$$

By using Eq. (9.5) and then letting $N-1-n=m$, $m=n$ the last summation in the above equation can be expressed as

$$\sum_{n=(N+1)/2}^{N-1} h(nT)e^{-j\omega nT} = \sum_{n=(N+1)/2}^{N-1} h[(N-1-n)T]e^{-j\omega nT}$$

$$= \sum_{n=0}^{(N-3)/2} h(nT)e^{-j\omega(N-1-n)T} \qquad (9.9)$$

Table 9.1 Frequency response of constant-delay nonrecursive filters

$h(nT)$	N	$H(e^{j\omega T})$
Symmetrical	Odd	$e^{-j\omega(N-1)T/2} \displaystyle\sum_{k=0}^{(N-1)/2} a_k \cos \omega k T$
	Even	$e^{-j\omega(N-1)T/2} \displaystyle\sum_{k=1}^{N/2} b_k \cos [\omega(k - \frac{1}{2})T]$
Antisymmetrical	Odd	$e^{-j[\omega(N-1)T/2 - \pi/2]} \displaystyle\sum_{k=1}^{(N-1)/2} a_k \sin \omega k T$
	Even	$e^{-j[\omega(N-1)T/2 - \pi/2]} \displaystyle\sum_{k=1}^{N/2} b_k \sin [\omega(k - \frac{1}{2})T]$

where $\quad a_0 = h\left[\dfrac{(N-1)T}{2}\right] \qquad a_k = 2h\left[\left(\dfrac{N-1}{2} - k\right)T\right] \qquad b_k = 2h\left[\left(\dfrac{N}{2} - k\right)T\right]$

Now from Eqs. (9.8) and (9.9)

$$H(e^{j\omega T}) = e^{-j\omega(N-1)T/2}\left\{ h\left[\dfrac{(N-1)T}{2}\right] \right.$$
$$\left. + \sum_{n=0}^{(N-3)/2} 2h(nT) \cos \left[\omega\left(\dfrac{N-1}{2} - n\right)T\right] \right\}$$

and hence with $(N-1)/2 - n = k$ we have

$$H(e^{j\omega T}) = e^{-j\omega(N-1)T/2} \sum_{k=0}^{(N-1)/2} a_k \cos \omega k T$$

where

$$a_0 = h\left[\dfrac{(N-1)T}{2}\right] \tag{9.10}$$

$$a_k = 2h\left[\left(\dfrac{N-1}{2} - k\right)T\right] \tag{9.11}$$

Similarly, the frequency responses for the case of symmetrical impulse response with N even and for the two cases of antisymmetrical response simplify to the expressions summarized in Table 9.1.

Location of Zeros

The impulse response constraints of Eqs. (9.5) and (9.7) impose certain restrictions on the zeros of $H(z)$. For odd N, Eqs. (9.1), (9.5), and (9.7) yield

$$H(z) = \dfrac{1}{z^{(N-1)/2}} \sum_{n=0}^{(N-3)/2} h(nT)(z^{(N-1)/2-n} \pm z^{-[(N-1)/2-n]})$$
$$+ \tfrac{1}{2}h\left[\dfrac{(N-1)T}{2}\right](z^0 \pm z^0) \tag{9.12}$$

where the negative sign applies to the case of antisymmetrical impulse response. With $(N - 1)/2 - n = k$ Eq. (9.12) can be put in the form

$$H(z) = \frac{N(z)}{D(z)} = \frac{1}{z^{(N-1)/2}} \sum_{k=0}^{(N-1)/2} \frac{a_k}{2}(z^k \pm z^{-k})$$

where a_0 and a_k are given by Eqs. (9.10) and (9.11).

The zeros of $H(z)$ are the roots of $N(z)$ given by

$$N(z) = \sum_{k=0}^{(N-1)/2} a_k(z^k \pm z^{-k})$$

If z is replaced by z^{-1}, we have

$$N(z^{-1}) = \sum_{k=0}^{(N-1)/2} a_k(z^{-k} \pm z^k)$$

$$= \pm \sum_{k=0}^{(N-1)/2} a_k(z^k \pm z^{-k}) = \pm N(z)$$

The same relation holds for even N, as can easily be shown, and therefore if $z_i = r_i e^{j\psi_i}$ is a zero of $H(z)$, then $z_i^{-1} = e^{-j\psi_i}/r_i$ must also be a zero of $H(z)$. This has the following implications on the zero locations:

1. An arbitrary number of zeros can be located at $z_i = \pm 1$ since $z_i^{-1} = \pm 1$.

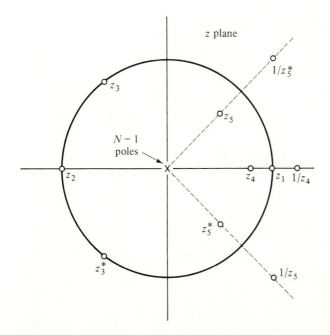

Figure 9.3 Typical zero-pole plot for a constant-delay nonrecursive filter.

2. An arbitrary number of complex-conjugate pairs of zeros can be located on the unit circle since

$$(z - z_i)(z - z_i^*) = (z - e^{j\psi_i})(z - e^{-j\psi_i}) = \left(z - \frac{1}{z_i^*}\right)\left(z - \frac{1}{z_i}\right)$$

3. Real zeros off the unit circle must occur in reciprocal pairs.
4. Complex zeros off the unit circle must occur in groups of four, namely z_i, z_i^*, and their reciprocals.

Polynomials with the above properties are often called *mirror-image polynomials*. A typical zero-pole plot for a constant-delay nonrecursive filter is shown in Fig. 9.3.

9.3 DESIGN USING THE FOURIER SERIES

As the frequency response of a nonrecursive filter is a periodic function of ω with period ω_s, it can be expressed as a Fourier series. We can write

$$H(e^{j\omega T}) = \sum_{n=-\infty}^{\infty} h(nT)e^{-j\omega nT} \tag{9.13}$$

where

$$h(nT) = \frac{1}{\omega_s} \int_{-\omega_s/2}^{\omega_s/2} H(e^{j\omega T})e^{j\omega nT} \, d\omega \tag{9.14}$$

and if $e^{j\omega T} = z$, Eq. (9.13) gives

$$H(z) = \sum_{n=-\infty}^{\infty} h(nT)z^{-n} \tag{9.15}$$

Hence with an analytic representation for the frequency response available, a corresponding transfer function can be readily derived. Unfortunately, however, this is noncausal and of infinite order. For a finite-order transfer function, the series of Eq. (9.15) can be truncated by assigning

$$h(nT) = 0 \qquad \text{for } |n| > \frac{N-1}{2}$$

in which case

$$H(z) = h(0) + \sum_{n=1}^{(N-1)/2} [h(-nT)z^n + h(nT)z^{-n}] \tag{9.16}$$

Causality can be brought about by multiplying $H(z)$ by $z^{-(N-1)/2}$ so that

$$H'(z) = z^{-(N-1)/2}H(z) \tag{9.17}$$

This modification is permissible since the amplitude response will remain unchanged and the group delay will be increased by a constant $(N-1)T/2$.

Example 9.1 Design a lowpass filter with a frequency response

$$H(e^{j\omega T}) \approx \begin{cases} 1 & \text{for} & |\omega| \le \omega_c \\ 0 & \text{for } \omega_c < |\omega| \le \dfrac{\omega_s}{2} \end{cases}$$

where ω_s is the sampling frequency.

SOLUTION From Eq. (9.14)

$$h(nT) = \frac{1}{\omega_s} \int_{-\omega_c}^{\omega_c} e^{j\omega nT} \, d\omega = \frac{1}{n\pi} \sin \omega_c nT$$

Hence Eqs. (9.16) and (9.17) yield

$$H(z) = z^{-(N-1)/2} \sum_{n=0}^{(N-1)/2} \frac{a_n}{2} (z^n + z^{-n})$$

where $$a_0 = h(0) \qquad a_n = 2h(nT)$$

The amplitude response of the preceding filter with ω_c and ω_s assumed to be 2 and 10 rad/s, respectively, is plotted in Fig. 9.4 for $N = 11, 21$, and 31. The passband and stopband oscillations observed are due to slow convergence in the Fourier series, which in turn, is caused by the discontinuity at the passband edge. These are known as *Gibbs' oscillations*. As N is increased, the frequency of these

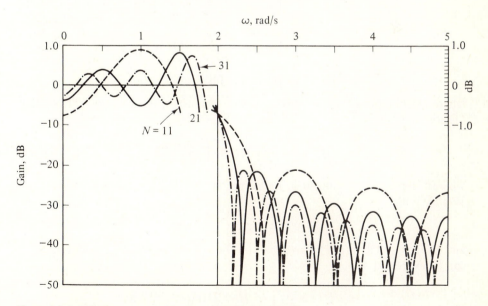

Figure 9.4 Amplitude response of lowpass filter (Example 9.1).

oscillations is seen to increase, and at both low and high frequencies their amplitude is decreased. The amplitude of the last passband ripple, however, and that of the first stopband ripple tend to remain virtually unchanged. This type of performance is often objectionable in practice, and ways must be sought for the reduction of Gibbs' oscillations.

A rudimentary method is to avoid discontinuities in the frequency response by introducing transition bands between passbands and stopbands [3]. For example, the response of the above lowpass filter could be redefined as

$$
H(e^{j\omega T}) \approx
\begin{cases}
1 & \text{for} \quad |\omega| \le \omega_p \\[2mm]
-\dfrac{\omega - \omega_a}{\omega_a - \omega_p} & \text{for} \quad \omega_p < |\omega| < \omega_a \\[2mm]
0 & \text{for} \quad \omega_a \le |\omega| \le \dfrac{\omega_s}{2}
\end{cases}
$$

9.4 USE OF WINDOW FUNCTIONS

An alternative and easy-to-apply technique for the reduction of Gibbs' oscillations is to precondition $h(nT)$ as given by Eq. (9.14) using a class of time-domain functions known as *window functions*.

Let

$$
H(z) = \mathscr{Z}[h(nT)] = \sum_{n=-\infty}^{\infty} h(nT)z^{-n} \tag{9.18}
$$

$$
W(z) = \mathscr{Z}[w(nT)] = \sum_{n=-\infty}^{\infty} w(nT)z^{-n} \tag{9.19}
$$

$$
H_w(z) = \mathscr{Z}[w(nT)h(nT)] \tag{9.20}
$$

where $w(nT)$ represents a window function. The use of the complex convolution (Theorem 2.7) gives

$$
H_w(z) = \frac{1}{2\pi j} \oint_\Gamma H(v)W\left(\frac{z}{v}\right)v^{-1}\, dv \tag{9.21}
$$

where Γ represents a contour in the common region of convergence of $H(v)$ and $W(z/v)$. With

$$
v = e^{j\Omega T} \quad \text{and} \quad z = e^{j\omega T}
$$

and $H(v)$ as well as $W(z/v)$ convergent on the unit circle of the v plane, Eq. (9.21) can be expressed as

$$
H_w(e^{j\omega T}) = \frac{T}{2\pi} \int_0^{2\pi/T} H(e^{j\Omega T})W(e^{j(\omega - \Omega)T})\, d\Omega \tag{9.22}
$$

For the sake of exposition let

$$H(e^{j\Omega T}) = \begin{cases} 1 & \text{for } 0 \le |\Omega| \le \omega_c \\ 0 & \text{for } \omega_c < |\Omega| \le \dfrac{\omega_s}{2} \end{cases}$$

Also let $W(e^{j\Omega T})$ be real with the form illustrated in Fig. 9.5b and assume that

$$W(e^{j\Omega T}) = 0 \qquad \text{for } \omega_m \le |\Omega| \le \frac{\omega_s}{2} \tag{9.23}$$

According to Eq. (9.22), $H_w(e^{j\omega T})$ can be formed by using the following graphical procedure:

1. Shift $W(e^{j\Omega T})$ to the right by ω, as in Fig. 9.5c.
2. Multiply $H(e^{j\Omega T})$ by $W(e^{j(\omega - \Omega)T})$, as in Fig. 9.5d.
3. Find the area in Fig. 9.5d.

As ω is varied in the range ω_1 to ω_5 through the discontinuity of $H(e^{j\Omega T})$, the successive values of $H_w(e^{j\omega T})$ can be determined as in Fig. 9.6. Evidently, with Eq. (9.23) satisfied and the area under the curve in Fig. 9.5b equal to unity, the derived function $H_w(e^{j\omega T})$ will be a close approximation for $H(e^{j\omega T})$, and furthermore it will be free of Gibbs' oscillations.

If $H(z)$ as given by Eq. (9.15) represents a constant-delay filter and $H_w(z)$ is to represent a finite-order constant-delay filter, $w(nT)$ must have the following time-domain properties: (1) it must be zero for $|n| > (N - 1)/2$; (2) for odd N, it must be symmetrical about sample $n = 0$. A typical window function is illustrated in Fig. 9.7.

In practice, the spectrum of $w(nT)$ is of the form shown in Fig. 9.8, where $k\omega_s/N$ is the width of the main lobe (k is a constant). This deviates from the ideal spectrum of Fig. 9.5b in that Eq. (9.23) is only approximately satisfied.

The effect of side lobes in $W(e^{j\Omega T})$ can be deduced from Fig. 9.5. As ω is varied in the passband of the filter, left- and right-hand side lobes will be swept in and out of the passband, respectively, and as their amplitudes vary with frequency, the area in Fig. 9.5d will oscillate about unity. Similarly, as ω is varied in the stopband, a number of left-hand-tail side lobes will be swept through the passband and the area in Fig. 9.5d will oscillate about zero. Clearly, side lobes in the spectrum of $w(nT)$ give rise to Gibbs' oscillations in the amplitude response of the filter. Therefore, for low passband ripple and large stopband attenuation the area under the side lobes should be a small proportion of that under the main lobe.

As ω is varied about the passband edge, $H_w(e^{j\omega T})$ will begin to decrease rapidly when

$$\omega = \omega_c - \frac{k\omega_s}{2N}$$

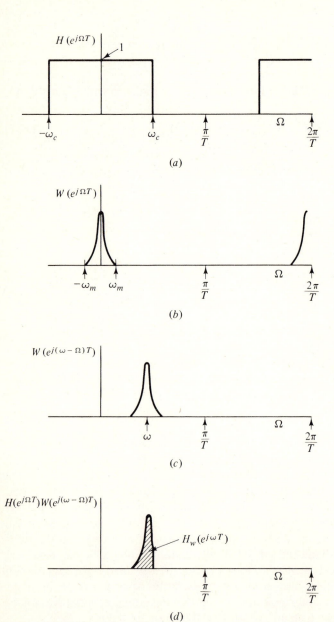

Figure 9.5 Complex convolution $(T = 2\pi)$.

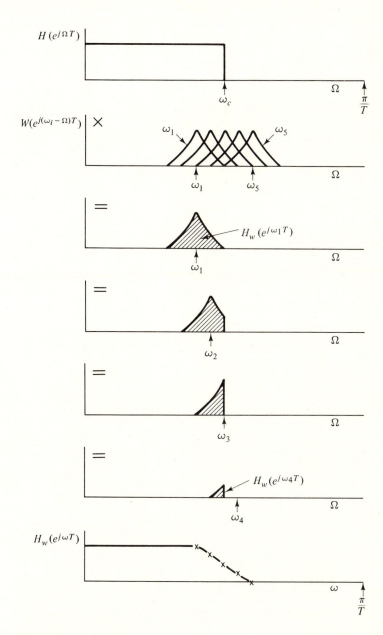

Figure 9.6 The effect of a window function.

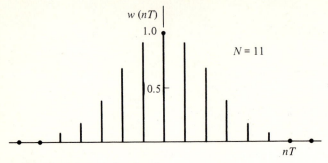

Figure 9.7 Typical window function.

and will slow down when

$$\omega = \omega_c + \frac{k\omega_s}{2N}$$

i.e., the main-lobe width determines the transition width of the resulting filter. For high selectivity, N and hence the order of the filter should be large, as may be expected, and k should be small.

The most frequently used window functions are [3, 4]:

1. Rectangular
2. Hann†
3. Hamming
4. Blackman
5. Kaiser

† Due to Julius von Hann and often referred to inaccurately as the Hanning window function.

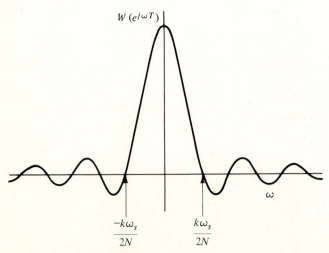

Figure 9.8 Typical spectrum of a window function.

Rectangular Window

The rectangular window is given by

$$w_R(nT) = \begin{cases} 1 & \text{for } |n| \le \dfrac{N-1}{2} \\ 0 & \text{otherwise} \end{cases} \tag{9.24}$$

This corresponds to the direct truncation of the Fourier series, and its effect on $H(e^{j\omega T})$ has been noted earlier.

The spectrum of $w_R(nT)$ can be deduced from Eqs. (9.19) and (9.24) as

$$W_R(e^{j\omega T}) = \sum_{n=-(N-1)/2}^{(N-1)/2} e^{-j\omega nT} = \frac{e^{j\omega(N-1)T/2} - e^{-j\omega(N+1)T/2}}{1 - e^{-j\omega T}}$$

$$= \frac{e^{j\omega NT/2} - e^{-j\omega NT/2}}{e^{j\omega T/2} - e^{-j\omega T/2}} = \frac{\sin(\omega NT/2)}{\sin(\omega T/2)} \tag{9.25}$$

This is plotted in Fig. 9.9 for $N = 11$ and $\omega_s = 10$ rad/s.

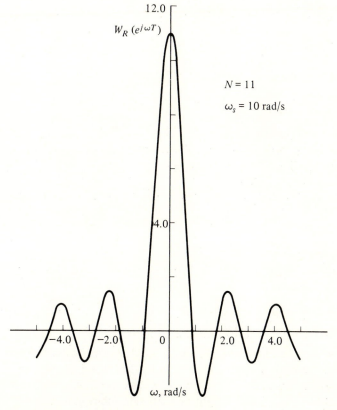

Figure 9.9 Spectrum of a rectangular window.

Since $W(e^{j\omega T}) = 0$ at $\omega = m\omega_s/N$ for $m = \pm 1, \pm 2, \ldots$, the main-lobe width is $2\omega_s/N$. The ripple ratio (RR) defined as

$$RR = \frac{100 \ (\text{maximum side-lobe amplitude})}{\text{main-lobe amplitude}} \%$$

is 22.3 percent for $N = 11$ and decreases slightly as N is increased.

Hann and Hamming Windows

The Hann and Hamming windows are given by

$$w_H(nT) = \begin{cases} \alpha + (1 - \alpha) \cos \dfrac{2\pi n}{N - 1} & \text{for } |n| \le \dfrac{N - 1}{2} \\ 0 & \text{otherwise} \end{cases} \tag{9.26}$$

The two differ in the choice of α. In the Hann window $\alpha = 0.5$, and in the Hamming window $\alpha = 0.54$.

The spectra of these windows can be related to that of the rectangular window. Equation (9.26) can be expressed as

$$w_H(nT) = w_R(nT)\left[\alpha + (1 - \alpha) \cos \frac{2\pi n}{N - 1}\right]$$

$$= \alpha w_R(nT) + \frac{1 - \alpha}{2} w_R(nT)(e^{j2\pi n/(N - 1)} + e^{-j2\pi n/(N - 1)})$$

and on using Theorem 2.4 we have

$$W_H(e^{j\omega T}) = \mathscr{Z}[w_H(nT)]_{z = e^{j\omega T}}$$

$$= \alpha W_R(e^{j\omega T}) + \frac{1 - \alpha}{2} W_R(e^{j[\omega T - 2\pi/(N - 1)]})$$

$$+ \frac{1 - \alpha}{2} W_R(e^{j[\omega T + 2\pi/(N - 1)]})$$

Now from Eq. (9.25)

$$W_H(e^{j\omega T}) = \frac{\alpha \sin (\omega NT/2)}{\sin (\omega T/2)} + \frac{1 - \alpha}{2} \frac{\sin [\omega NT/2 - N\pi/(N - 1)]}{\sin [\omega T/2 - \pi/(N - 1)]}$$

$$+ \frac{1 - \alpha}{2} \frac{\sin [\omega NT/2 + N\pi/(N - 1)]}{\sin [\omega T/2 + \pi/(N - 1)]} \tag{9.27}$$

Consequently, the spectra for the Hann and Hamming windows can be formed by shifting $W_R(e^{j\omega T})$ first to the right and then to the left by $2\pi/(N - 1)T$ and subsequently forming the sum in Eq. (9.27), as in Fig. 9.10. As can be observed, the second and third terms tend to cancel the right and left side lobes in $\alpha W_R(e^{j\omega T})$, and as a result both the Hann and Hamming windows have reduced side lobes

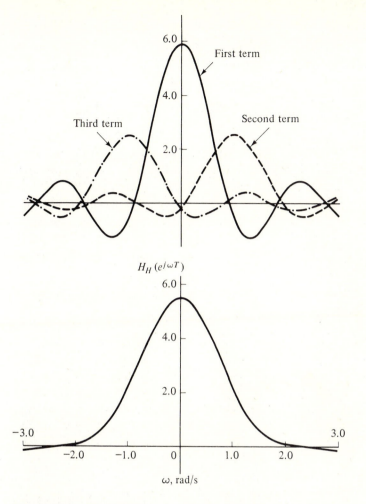

Figure 9.10 Spectrum of the Hann or Hamming window.

compared with those of the rectangular window. For $N = 11$ and $\omega_s = 10$ the ripple ratios for the two windows are 2.67 percent and 0.93 percent, respectively.

The first term in Eq. (9.27) is zero at

$$\omega = \frac{m\omega_s}{N}$$

and, similarly, the second and third terms are zero at

$$\omega = \left(m + \frac{N}{N-1}\right)\frac{\omega_s}{N} \quad \text{and} \quad \omega = \left(m - \frac{N}{N-1}\right)\frac{\omega_s}{N}$$

respectively, for $m = \pm 1, \pm 2, \ldots$. If $N \gg 1$, all three terms in Eq. (9.27) have their first common zero at $|\omega| \approx 2\omega_s/N$, and hence the main-lobe width for the Hann and Hamming windows is approximately $4\omega_s/N$.

Blackman Window

The Blackman window is similar to the preceding two and is given by

$$w_B(nT) = \begin{cases} 0.42 + 0.5 \cos \dfrac{2\pi n}{N-1} + 0.08 \cos \dfrac{4\pi n}{N-1} & \text{for } |n| \leq \dfrac{N-1}{2} \\ 0 & \text{otherwise} \end{cases}$$

The additional cosine term leads to a further reduction in the amplitude of Gibbs' oscillations. The ripple ratio for $N = 11$ and $\omega_s = 10$ is 0.124 percent. The main-lobe width, however, is increased to about $6\omega_s/N$.

The important parameters of the windows considered so far are summarized in Table 9.2.

Example 9.2 Redesign the lowpass filter of Example 9.1 using the Hann, Hamming, and Blackman windows.

SOLUTION The impulse response is the same as in Example 9.1, that is,

$$h(nT) = \frac{1}{n\pi} \sin \omega_c nT$$

On multiplying $h(nT)$ by the appropriate window function and then using Eqs. (9.20) and (9.17) we obtain

$$H'_w(z) = z^{-(N-1)/2} \sum_{n=0}^{(N-1)/2} \frac{a'_n}{2} (z^n + z^{-n})$$

Table 9.2 Summary of window parameters

Type of window	Main-lobe width	Ripple ratio, %		
		$N = 11$	$N = 21$	$N = 31$
Rectangular	$\dfrac{2\omega_s}{N}$	22.34	21.89	21.80
Hann	$\dfrac{4\omega_s}{N}$	2.62	2.67	2.67
Hamming	$\dfrac{4\omega_s}{N}$	1.47	0.93	0.82
Blackman	$\dfrac{6\omega_s}{N}$	0.08	0.12	0.12

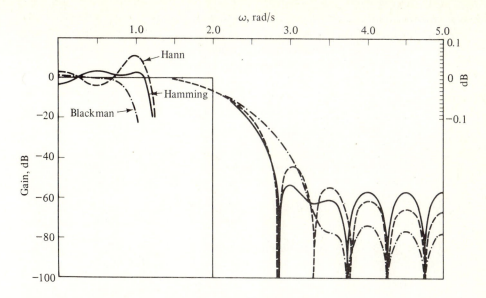

Figure 9.11 Amplitude response of lowpass filter (Example 9.2).

where $\qquad a_0' = w(0)h(0) \qquad a_n' = 2w(nT)h(nT)$

The amplitude responses for the three filters are given by

$$M(\omega) = \left| \sum_{n=0}^{(N-1)/2} a_n' \cos \omega nT \right|$$

These are plotted in Fig. 9.11 for $N = 21$ and $\omega_s = 10$. As expected, the passband ripple is reduced, and the minimum stopband attenuation as well as the transition width are increased progressively from the Hann to the Hamming to the Blackman window.

Kaiser Window

As can be seen in Table 9.2, a trade-off exists between ripple ratio and main-lobe width; i.e., as the ripple ratio is decreased from window to window, the main-lobe width is increased. The latter parameter can be adjusted by varying N. The ripple ratio, however, is approximately constant for a given window. Thus in order to achieve prescribed minimum stopband attenuation and passband ripple, the designer should first select a window with an appropriate ripple ratio and then choose N to achieve the prescribed transition width. As the number of possible ripple ratios is limited by the number of available windows, the designer may often have to settle for a window with an unnecessarily low ripple ratio, which will of course have an unnecessarily high main-lobe width. Subsequently, to achieve the desired transition width the value of N, and hence the order of the filter, will have to be increased to an unnecessarily high value. A window which overcomes this problem is that due to Kaiser. This window is given by

$$w_K(nT) = \begin{cases} \dfrac{I_0(\beta)}{I_0(\alpha)} & \text{for } |n| \le \dfrac{N-1}{2} \\ 0 & \text{otherwise} \end{cases} \qquad (9.28)$$

where α is an independent parameter and

$$\beta = \alpha \sqrt{1 - \left(\frac{2n}{N-1}\right)^2}$$

$I_0(x)$ is the zeroth-order Bessel function of the first kind. This can be evaluated to any desired degree of accuracy by using the rapidly converging series

$$I_0(x) = 1 + \sum_{k=1}^{\infty} \left[\frac{1}{k!}\left(\frac{x}{2}\right)^k\right]^2$$

The spectrum of $w_K(nT)$ can be readily obtained from Eq. (9.19) as

$$W_K(e^{j\omega T}) = w_K(0) + 2 \sum_{n=1}^{(N-1)/2} w_K(nT) \cos \omega nT$$

The continuous-time counterpart of the Kaiser window is

$$w_K(t) = \begin{cases} \dfrac{I_0(\beta)}{I_0(\alpha)} & \text{for } |t| \le \tau \\ 0 & \text{otherwise} \end{cases}$$

where $\qquad \beta = \alpha \sqrt{1 - \left(\dfrac{t}{\tau}\right)^2} \qquad$ and $\qquad \tau = \dfrac{(N-1)T}{2}$

The spectrum of $w_K(t)$ can be shown to be [5]

$$W_K(j\omega) = \frac{2}{I_0(\alpha)} \frac{\sin(\tau\sqrt{\omega^2 - \omega_a^2})}{\sqrt{\omega^2 - \omega_a^2}} \qquad (9.29)$$

where $\qquad \omega_a = \dfrac{\alpha}{\tau}$

If

$$W_K(j\omega) \approx 0 \qquad \text{for } |\omega| \ge \frac{\omega_s}{2}$$

the spectrum of sampled signal $w_K^*(t)$ or equivalently the spectrum of $w_K(nT)$ can be expressed as

$$W_K^*(j\omega) = W_K(e^{j\omega T}) \approx \frac{1}{T} W_K(j\omega) \qquad \text{for } 0 \le |\omega| < \frac{\omega_s}{2} \qquad (9.30)$$

according to Eq. (6.14). Hence from Eqs. (9.29) and (9.30) a closed-form but approximate expression for $W_K(e^{j\omega T})$ can be deduced as

$$W_K(e^{j\omega T}) \approx \frac{N-1}{\alpha I_0(\alpha)} \frac{\sin \left[\alpha \sqrt{(\omega/\omega_a)^2 - 1} \right]}{\sqrt{(\omega/\omega_a)^2 - 1}}$$

The attractive property of the Kaiser window is that the ripple ratio can be varied continuously from the low value in the Blackman window to the high value in the rectangular window by simply varying the parameter α. Also, as in the other windows, the main-lobe width can be adjusted by varying N. The effect of α on the ripple ratio and the main-lobe width B_m is illustrated in Fig. 9.12a and b for $N = 21$ and $\omega_s = 10$ rad/s. In a sense, the Kaiser window is a variable one which can be adjusted to suit almost any set of prescribed filter specifications.

Consider the lowpass-filter specification of Fig. 9.13a. The passband ripple and minimum stopband attenuation in decibels are given by

$$A_p = 20 \log \frac{1 + \delta}{1 - \delta} \qquad (9.31)$$

and

$$A_a = -20 \log \delta \qquad (9.32)$$

respectively, and the transition width in radians per second is

$$B_t = \omega_a - \omega_p$$

A filter with a passband ripple equal to or less than A_p', a minimum stopband attenuation equal to or greater than A_a', and a transition width B_t can readily be designed using the following procedure:

1. Determine $h(nT)$ using the Fourier-series approach of Sec. 9.3 assuming an idealized frequency response

$$H(e^{j\omega T}) = \begin{cases} 1 & \text{for } |\omega| \le \omega_c \\ 0 & \text{for } \omega_c < |\omega| \le \dfrac{\omega_s}{2} \end{cases}$$

(dotted line in Fig. 9.13a), where

$$\omega_c = \tfrac{1}{2}(\omega_p + \omega_a)$$

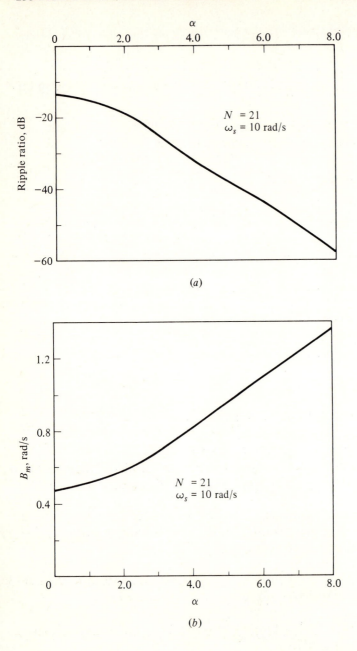

Figure 9.12 Kaiser window: (a) ripple ratio versus α, (b) main-lobe width versus α.

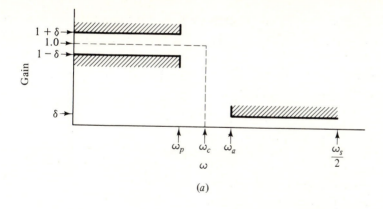

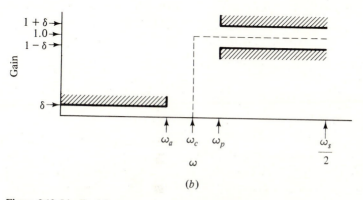

Figure 9.13 Idealized frequency responses: (a) lowpass filter. (b) highpass filter.

2. Choose δ in Eqs. (9.31) and (9.32) such that $A_p \leq A'_p$ and $A_a \geq A'_a$. A suitable value is

$$\delta = \min(\delta_1, \delta_2)$$

where

$$\delta_1 = 10^{-0.05A_{a'}} \qquad \delta_2 = \frac{10^{0.05A_{p'}} - 1}{10^{0.05A_{p'}} + 1}$$

3. Calculate A_a using Eq. (9.32).
4. Choose parameter α as follows:

$$\alpha = \begin{cases} 0 & \text{for} \quad A_a \leq 21 \\ 0.5842(A_a - 21)^{0.4} + 0.07886(A_a - 21) & \text{for } 21 < A_a \leq 50 \\ 0.1102(A_a - 8.7) & \text{for} \quad A_a > 50 \end{cases}$$

5. Choose parameter D as follows:

$$D = \begin{cases} 0.9222 & \text{for } A_a \leq 21 \\ \dfrac{A_a - 7.95}{14.36} & \text{for } A_a > 21 \end{cases}$$

Then select the lowest odd value of N satisfying the inequality

$$N \geq \frac{\omega_s D}{B_t} + 1$$

6. Form $w_K(nT)$ using Eq. (9.28).
7. Form

$$H'_w(z) = z^{-(N-1)/2} H_w(z) \qquad \text{where } H_w(z) = \mathcal{Z}[w_K(nT)h(nT)]$$

The above expressions for α and D are empirical relations developed by Kaiser [5].

Example 9.3 Design a lowpass filter satisfying the following specifications:

Passband ripple in frequency range 0 to 1.5 rad/s ≤ 0.1 dB
Minimum stopband attenuation in frequency range 2.5 to 5.0 rad/s ≥ 40 dB
Sampling frequency: 10 rad/s

SOLUTION From step 1 and Example 9.1

$$h(nT) = \frac{1}{n\pi} \sin \omega_c nT \qquad \text{where } \omega_c = \tfrac{1}{2}(1.5 + 2.5) = 2.0 \text{ rad/s}$$

Step 2 gives

$$\delta_1 = 10^{-0.05(40)} = 0.01$$

$$\delta_2 = \frac{10^{0.05(0.1)} - 1}{10^{0.05(0.1)} + 1} = 5.7564 \times 10^{-3}$$

Hence

$$\delta = 5.7564 \times 10^{-3}$$

and from step 3

$$A_a = 44.797 \text{ dB}$$

Steps 4 and 5 yield

$$\alpha = 3.9524 \qquad D = 2.5660$$

Hence

$$N \geq \frac{10(2.566)}{1} + 1 = 26.66$$

or

$$N = 27$$

Finally steps 6 and 7 give

$$H'_w(z) = z^{-(N-1)/2} \sum_{n=0}^{(N-1)/2} \frac{a'_n}{2} (z^n + z^{-n})$$

where

$$a'_0 = w_K(0)h(0) \qquad a'_n = 2w_K(nT)h(nT)$$

The numerical values of $h(nT)$ and $w_K(nT)h(nT)$ are given in Table 9.3, and the amplitude response achieved is plotted in Fig. 9.14. This satisfies the prescribed specifications.

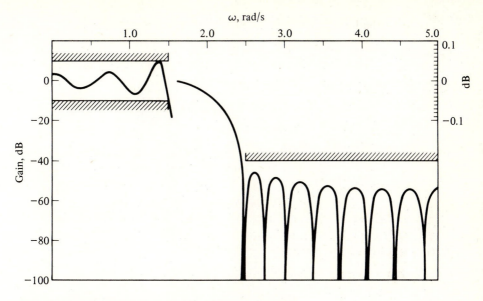

Figure 9.14 Amplitude response of lowpass filter (Example 9.3).

The above procedure can readily be applied to the design of highpass, bandpass, and bandstop filters. For the highpass specification of Fig. 9.13b, the transition width and idealized frequency response in step 1 can be taken as

$$B_t = \omega_p - \omega_a \quad \text{and} \quad H(e^{j\omega T}) = \begin{cases} 0 & \text{for} & |\omega| < \omega_c \\ 1 & \text{for } \omega_c \le |\omega| \le \dfrac{\omega_s}{2} \end{cases}$$

Table 9.3 Numerical values of $h(nT)$ and $w_K(nT)h(nT)$ (Example 9.3)

n	$h(nT)$	$w_K(nT)h(nT)$
0	4.00000×10^{-1}	4.00000×10^{-1}
1	3.02731×10^{-1}	2.99692×10^{-1}
2	9.35489×10^{-2}	8.98359×10^{-2}
3	-6.23660×10^{-2}	-5.69018×10^{-2}
4	-7.56827×10^{-2}	-6.42052×10^{-2}
5	0.0	0.0
6	5.04551×10^{-2}	3.45003×10^{-2}
7	2.67283×10^{-2}	1.57769×10^{-2}
8	-2.33872×10^{-2}	-1.15598×10^{-2}
9	-3.36367×10^{-2}	-1.34373×10^{-2}
10	0.0	0.0
11	2.75210×10^{-2}	6.23505×10^{-3}
12	1.55915×10^{-2}	2.39574×10^{-3}
13	-1.43921×10^{-2}	-1.32685×10^{-3}

where
$$\omega_c = \tfrac{1}{2}(\omega_a + \omega_p)$$

The remaining steps apply without modification.

For the bandpass-filter specification of Fig. 9.15a, the design must be based on the narrower of the two transition bands, i.e.,

$$B_t = \min\left[(\omega_{p1} - \omega_{a1}), (\omega_{a2} - \omega_{p2})\right] \tag{9.33}$$

Hence
$$H(e^{j\omega T}) = \begin{cases} 0 & \text{for} \quad 0 \le |\omega| < \omega_{c1} \\ 1 & \text{for } \omega_{c1} \le |\omega| \le \omega_{c2} \\ 0 & \text{for } \omega_{c2} < |\omega| \le \dfrac{\omega_s}{2} \end{cases} \tag{9.34}$$

where
$$\omega_{c1} = \omega_{p1} - \frac{B_t}{2} \qquad \omega_{c2} = \omega_{p2} + \frac{B_t}{2} \tag{9.35}$$

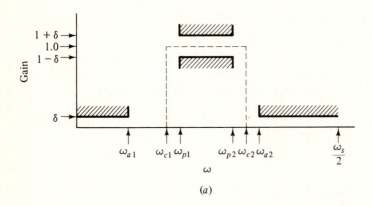

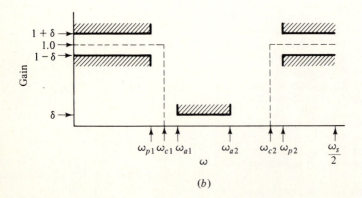

Figure 9.15 Idealized frequency responses: (a) bandpass filter. (b) bandstop filter.

Similarly, for the bandstop specification of Fig. 9.15b

$$B_t = \min \left[(\omega_{a1} - \omega_{p1}), (\omega_{p2} - \omega_{a2})\right]$$

and

$$H(e^{j\omega T}) = \begin{cases} 1 & \text{for} \quad 0 \le |\omega| \le \omega_{c1} \\ 0 & \text{for} \quad \omega_{c1} < |\omega| < \omega_{c2} \\ 1 & \text{for} \quad \omega_{c2} \le |\omega| < \dfrac{\omega_s}{2} \end{cases}$$

where

$$\omega_{c1} = \omega_{p1} + \frac{B_t}{2} \qquad \omega_{c2} = \omega_{p2} - \frac{B_t}{2}$$

Example 9.4 Design a bandpass filter satisfying the following specifications:

Minimum attenuation for $0 \le \omega \le 200$: 45 dB
Passband ripple for $400 < \omega < 600$: 0.2 dB
Minimum attenuation for $700 \le \omega \le 1000$: 45 dB
Sampling frequency: 2000 rad/s

SOLUTION From Eq. (9.33)

$$B_t = \min \left[(400 - 200), (700 - 600)\right] = 100$$

Hence from Eq. (9.35)

$$\omega_{c1} = 400 - 50 = 350 \qquad \omega_{c2} = 600 + 50 = 650$$

Step 1 of the design procedure yields

$$h(nT) = \frac{1}{\omega_s} \int_{-\omega_s/2}^{\omega_s/2} H(e^{j\omega T}) e^{j\omega nT} \, d\omega$$

$$= \frac{1}{\omega_s} \int_{0}^{\omega_s/2} \left[H(e^{j\omega T}) e^{j\omega nT} + H(e^{-j\omega T}) e^{-j\omega nT} \right] d\omega$$

and from Eq. (9.34)

$$h(nT) = \frac{1}{\omega_s} \int_{\omega_{c1}}^{\omega_{c2}} 2 \cos (\omega nT) \, d\omega$$

or

$$h(nT) = \frac{1}{\pi n} (\sin \omega_{c2} nT - \sin \omega_{c1} nT)$$

Now according to step 2,

$$\delta_1 = 10^{-0.05(45)} = 5.6234 \times 10^{-3}$$

$$\delta_2 = \frac{10^{0.05(0.2)} - 1}{10^{0.05(0.2)} + 1} = 1.1512 \times 10^{-2}$$

and

$$\delta = 5.6234 \times 10^{-3} \qquad \text{or} \quad A_a = 45 \text{ dB}$$

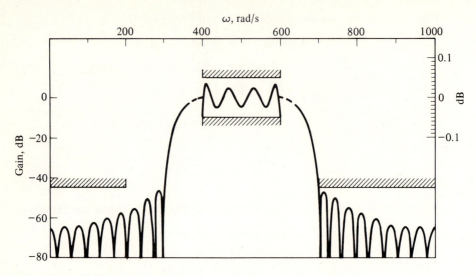

Figure 9.16 Amplitude response of bandpass filter (Example 9.4).

The design can be completed as in Example 9.3. The resulting values for α, D, and N are

$$\alpha = 3.9754 \qquad D = 2.580 \qquad N = 53$$

The amplitude response achieved is plotted in Fig. 9.16.

The above expression for $h(nT)$ can also be used to design lowpass or highpass filters by letting $\omega_{c1} = 0$ and $\omega_{c2} = \omega_c$ or $\omega_{c1} = \omega_c$ and $\omega_{c2} = \omega_s/2$.

The design of nonrecursive filters satisfying prescribed specifications can be carried out by using Program B.15.

9.5 DESIGN BASED ON NUMERICAL-ANALYSIS FORMULAS

A signal $x(t)$ whose values are known at $t = nT$ for $n = 0, 1, 2, \ldots$ can be interpolated, differentiated, or integrated by using the many available numerical-analysis formulas [6, 7]. The most basic of these are the interpolation formulas, which are derived directly from the Taylor series.

The commonly used interpolation formulas are the Gregory-Newton forward- and backward-difference formulas and the Bessel, Everett, and Stirling central-difference formulas. The value of $x(t)$ at $t = nT + pT$, where $0 \le p < 1$, is given by the Gregory-Newton formulas as

$$x(nT + pT) = (1 + \Delta)^p x(nT) = \left[1 + p\Delta + \frac{p(p-1)}{2!}\Delta^2 + \cdots\right]x(nT)$$

or $\quad x(nT + pT) = (1 - \nabla)^{-p}x(nT) = \left[1 + p\nabla + \dfrac{p(p + 1)}{2!}\nabla^2 + \cdots\right]x(nT)$

where $\quad \Delta x(nT) = x(nT + T) - x(nT) \qquad$ and $\qquad \nabla x(nT) = x(nT) - x(nT - T)$

On the other hand, the Stirling formula yields

$$x(nT + pT) = \left[1 + \frac{p^2}{2!}\delta^2 + \frac{p^2(p^2 - 1)}{4!}\delta^4 + \cdots\right]x(nT)$$

$$+ \frac{p}{2}[\delta x(nT - \tfrac{1}{2}T) + \delta x(nT + \tfrac{1}{2}T)]$$

$$+ \frac{p(p^2 - 1)}{2(3!)}[\delta^3 x(nT - \tfrac{1}{2}T) + \delta^3 x(nT + \tfrac{1}{2}T)]$$

$$+ \frac{p(p^2 - 1)(p^2 - 2^2)}{2(5!)}[\delta^5 x(nT - \tfrac{1}{2}T) + \delta^5 x(nT + \tfrac{1}{2}T)] + \cdots$$

$$(9.36)$$

where $\qquad\qquad \delta x(nT + \tfrac{1}{2}T) = x(nT + T) - x(nT) \qquad\qquad (9.37)$

The differential of $x(t)$ at $t = nT + pT$ can be expressed as

$$\frac{dx(t)}{dt}\bigg|_{t = nT + pT} = \frac{dx(nT + pT)}{dp} \times \frac{dp}{dt}$$

$$= \frac{1}{T}\frac{dx(nT + pT)}{dp} \qquad\qquad (9.38)$$

and therefore the above interpolation formulas lead directly to corresponding differentiation formulas. Similarly, integration formulas can be derived by writing

$$\int_{nT}^{t_2} x(t)\, dt = T \int_0^{p_2} x(nT + pT)\, dp$$

where $\qquad\qquad nT < t_2 \leq nT + T \qquad$ and $\qquad p_2 = \dfrac{t_2 - nT}{T}$

that is, $0 < p_2 \leq 1$.

Formulas like the above can be used to design nonrecursive filters which will perform interpolation, differentiation, or integration. Let $x(nT)$ and $y(nT)$ be the input and output in a nonrecursive filter and assume that $y(nT)$ is equal to the desired function of $x(t)$, that is,

$$y(nT) = f[x(t)] \qquad\qquad (9.39)$$

For example, if the differential of $x(t)$ at $t = nT + pT$ is required, let

$$y(nT) = \frac{dx(t)}{dt}\bigg|_{t = nT + pT} \qquad\qquad (9.40)$$

By choosing an appropriate formula for $f[x(t)]$ and then eliminating operators using their definitions, Eq. (9.39) can be put in the form

$$y(nT) = \sum_{i=-k}^{m} a_i x(nT - iT)$$

Thus the desired transfer function is obtained as

$$H(z) = \sum_{n=-k}^{m} h(nT)z^{-n}$$

For the case of a forward- or central-difference formula, $H(z)$ is noncausal. Hence for real-time applications it will be necessary to multiply $H(z)$ by an appropriate negative power of z.

Example 9.5 A signal $x(t)$ is sampled at a rate $1/T$ Hz. Design a sixth-order differentiating filter with a time-domain response

$$y(nT) = \left. \frac{dx(t)}{dt} \right|_{t=nT}$$

Use the Stirling formula.

SOLUTION From Eqs. (9.36) and (9.38)

$$y(nT) = \left. \frac{dx(t)}{dt} \right|_{t=nT} = \frac{1}{2T}[\delta x(nT - \tfrac{1}{2}T) + \delta x(nT + \tfrac{1}{2}T)]$$

$$- \frac{1}{12T}[\delta^3 x(nT - \tfrac{1}{2}T) + \delta^3 x(nT + \tfrac{1}{2}T)]$$

$$+ \frac{1}{60T}[\delta^5 x(nT - \tfrac{1}{2}T) + \delta^5 x(nT + \tfrac{1}{2}T)] + \cdots$$

Now on using Eq. (9.37)

$$\delta x(nT - \tfrac{1}{2}T) + \delta x(nT + \tfrac{1}{2}T) = x(nT + T) - x(nT - T)$$

$$\delta^3 x(nT - \tfrac{1}{2}T) + \delta^3 x(nT + \tfrac{1}{2}T) = x(nT + 2T) - 2x(nT + T)$$

$$+ 2x(nT - T) - x(nT - 2T)$$

$$\delta^5 x(nT - \tfrac{1}{2}T) + \delta^5 x(nT + \tfrac{1}{2}T) = x(nT + 3T) - 4x(nT + 2T)$$

$$+ 5x(nT + T) - 5x(nT - T)$$

$$+ 4x(nT - 2T) - x(nT - 3T)$$

Hence

$$y(nT) = \frac{1}{60T}[x(nT + 3T) - 9x(nT + 2T) + 45x(nT + T)$$

$$- 45x(nT - T) + 9x(nT - 2T) - x(nT - 3T)]$$

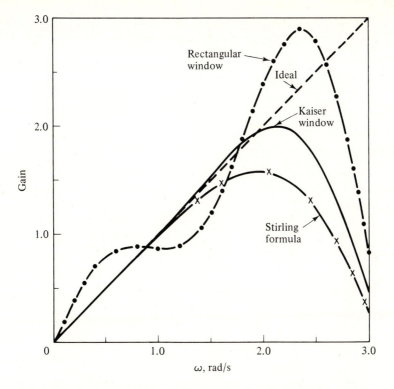

Figure 9.17 Amplitude response of differentiating filter (Examples 9.5 and 9.6).

and therefore

$$H(z) = \frac{1}{60T}(z^3 - 9z^2 + 45z - 45z^{-1} + 9z^{-2} - z^{-3})$$

The filter has an antisymmetrical impulse response and is noncausal. A causal filter can be obtained by multiplying $H(z)$ by z^{-3}. The amplitude response of the filter is plotted in Fig. 9.17 for $\omega_s = 2\pi$.

Differentiating filters can also be designed by employing the Fourier series method of Sec. 9.3. An analog differentiator is characterized by

$$H(s) = s$$

Hence a corresponding digital differentiator can be designed by assigning

$$H(e^{j\omega T}) = j\omega \qquad \text{for } 0 \le |\omega| < \frac{\omega_s}{2} \tag{9.41}$$

Then on assuming a periodic frequency response, the appropriate impulse response can be determined from Eq. (9.14). Gibbs' oscillations due to the transition in $H(e^{j\omega T})$ at $\omega = \omega_s/2$ can be reduced as before, by using the window technique.

Example 9.6 Redesign the differentiating filter of Example 9.5 employing the Fourier-series method. Use (a) a rectangular window and (b) the Kaiser window with $\alpha = 3.0$.

SOLUTION (a) From Eqs. (9.41) and (9.14)

$$h(nT) = \frac{1}{\omega_s} \int_{-\omega_s/2}^{\omega_s/2} j\omega e^{j\omega nT}\, d\omega = \frac{-1}{\omega_s} \int_{0}^{\omega_s/2} 2\omega \sin (\omega nT)\, d\omega$$

On integrating by parts

$$h(nT) = \frac{1}{nT} \cos \pi n - \frac{1}{n^2 \pi T} \sin \pi n$$

or

$$h(nT) = \begin{cases} 0 & \text{for } n = 0 \\ \dfrac{1}{nT} \cos \pi n & \text{otherwise} \end{cases}$$

Now by using the rectangular window with $N = 7$, we deduce

$$H_w(z) = \frac{1}{6T} (2z^3 - 3z^2 + 6z - 6z^{-1} + 3z^{-2} - 2z^{-3})$$

(b) Similarly, the Kaiser window yields

$$H_w(z) = \sum_{n=-3}^{3} w_K(nT) h(nT) z^{-n}$$

where $w_K(nT)$ can be computed from Eq. (9.28).

The amplitude responses for the two filters are compared in Fig. 9.17 with the corresponding response obtained in Example 9.5.

As before, the parameter α in the Kaiser window can be increased to increase the in-band accuracy or decreased to increase the bandwidth. Thus the Kaiser differentiator has the important advantage that it can be adjusted to suit the application.

9.6 COMPARISON BETWEEN RECURSIVE AND NONRECURSIVE DESIGNS

Before a solution is sought for the approximation problem, a choice must be made between a recursive and a nonrecursive design. In recursive filters the poles of the transfer function can be placed anywhere inside the unit circle. A consequence of this degree of freedom is that high selectivity can easily be achieved with low-order transfer functions. In nonrecursive filters, on the other hand, with the poles fixed at the origin, high selectivity can be achieved only by using a relatively high order for the transfer function. For the same filter specification the required order in a nonrecursive design can be as high as 5 to 10 times that in a recursive design. For example, the bandpass-filter specification in Example 9.4 can be met using a nonrecursive filter of order 52 or a recursive elliptic filter of order 8. In practice, the cost of a digital filter tends to increase and its speed tends to decrease as the order of the transfer function is increased. Hence, for high-selectivity applications where the delay characteristic is of secondary importance, the choice is expected to be a recursive design.

For certain applications constant group delay is mandatory, e.g., in data transmission. Although it may sometimes be possible to find a satisfactory recursive approximation, usually much computation may be required for equalization. By contrast, a nonrecursive constant-delay filter can readily be designed using the techniques of this chapter.

The nonrecursive filter is naturally suited for certain specific applications, e.g., to perform numerical differentiation or integration or to simulate a differential equation; it is also suited for applications where the prescribed specifications cannot be met by conventional Butterworth, Tschebyscheff, or elliptic approximations, e.g., if a triangular amplitude response is required for some reason.

A very important advantage of the nonrecursive filter is that it can be implemented by using the fast Fourier-transform method. This possibility is considered in Sec. 13.12.

REFERENCES

1. L. R. Rabiner and B. Gold, "Theory and Application of Digital Signal Processing," Prentice-Hall, Englewood Cliffs, N.J., 1975.
2. A. V. Oppenheim and R. W. Schafer, "Digital Signal Processing," Prentice-Hall, Englewood Cliffs, N.J., 1975.
3. F. F. Kuo and J. F. Kaiser, "System Analysis by Digital Computer," chap. 7, Wiley, New York, 1966.
4. R. B. Blackman, "Data Smoothing and Prediction," Addison-Wesley, Reading, Mass., 1965.
5. J. F. Kaiser, Nonrecursive Digital Filter Design Using the I_0-sinh Window Function, *Proc. 1974 IEEE Int. Symp. Circuit Theory*, pp. 20–23.
6. R. Butler and E. Kerr, "An Introduction to Numerical Methods," Pitman, London, 1962.
7. C. E. Fröberg, "Introduction to Numerical Analysis," Addison-Wesley, Reading, Mass., 1965.

ADDITIONAL REFERENCES

Constantinides, A. G.: The Design of Linear Phase Nonrecursive Lowpass Digital Filters Having Equiripple Passbands, *Proc. Imp. Coll. Symp. Digital Filtering*, September 1971.
Crochiere, R. E., and L. R. Rabiner: Optimum FIR Digital Filter Implementations for Decimation, Interpolation, and Narrow-Band Filtering, *IEEE Trans. Acoust., Speech, Signal Process.*, vol. ASSP-23, pp. 444–456, October 1975.
McClellan, J. H., T. W. Parks, and L. R. Rabiner: A Computer Program for Designing Optimum FIR Linear Phase Digital Filters, *IEEE Trans. Audio Electroacoust.*, vol. AU-21, pp. 506–526, December 1973.

Rabiner, L. R.: Approximate Design Relationships for Low-Pass FIR Digital Filters, *IEEE Trans. Audio Electroacoust.*, vol. AU-21, pp. 456–460, October 1973.

Rabiner, L. R., and O. Herrmann: On the Design of Optimum FIR Low-Pass Filters with Even Impulse Response Duration, *IEEE Trans. Audio Electroacoust.*, vol. AU-21, pp. 329–336, August 1973.

———, J. F. Kaiser, O. Herrmann, and M. T. Dolan: Some Comparisons between FIR and IIR Digital Filters, *Bell Syst. Tech. J.*, vol. 53, pp. 305–331, February 1974.

———, J. H. McClellan, and T. W. Parks: FIR Digital Filter Design Techniques Using Weighted Chebyshev Approximation, *Proc. IEEE*, vol. 63, pp. 595–610, April 1975.

Schafer, R. W., and L. R. Rabiner: A Digital Signal Processing Approach to Interpolation, *Proc. IEEE*, vol. 61, pp. 692–702, June 1973.

PROBLEMS

9.1 (*a*) A nonrecursive filter is characterized by the transfer function

$$H(z) = \frac{1 + 2z + 3z^2 + 4z^3 + 3z^4 + 2z^5 + z^6}{z^6}$$

Find the group delay of the filter.

(*b*) Repeat part (*a*) if

$$H(z) = \frac{1 - 2z + 3z^2 - 4z^3 + 3z^4 - 2z^5 + z^6}{z^6}$$

9.2 Figure P9.2 shows the zero-pole plots of two nonrecursive filters. Check each filter for phase-response linearity.

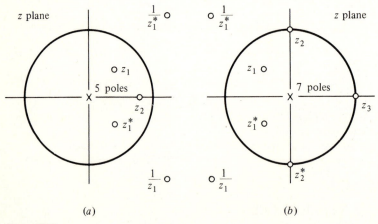

(*a*)   (*b*)

Figure P9.2

9.3 Design a nonrecursive bandstop filter assuming an idealized frequency response

$$H(e^{j\omega T}) = \begin{cases} 1 & \text{for} \quad |\omega| \le \omega_{c1} \\ 0 & \text{for } \omega_{c1} < |\omega| < \omega_{c2} \\ 1 & \text{for } \omega_{c2} \le |\omega| \le \dfrac{\omega_s}{2} \end{cases}$$

9.4 A digital filter is required with a frequency response

$$H(e^{j\omega T}) \approx \begin{cases} 0 & \text{for} & |\omega| < \omega_{c1} \\ 1 & \text{for } \omega_{c1} \leq |\omega| \leq \omega_{c2} \\ 0 & \text{for } \omega_{c2} < |\omega| < \omega_{c3} \\ 1 & \text{for } \omega_{c3} \leq |\omega| \leq \omega_{c4} \\ 0 & \text{for } \omega_{c4} < |\omega| \leq \dfrac{\omega_s}{2} \end{cases}$$

Obtain a causal transfer function by using the Fourier-series method.

9.5 (*a*) Derive an exact expression for the spectrum of the Blackman window.

(*b*) By using the result in part (*a*) and assuming that $N \gg 1$, show that the main-lobe width for the Blackman window is approximately $6\omega_s/N$.

9.6 The Bartlett (or triangular) window is given by

$$w_{BA}(nT) = \begin{cases} 1 - \dfrac{2|n|}{N-1} & \text{for } |n| \leq \dfrac{N-1}{2} \\ 0 & \text{otherwise} \end{cases}$$

(*a*) Assuming that $w_{BA}(t)$ is bandlimited, obtain an approximate expression for $W_{BA}(e^{j\omega T})$.

(*b*) Estimate the main-lobe width if $N \gg 1$.

(*c*) Estimate the ripple ratio if $N \gg 1$.

Hint: See Prob. 6.11.

9.7 Show that the Kaiser window includes the rectangular window as a special case.

9.8 (*a*) Design a nonrecursive highpass filter in which

$$H(e^{j\omega T}) \approx \begin{cases} 0 & \text{for} & |\omega| < 2.5 \text{ rad/s} \\ 1 & \text{for } 2.5 \leq |\omega| \leq 5.0 \text{ rad/s} \end{cases}$$

Use the rectangular window and assume that $\omega_s = 10$ rad/s and $N = 11$.

(*b*) Repeat part (*a*) with $N = 21$ and $N = 31$. Compare the three designs.

9.9 Redesign the filter of Prob. 9.8 using the Hann, Hamming, and Blackman windows in turn. Assume that $N = 21$. Compare the three designs.

9.10 Design a nonrecursive bandpass filter in which

$$H(e^{j\omega T}) \approx \begin{cases} 0 & \text{for} & |\omega| < 400 \text{ rad/s} \\ 1 & \text{for } 400 \leq |\omega| \leq 600 \text{ rad/s} \\ 0 & \text{for } 600 < |\omega| \leq 1000 \text{ rad/s} \end{cases}$$

Use the Hann window and assume that $\omega_s = 2000$ rad/s and $N = 21$.

9.11 Design a nonrecursive bandstop filter with a frequency response

$$H(e^{j\omega T}) \approx \begin{cases} 1 & \text{for} & |\omega| \leq 300 \text{ rad/s} \\ 0 & \text{for } 300 < |\omega| < 700 \text{ rad/s} \\ 1 & \text{for } 700 \leq |\omega| \leq 1000 \text{ rad/s} \end{cases}$$

Use the Hamming window and assume that $\omega_s = 2000$ rad/s and $N = 21$.

9.12 A digital filter is required with a frequency response like that depicted in Fig. P9.12. Obtain a nonrecursive design using the rectangular window. Assume that $\omega_s = 10$ rad/s and $N = 21$.

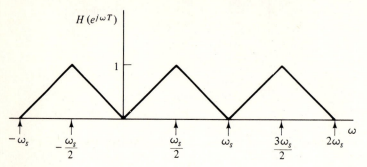

Figure P9.12

9.13 Design a nonrecursive filter with a frequency response like that depicted in Fig. P9.13. Use a sampling frequency $\omega_s = 10$ rad/s and assume that $N = 21$.

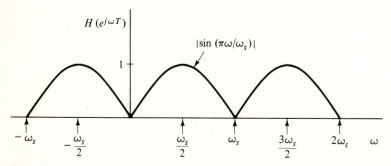

Figure P9.13

9.14 Design a nonrecursive lowpass filter satisfying the following specifications:

$$A_p \text{ (passband ripple)} \leq 0.1 \text{ dB} \qquad A_a \geq 44.0 \text{ dB}$$

$$\omega_p = 20 \text{ rad/s} \qquad \omega_a = 30 \text{ rad/s} \qquad \omega_s = 100 \text{ rad/s}$$

9.15 Design a nonrecursive highpass filter satisfying the following specifications:

$$A_p \leq 0.3 \text{ dB} \qquad A_a \geq 45.0 \text{ dB} \qquad \omega_p = 3 \text{ rad/s} \qquad \omega_a = 2 \text{ rad/s} \qquad \omega_s = 10 \text{ rad/s}$$

9.16 Design a nonrecursive bandpass filter satisfying the following specifications:

$$A_p \leq 0.5 \text{ dB} \qquad A_a \geq 35.0 \text{ dB} \qquad \omega_{p1} = 40 \text{ rad/s} \qquad \omega_{p2} = 60 \text{ rad/s}$$

$$\omega_{a1} = 20 \text{ rad/s} \qquad \omega_{a2} = 80 \text{ rad/s} \qquad \omega_s = 200 \text{ rad/s}$$

9.17 Design a nonrecursive bandstop filter satisfying the following specifications:

$$A_p \leq 0.2 \text{ dB} \qquad A_a \geq 40 \text{ dB} \qquad \omega_{p1} = 1000 \text{ rad/s} \qquad \omega_{p2} = 4000 \text{ rad/s}$$

$$\omega_{a1} = 2000 \text{ rad/s} \qquad \omega_{a2} = 3000 \text{ rad/s} \qquad \omega_s = 10,000 \text{ rad/s}$$

9.18 Redesign the filter of Prob. 9.16 as a recursive filter. Use an elliptic approximation. Compare this design with the nonrecursive one.

9.19 (*a*) Show that

$$\mathscr{Z} \nabla^k x(nT) = (1 - z^{-1})^k X(z)$$

(*b*) A signal $x(t)$ is sampled at a rate of 2π rad/s. Design a sixth-order differentiating filter in which

$$y(nT) \approx \left. \frac{dx(t)}{dt} \right|_{t=nT}$$

Use the Gregory-Newton backward-difference formula.

9.20 The phase response $\theta(\omega)$ of an analog filter is sampled at $\omega = n\Omega$ for $n = 0, 1, 2, \ldots$. Design a sixth-order digital filter which can be used to generate the group delay of the analog filter. Use the Stirling formula.

9.21 A signal $x(t)$ is sampled at a rate of 2π rad/s. Design a sixth-order integrating filter in which

$$y(nT) \approx \int_{nT}^{(n+1)T} x(t) \, dt$$

Use the Gregory-Newton backward-difference formula.

9.22 Two digital filters are to be cascaded. The sampling frequency in the first filter is 2π rad/s, and that in the second is 4π rad/s. Design a sixth-order interface using the Gregory-Newton backward-difference formula. *Hint:* Design an interpolating filter.

# TEN

RANDOM SIGNALS

10.1 INTRODUCTION

The methods of analysis considered so far assume deterministic signals. Frequently in digital filters and communication systems in general random signals are encountered, e.g., the noise generated by an A/D converter or the noise generated by an amplifier. Signals of this type can assume an infinite number of waveforms, and measurement will at best yield a set of typical waveforms. Despite the lack of a complete description, many statistical attributes of a random signal can be determined from a statistical description of the signal.

The time- and frequency-domain statistical attributes of random signals as well as the effect of filtering on such signals can be studied by using the concept of a random process.

This chapter provides a brief description of random processes. The main results are presented in terms of continuous-time random signals and are then extended to discrete-time signals by using the interrelation between the Fourier and z transforms. The chapter begins with a brief summary of the essential features of random variables. Detailed discussions of random variables and processes can be found in Refs. 1 to 4.

10.2 RANDOM VARIABLES

Definition

Consider an experiment which may have a finite or infinite number of random outcomes, and let $\zeta_1, \zeta_2, \ldots$ be the possible outcomes. A set S comprising all ζ can be constructed, and a number $\mathbf{x}(\zeta)$ can be assigned to each ζ according to some rule. The function $\mathbf{x}(\zeta)$ or simply $\mathbf{x}$ whose domain is set S and whose range is a set of numbers is called a *random variable*.

Probability Distribution Function

A random variable $\mathbf{x}$ may assume values in a certain range (x_1, x_2) where x_1 can be as low as $-\infty$ and x_2 as high as $+\infty$. The probability of observing random variable $\mathbf{x}$ below value x is referred to as the *probability distribution function* of $\mathbf{x}$ and is denoted by

$$P_{\mathbf{x}}(x) = \Pr\,[\mathbf{x} \leq x]$$

Probability-Density Function

The derivative of $P_{\mathbf{x}}(x)$ with respect to x is called the *probability-density function* of $\mathbf{x}$ and is denoted by

$$p_{\mathbf{x}}(x) = \frac{dP_{\mathbf{x}}(x)}{dx}$$

A fundamental property of $p_{\mathbf{x}}(x)$ is

$$\int_{-\infty}^{\infty} p_{\mathbf{x}}(x)\,dx = 1$$

since range $(-\infty, +\infty)$ must necessarily include the value of $\mathbf{x}$. Also

$$\Pr\,[x_1 \leq \mathbf{x} \leq x_2] = \int_{x_1}^{x_2} p_{\mathbf{x}}(x)\,dx$$

Uniform Probability Density

In many situations there is no preferred range for the random variable. If such is the case, the probability density is said to be *uniform* and is given by

$$p_{\mathbf{x}}(x) = \begin{cases} \dfrac{1}{x_2 - x_1} & \text{for } x_1 \leq x \leq x_2 \\ 0 & \text{otherwise} \end{cases}$$

Gaussian Probability Density

Very common in nature is the gaussian probability density, given by

$$p_{\mathbf{x}}(x) = \frac{1}{\sigma\sqrt{2\pi}}e^{-(x-\eta)^2/2\sigma^2} \qquad -\infty \leq x \leq \infty \tag{10.1}$$

The parameters σ and η are constants.

There are many other important probability-density functions, e.g., binomial, Poisson, and Rayleigh [1].

Joint Distributions

An experiment may have two sets of random outcomes, say $\zeta_{x1}, \zeta_{x2}, \ldots$ and $\zeta_{y1}, \zeta_{y2}, \ldots$. For example, in an experiment of target practice, the hit position can be described in terms of two coordinates. Experiments of this type necessitate two random variables, say $\mathbf{x}$ and $\mathbf{y}$. The probability of observing $\mathbf{x}$ and $\mathbf{y}$ below x and y, respectively, is said to be the *joint distribution function* of $\mathbf{x}$ and $\mathbf{y}$ and is denoted by

$$P_{\mathbf{xy}}(x, y) = \Pr\left[\mathbf{x} \leq x, \mathbf{y} \leq y\right]$$

The joint probability-density function of $\mathbf{x}$ and $\mathbf{y}$ is denoted by

$$p_{\mathbf{xy}}(x, y) = \frac{\partial^2 P_{\mathbf{xy}}(x, y)}{\partial x \, \partial y}$$

The range $(-\infty, \infty)$ must include $\mathbf{x}$ and $\mathbf{y}$, and hence

$$\int_{-\infty}^{\infty} \int_{-\infty}^{\infty} p_{\mathbf{xy}}(x, y) \, dx \, dy = 1$$

Two random variables $\mathbf{x}$ and $\mathbf{y}$ representing outcomes $\zeta_{x1}, \zeta_{x2}, \ldots$ and $\zeta_{y1}, \zeta_{y2}, \ldots$ of an experiment are said to be *statistically independent* if the occurrence of any outcome ζ_x does not influence the occurrence of any outcome ζ_y and vice versa. A necessary and sufficient condition for statistical independence is

$$p_{\mathbf{xy}}(x, y) = p_{\mathbf{x}}(x)p_{\mathbf{y}}(y) \tag{10.2}$$

Mean Values and Moments

The *mean* or *expected value* of random variable $\mathbf{x}$ is defined as

$$E\{\mathbf{x}\} = \int_{-\infty}^{\infty} x p_{\mathbf{x}}(x) \, dx$$

Similarly, if

$$\mathbf{z} = f(\mathbf{x}, \mathbf{y})$$

then

$$E\{\mathbf{z}\} = \int_{-\infty}^{\infty} z p_{\mathbf{z}}(z) \, dz \tag{10.3}$$

If $\mathbf{z}$ is a single-valued function of $\mathbf{x}$ and $\mathbf{y}$ and $x \leq \mathbf{x} \leq x + dx$, $y \leq \mathbf{y} \leq y + dy$, then $z \leq \mathbf{z} \leq z + dz$. Hence

$$\Pr\left[z \leq \mathbf{z} \leq z + dz\right] = \Pr\left[x \leq \mathbf{x} \leq x + dx, y \leq \mathbf{y} \leq y + dy\right]$$

or

$$p_{\mathbf{z}}(z) \, dz = p_{\mathbf{xy}}(x, y) \, dx \, dy$$

and from Eq. (10.3)

$$E\{\mathbf{z}\} = \int_{-\infty}^{\infty} \int_{-\infty}^{\infty} f(x, y)p_{\mathbf{xy}}(x, y) \, dx \, dy$$

Actually this is a general relation which holds for multivalued functions as well [1].

For

$$\mathbf{z} = \mathbf{xy}$$

we have

$$E\{\mathbf{xy}\} = \int_{-\infty}^{\infty} \int_{-\infty}^{\infty} xy p_{\mathbf{xy}}(x, y)\, dx\, dy$$

and if $\mathbf{x}$ and $\mathbf{y}$ are independent variables, the use of Eq. (10.2) yields

$$E\{\mathbf{xy}\} = \int_{-\infty}^{\infty} x p_{\mathbf{x}}(x)\, dx \int_{-\infty}^{\infty} y p_{\mathbf{y}}(y)\, dy = E\{\mathbf{x}\}E\{\mathbf{y}\} \tag{10.4}$$

The nth moment of $\mathbf{x}$ is defined as

$$E\{\mathbf{x}^n\} = \int_{-\infty}^{\infty} x^n p_{\mathbf{x}}(x)\, dx$$

Similarly, the nth central moment of $\mathbf{x}$ is

$$E\{(\mathbf{x} - E\{\mathbf{x}\})^n\} = \int_{-\infty}^{\infty} (x - E\{\mathbf{x}\})^n p_{\mathbf{x}}(x)\, dx \tag{10.5}$$

The second central moment is also known as the *variance* and is given by

$$\sigma_{\mathbf{x}}^2 = E\{(\mathbf{x} - E\{\mathbf{x}\})^2\} = E\{\mathbf{x}^2\} - E^2\{\mathbf{x}\} \tag{10.6}$$

If

$$\mathbf{z} = a_1 \mathbf{x}_1 + a_2 \mathbf{x}_2$$

where a_1, a_2 are constants and $\mathbf{x}_1$, $\mathbf{x}_2$ are independent random variables, then from Eqs. (10.4) and (10.5)

$$\sigma_{\mathbf{z}}^2 = a_1^2 \sigma_{\mathbf{x}_1}^2 + a_2^2 \sigma_{\mathbf{x}_2}^2$$

In general, if

$$\mathbf{z} = \sum_{i=1}^{n} a_i \mathbf{x}_i$$

and $\mathbf{x}_1, \mathbf{x}_2, \ldots, \mathbf{x}_n$ are statistically independent random variables, then

$$\sigma_{\mathbf{z}}^2 = \sum_{i=1}^{n} a_i^2 \sigma_{\mathbf{x}_i}^2 \tag{10.7}$$

Example 10.1 (*a*) Find the mean and variance for the case of uniform probability density:

$$p_{\mathbf{x}}(x) = \begin{cases} \dfrac{1}{x_2 - x_1} & \text{for } x_1 \leq x \leq x_2 \\ 0 & \text{otherwise} \end{cases}$$

(b) Repeat part (a) for the gaussian density

$$p_\mathbf{x}(x) = \frac{1}{\sigma\sqrt{2\pi}} e^{-(x-\eta)^2/2\sigma^2} \qquad -\infty \le x \le \infty$$

SOLUTION (a)

$$E\{\mathbf{x}\} = \int_{x_1}^{x_2} \frac{x}{x_2 - x_1} \, dx = \tfrac{1}{2}(x_1 + x_2) \tag{10.8}$$

$$E\{\mathbf{x}^2\} = \int_{x_1}^{x_2} \frac{x^2}{x_2 - x_1} \, dx = \frac{x_2^3 - x_1^3}{3(x_2 - x_1)} \tag{10.9}$$

Hence from Eq. (10.6)

$$\sigma_\mathbf{x}^2 = \frac{(x_2 - x_1)^2}{12} \tag{10.10}$$

(b) $$E\{\mathbf{x}\} = \frac{1}{\sigma\sqrt{2\pi}} \int_{-\infty}^{\infty} x e^{-(x-\eta)^2/2\sigma^2} \, dx$$

With $x = y + \eta$

$$E\{\mathbf{x}\} = \frac{1}{\sigma\sqrt{2\pi}} \left(\int_{-\infty}^{\infty} y e^{-y^2/2\sigma^2} \, dy + \eta \int_{-\infty}^{\infty} e^{-y^2/2\sigma^2} \, dy \right)$$

The first integral is zero because the integrand is an odd function of y, whereas the second integral is $\sigma\sqrt{2\pi}$, according to standard tables. Hence

$$E\{\mathbf{x}\} = \eta$$

Now

$$E\{\mathbf{x}^2\} = \frac{1}{\sigma\sqrt{2\pi}} \int_{-\infty}^{\infty} x^2 e^{-(x-\eta)^2/2\sigma^2} \, dx$$

and, as before,

$$E\{\mathbf{x}^2\} = \sigma^2 + \eta^2 \qquad \text{or} \qquad \sigma_\mathbf{x}^2 = \sigma^2$$

10.3 RANDOM PROCESSES

A random process is an extension of the concept of a random variable.

Definition

Consider an experiment with possible random outcomes $\zeta_1, \zeta_2, \ldots$. A set S comprising all ζ can be constructed and a waveform $\mathbf{x}(t, \zeta)$ can be assigned to each ζ according to some rule. The set of waveforms obtained is called an *ensemble*, and

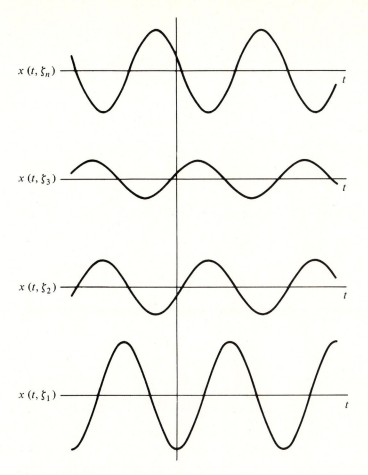

$x(t, \zeta_n)$

$x(t, \zeta_3)$

$x(t, \zeta_2)$

$x(t, \zeta_1)$

Figure 10.1 A random process.

each individual waveform is said to be a *sample function*. Set S, the ensemble, and the probability description associated with S constitute a *random process*.

The concept of a random process can be illustrated by an example. Suppose that a large number of radio receivers of one model are receiving a carrier signal transmitted by a station. With the receivers located at different distances from the station, the amplitude and phase of the received carrier will be different at each receiver. As a result, the set of the received waveforms, illustrated in Fig. 10.1, can be described by

$$\mathbf{x}(t, \zeta) = \mathbf{z} \cos (\omega_c t + \mathbf{y})$$

where $\mathbf{z}$ and $\mathbf{y}$ are random variables and $\zeta = \zeta_1, \zeta_2, \ldots.$ The set of all possible waveforms that might be received constitutes an ensemble and the ensemble together with the probability densities of $\mathbf{z}$ and $\mathbf{y}$ constitute a random process.

Notation

A random process can be represented by $\mathbf{x}(t, \zeta)$ or in a simplified notation by $\mathbf{x}(t)$. Depending on the circumstances, $\mathbf{x}(t, \zeta)$ can represent one of four things:

1. The ensemble, if t and ζ are variables.
2. A sample function, if t is variable and ζ is fixed.
3. A random variable, if t is fixed and ζ is variable.
4. A single number, if t and ζ are fixed.

10.4 FIRST- AND SECOND-ORDER STATISTICS

For a fixed value of t, $\mathbf{x}(t)$ is a random variable representing the instantaneous values of the various sample functions over the ensemble. The probability distribution and probability density of $\mathbf{x}(t)$ are denoted by

$$P(x; t) = \text{Pr}\ [\mathbf{x}(t) \leq x] \quad \text{and} \quad p(x; t) = \frac{\partial P(x; t)}{\partial x}$$

respectively. These two equations constitute the first-order statistics of the random process.

At any two instants t_1 and t_2, $\mathbf{x}(t_1)$ and $\mathbf{x}(t_2)$ are distinct random variables. Their joint probability distribution and joint probability density depend on t_1 and t_2 in general, and are denoted by

$$P(x_1, x_2; t_1, t_2) = \text{Pr}\ [\mathbf{x}(t_1) \leq x_1, \mathbf{x}(t_2) \leq x_2]$$

and
$$p(x_1, x_2; t_1, t_2) = \frac{\partial^2 P(x_1, x_2; t_1, t_2)}{\partial x_1\, \partial x_2}$$

respectively. These two equations constitute the second-order statistics of the random process.

Higher-order statistics can be similarly defined.

Example 10.2 Find the first-order probability density $p(x; t)$ for random process

$$\mathbf{x}(t) = \mathbf{y}t - 2$$

where $\mathbf{y}$ is a random variable with a probability density

$$p_y(y) = \frac{1}{\sqrt{2\pi}} e^{-y^2/2} \qquad -\infty \leq y \leq \infty$$

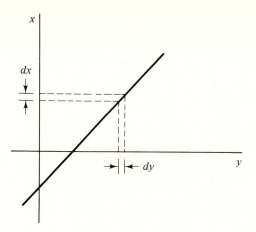

Figure 10.2 Function $x = yt - 2$ Example 10.2).

SOLUTION If x and y are possible values of $\mathbf{x}(t)$ and $\mathbf{y}$, then

$$x = yt - 2 \quad \text{or} \quad y = \frac{1}{t}(x + 2)$$

From Fig. 10.2

$$\Pr\left[x \le \mathbf{x} \le x + |dx|\right] = \Pr\left[y \le \mathbf{y} \le y + |dy|\right]$$

that is,

$$p_{\mathbf{x}}(x)|dx| = p_{\mathbf{y}}(y)|dy| \quad \text{or} \quad p_{\mathbf{x}}(x) = \frac{p_{\mathbf{y}}(y)}{|dx/dy|}$$

Since

$$\frac{dx}{dy} = t$$

we obtain

$$p(x; t) = p_{\mathbf{x}}(x) = \frac{1}{|t|\sqrt{2\pi}} e^{-(x+2)^2/2t^2} \qquad -\infty \le x \le \infty$$

Example 10.3 Find the first-order probability density $p(x; t)$ of the random process

$$\mathbf{x}(t) = \cos\left(\omega_c t + \mathbf{y}\right)$$

where $\mathbf{y}$ is a random variable with probability density

$$p_{\mathbf{y}}(y) = \begin{cases} \dfrac{1}{2\pi} & \text{for } 0 \le y \le 2\pi \\ 0 & \text{otherwise} \end{cases}$$

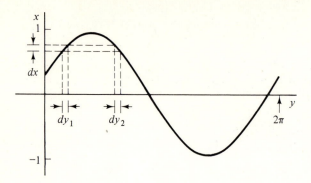

Figure 10.3 Function $x = \cos(\omega_c t + y)$ (Example 10.3).

SOLUTION If x and y are possible values of $\mathbf{x}(t)$ and $\mathbf{y}$, then

$$x = \cos(\omega_c t + y)$$

and from Fig. 10.3

$$\Pr\left[x \leq \mathbf{x} \leq x + |dx|\right] = \Pr\left[y_1 \leq \mathbf{y} \leq y_1 + |dy_1|\right]$$
$$+ \Pr\left[y_2 \leq \mathbf{y} \leq y_2 + |dy_2|\right]$$

or

$$p_x(x)|dx| = p_y(y_1)|dy_1| + p_y(y_2)|dy_2|$$

Hence

$$p_x(x) = \frac{p_y(y_1)}{|x'(y_1)|} + \frac{p_y(y_2)}{|x'(y_2)|}$$

where

$$x'(y) = \frac{dx}{dy} = -\sin(\omega_c t + y) = -\sqrt{1 - x^2}$$

Since

$$|x'(y_1)| = |x'(y_2)|$$

we deduce

$$p(x; t) = p_x(t) = \begin{cases} \dfrac{1}{\pi\sqrt{1 - x^2}} & \text{for } |x| < 1 \\ 0 & \text{otherwise} \end{cases}$$

10.5 MOMENTS AND AUTOCORRELATION

The first-order statistics give the mean, mean square, and other moments of a random process at any instant t. From Sec. 10.2

$$E\{\mathbf{x}(t)\} = \int_{-\infty}^{\infty} x p(x; t)\, dx$$

$$E\{\mathbf{x}^2(t)\} = \int_{-\infty}^{\infty} x^2 p(x; t)\, dx$$

The second-order statistics give the *autocorrelation function* of a random process, which is defined as

$$r_x(t_1, t_2) = E\{x(t_1)x(t_2)\} = \int_{-\infty}^{\infty} \int_{-\infty}^{\infty} x_1 x_2 \, p(x_1, x_2; t_1, t_2) \, dx_1 \, dx_2$$

The autocorrelation is a measure of the interdependence between the instantaneous signal values at $t = t_1$ and those at $t = t_2$. This is the most important attribute of a random process, as it leads to a frequency-domain description of the process.

Example 10.4 (*a*) Find the mean, mean square, and autocorrelation for the process in Example 10.2. (*b*) Repeat part (*a*) for the process in Example 10.3.

SOLUTION (*a*) From part (*b*) of Example 10.1 and from Example 10.2

$$E\{x(t)\} = -2 \qquad E\{x^2(t)\} = t^2 + 4$$

$$r_x(t_1, t_2) = E\{(yt_1 - 2)(yt_2 - 2)\} = t_1 t_2 E\{y^2\} - 2(t_1 + t_2)E\{y\} + 4$$

and since y is gaussian with

$$E\{y\} = 0 \qquad E\{y^2\} = 1$$

we have

$$r_x(t_1, t_2) = t_1 t_2 + 4$$

(*b*) From Example 10.3

$$E\{x(t)\} = \frac{1}{\pi} \int_{-1}^{1} \frac{x}{\sqrt{1 - x^2}} \, dx = 0 \tag{10.11}$$

$$E\{x^2(t)\} = \frac{1}{\pi} \int_{-1}^{1} \frac{x^2}{\sqrt{1 - x^2}} \, dx = \tfrac{1}{2}$$

$$r_x(t_1, t_2) = E\{\cos(\omega_c t_1 + y) \cos(\omega_c t_2 + y)\}$$
$$= \tfrac{1}{2} \cos(\omega_c t_1 - \omega_c t_2) - \tfrac{1}{2}E\{\cos(\omega_c t_1 + \omega_c t_2 + 2y)\}$$
$$= \tfrac{1}{2} \cos[\omega_c(t_1 - t_2)] \tag{10.12}$$

10.6 STATIONARY PROCESSES

A random process is said to be *strictly stationary* if $x(t)$ and $x(t + T)$ have the same statistics (all orders) for any value of T. If the mean of $x(t)$ is constant and its autocorrelation depends only on $t_2 - t_1$, that is,

$$E\{x(t)\} = \text{const.} \qquad E\{x(t_1)x(t_2)\} = r_x(t_2 - t_1)$$

the process is called *wide-sense stationary*. A strictly stationary process is also stationary in the wide sense; however, the converse is not necessarily true. The process of part (*b*) Example 10.4 is wide-sense stationary, according to Eqs. (10.11) and (10.12); that of part (*a*) of Example 10.4, however, is not stationary.

10.7 FREQUENCY-DOMAIN REPRESENTATION

The frequency-domain representation of deterministic signals is normally in terms of amplitude, phase, and energy-density spectra (see Chap. 6). Although such representations are possible for random processes [1], they are avoided in practice because of the mathematical difficulties associated with infinite-energy signals (see Sec. 6.3). Usually, random processes are represented in terms of power-density spectra.

Consider a signal $x(t)$, and let

$$x_T(t) = \begin{cases} x(t) & \text{for } |t| \leq T \\ 0 & \text{otherwise} \end{cases}$$

The average power of $x(t)$ over interval $(-T, T)$ is

$$P_T = \frac{1}{2T} \int_{-T}^{T} x^2(t) \, dt = \frac{1}{2T} \int_{-\infty}^{\infty} x_T^2(t) \, dt$$

and by virtue of Parseval's formula (Theorem 6.8)

$$P_T = \int_{-\infty}^{\infty} \frac{|X_T(j\omega)|^2}{2T} \frac{d\omega}{2\pi} \qquad \text{where } X_T(j\omega) = \mathscr{F} x_T(t)$$

Clearly, the quantity

$$\frac{|X_T(j\omega)|^2}{2T}$$

represents the average power per unit bandwidth (in hertz) and can be referred to as the *power spectral density* (PSD) of $x_T(t)$. If $x(t)$ is a sample function of random process $\mathbf{x}(t)$, we can define

$$\text{PSD of } \mathbf{x}_T(t) = E\left\{ \frac{|X_T(j\omega)|^2}{2T} \right\}$$

and since $\mathbf{x}_T(t) \rightarrow \mathbf{x}(t)$ as $T \rightarrow \infty$, we can define

$$\text{PSD of } \mathbf{x}(t) = S_\mathbf{x}(\omega) = \lim_{T \to \infty} E\left\{ \frac{|X_T(j\omega)|^2}{2T} \right\} \qquad (10.13)$$

The plot of $S_\mathbf{x}(\omega)$ versus frequency is said to be the *power-density spectrum* of the process.

For stationary processes, the PSD is the Fourier transform of the autocorrelation function, as will now be shown. From Eq. (10.13)

$$S_x(\omega) = \lim_{T \to \infty} E\left\{\frac{X_T(j\omega)X_T^*(j\omega)}{2T}\right\}$$

$$= \lim_{T \to \infty} \frac{1}{2T} E\left\{\int_{-T}^{T} x(t_2)e^{-j\omega t_2}\, dt_2 \int_{-T}^{T} x(t_1)e^{j\omega t_1}\, dt_1\right\}$$

$$= \lim_{T \to \infty} \frac{1}{2T} \int_{-T}^{T}\int_{-T}^{T} E\{x(t_1)x(t_2)\}e^{-j\omega(t_2 - t_1)}\, dt_1\, dt_2$$

For a wide-sense-stationary process

$$E\{x(t_1)x(t_2)\} = r_x(t_2 - t_1)$$

and hence we can write

$$S_x(\omega) = \lim_{T \to \infty} \frac{1}{2T}\int_{-T}^{T}\int_{-T}^{T} f(t_2 - t_1)\, dt_1\, dt_2 \tag{10.14}$$

where

$$f(t_2 - t_1) = r_x(t_2 - t_1)e^{-j\omega(t_2 - t_1)} \tag{10.15}$$

The preceding double integral represents the volume under the surface $y = f(t_2 - t_1)$ and above the square region in Fig. 10.4. Since $f(t_2 - t_1)$ is constant on any line of the form

$$t_2 = t_1 + c$$

the volume over the elemental area bounded by the square region and the lines

$$t_2 = t_1 + \tau \qquad t_2 = t_1 + \tau + d\tau$$

is approximately constant. From the geometry of Fig. 10.4

$$dA \approx \begin{cases} (2T - \tau)\, d\tau & \text{for } \tau \geq 0 \\ (2T + \tau)\, d\tau & \text{for } \tau < 0 \end{cases}$$

or in general

$$dA = (2T - |\tau|)\, d\tau$$

Hence the elemental volume for $t_2 - t_1 = \tau$ is

$$dV = f(\tau)(2T - |\tau|)\, d\tau$$

and from Eq. (10.14)

$$S_x(\omega) = \lim_{T \to \infty} \frac{1}{2T}\int_{-2T}^{2T} f(\tau)(2T - |\tau|)\, d\tau$$

$$= \int_{-\infty}^{\infty} f(\tau) \lim_{T \to \infty}\left(1 - \frac{|\tau|}{2T}\right) d\tau = \int_{-\infty}^{\infty} f(\tau)\, d\tau$$

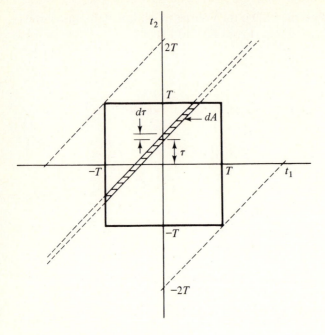

Figure 10.4 Domain of $y = f(t_2 - t_1)$.

Therefore, from Eq. (10.15)

$$S_x(\omega) = \int_{-\infty}^{\infty} r_x(\tau)e^{-j\omega\tau}\, d\tau \qquad (10.16)$$

and if

$$\int_{-\infty}^{\infty} |r_x(\tau)|\, d\tau < \infty$$

we can write

$$r_x(\tau) = E\{x(t)x(t+\tau)\} = \frac{1}{2\pi}\int_{-\infty}^{\infty} S_x(\omega)e^{j\omega\tau}\, d\omega \qquad (10.17)$$

that is,

$$r_x(\tau) \leftrightarrow S_x(\omega)$$

by virtue of Theorem 6.1.

Example 10.5 Find the PSD of the process in Example 10.3.

SOLUTION From part (b) of Example 10.4 [Eq. (10.12)] the autocorrelation of the process can be expressed as

$$r(\tau) = \tfrac{1}{2}\cos\omega_c\tau$$

Hence from Eq. (10.16) and Table 6.1

$$S_x(\omega) = \frac{\pi}{2}[\delta(\omega - \omega_c) + \delta(\omega + \omega_c)]$$

The autocorrelation is an even function of τ, that is,

$$r_x(\tau) = r_x(-\tau)$$

as can be easily shown, and $S_x(\omega)$ is an even function of ω by definition. Equations (10.16) and (10.17) can thus be written as

$$S_x(\omega) = \int_{-\infty}^{\infty} r_x(\tau) \cos{(\omega\tau)} \, d\tau$$

$$r_x(\tau) = \frac{1}{2\pi} \int_{-\infty}^{\infty} S_x(\omega) \cos{(\omega\tau)} \, d\omega$$

If $\omega = 0$,

$$S_x(0) = \int_{-\infty}^{\infty} r_x(\tau) \, d\tau$$

i.e., the total area under the autocorrelation function equals the PSD at zero frequency. The average power of $x(t)$ is given by

$$\text{Average power} = E\{x^2(t)\} = r_x(0) = \int_{-\infty}^{\infty} S_x(\omega) \frac{d\omega}{2\pi}$$

as is to be expected.

A random process whose PSD is constant at all frequencies is said to be a *white-noise process*. If

$$S_x(\omega) = K$$

we have

$$r_x(\tau) = K\delta(\tau)$$

i.e., the autocorrelation of a white-noise process is an impulse at the origin.

10.8 DISCRETE-TIME RANDOM PROCESSES

The concept of a random process can readily be extended to discrete-time random signals by simply assigning discrete-time waveforms to the possible outcomes of the experiment. The mean, mean square, and autocorrelation of a discrete-time process $x(nT)$ can be expressed as

$$E\{x(nT)\} = \int_{-\infty}^{\infty} xp(x; nT) \, dx$$

$$E\{x^2(nT)\} = \int_{-\infty}^{\infty} x^2 p(x; nT) \, dx$$

$$r_x(kT) = E\{x(nT)x(nT + kT)\}$$

A frequency-domain representation for a discrete-time process can be deduced by using the interrelation between the two-sided z transform and the Fourier transform (see Sec. 6.4). We can write

$$\mathscr{Z}r_x(kT) = \sum_{k=-\infty}^{\infty} r(kT)z^{-k} = R_x(z)$$

and from Eq. (6.11)

$$R_x(e^{j\omega T}) = \mathscr{F}r_x^*(\tau) = S_x^*(\omega) \tag{10.18}$$

where

$$r_x^*(\tau) = E\{\mathbf{x}^*(t)\mathbf{x}^*(t+\tau)\}$$

$$\mathbf{x}^*(t) = \sum_{n=-\infty}^{\infty} \mathbf{x}(nT)\delta(t-nT)$$

$$\tau = kT$$

Therefore, from Eqs. (10.13) and (10.18)

$$R_x(e^{j\omega T}) = \lim_{T_0 \to \infty} E\left\{\frac{|X_{T_0}^*(j\omega)|^2}{2T_0}\right\}$$

where

$$X_{T_0}^*(j\omega) = \mathscr{F}\mathbf{x}_{T_0}^*(t)$$

and

$$\mathbf{x}_{T_0}^*(t) = \begin{cases} \mathbf{x}^*(t) & \text{for } |t| \le T_0 \\ 0 & \text{otherwise} \end{cases}$$

In effect, the z transform of the autocorrelation of discrete-time process $\mathbf{x}(nT)$ evaluated on the unit circle $|z| = 1$ is numerically equal to the PSD of sampled process $\mathbf{x}^*(t)$. This can thus be referred to as the PSD of discrete-time process $\mathbf{x}(nT)$. Consequently, we can write

$$\mathscr{Z}r_x(kT) = S_x(z)$$

where

$$r_x(kT) = \frac{1}{2\pi j}\oint_\Gamma S_x(z)z^{k-1}\,dz \tag{10.19}$$

by virtue of Eq. (2.4).

10.9 FILTERING OF DISCRETE-TIME RANDOM SIGNALS

If a discrete-time random signal is passed through a digital filter, we expect the PSD of the output signal to be related to that of the input signal. This indeed is the case, as will now be shown.

Consider a filter characterized by $H(z)$, and let $\mathbf{x}(n)$ and $\mathbf{y}(n)$ be the input and output processes, respectively. From the convolution summation (see Sec. 1.7)

$$\mathbf{y}(i) = \sum_{p=-\infty}^{\infty} h(p)\mathbf{x}(i-p) \qquad \mathbf{y}(j) = \sum_{q=-\infty}^{\infty} h(q)\mathbf{x}(j-q)$$

and hence
$$E\{\mathbf{y}(i)\mathbf{y}(j)\} = E\left\{\sum_{q=-\infty}^{\infty} \sum_{p=-\infty}^{\infty} h(p)h(q)\mathbf{x}(i-p)\mathbf{x}(j-q)\right\}$$

With $j = i + k$ and $q = p + n$ we have
$$r_y(k) = \sum_{n=-\infty}^{\infty} \sum_{p=-\infty}^{\infty} h(p)h(p+n)E\{\mathbf{x}(i-p)\mathbf{x}(i-p+k-n)\}$$

or
$$r_y(k) = \sum_{n=-\infty}^{\infty} g(n)r_x(k-n)$$

where
$$g(n) = \sum_{p=-\infty}^{\infty} h(p)h(p+n)$$

The use of Theorem 2.6(a) gives
$$S_y(z) = \mathcal{Z}r_y(k) = \mathcal{Z}g(k)\mathcal{Z}r_x(k) = G(z)S_x(z) \tag{10.20}$$

Now
$$G(z) = \mathcal{Z}\sum_{p=-\infty}^{\infty} h(p)h(p+n) = \sum_{n=-\infty}^{\infty} \sum_{p=-\infty}^{\infty} h(p)h(p+n)z^{-n}$$

and with $n = k - p$
$$G(z) = \sum_{k=-\infty}^{\infty} h(k)z^{-k} \sum_{p=-\infty}^{\infty} h(p)(z^{-1})^{-p} = H(z)H(z^{-1}) \tag{10.21}$$

Therefore, from Eqs. (10.20) and (10.21)
$$S_y(z) = H(z)H(z^{-1})S_x(z) \tag{10.22}$$

or
$$S_y(e^{j\omega T}) = |H(e^{j\omega T})|^2 S_x(e^{j\omega T})$$

i.e., the PSD of the output process is equal to the squared amplitude response of the filter times the PSD of the input process.

Example 10.6 The output of a digital filter is given by
$$y(n) = x(n) + 0.8y(n-1)$$

The input of the filter is a random signal with zero mean and variance σ_x^2; successive values of $x(n)$ are statistically independent. (a) Find the output PSD of the filter. (b) Obtain an expression for the average output power.

SOLUTION (a)
$$r_x(k) = E\{\mathbf{x}(n)\mathbf{x}(n+k)\}$$

For $k = 0$
$$r_x(k) = E\{\mathbf{x}^2(n)\} = \sigma_x^2$$

For $k \neq 0$ the use of Eq. (10.4) gives

$$r_x(k) = E\{x(n)\}E\{x(n+k)\} = 0$$

Hence $\quad\quad\quad r_x(k) = \sigma_x^2 \delta(k) \quad\quad$ and $\quad\quad S_x(z) = \sigma_x^2$

Now from Eq. (10.22)

$$S_y(z) = \sigma_x^2 H(z)H(z^{-1})$$

where $\quad\quad\quad\quad\quad\quad H(z) = \dfrac{z}{z - 0.8}$

(b) From Eq. (10.19)

$$\text{Output power} = E\{y^2(n)\} = r_y(0) = \frac{1}{2\pi j}\oint_{\Gamma} \sigma_x^2 H(z)H(z^{-1})z^{-1}\,dz$$

and if Γ is taken to be the unit circle $|z| = 1$, we can let $z = e^{j\omega T}$, in which case

$$\text{Output power} = \frac{1}{\omega_s}\int_0^{\omega_s} \sigma_x^2 H(e^{j\omega T})H(e^{-j\omega T})\,d\omega$$

REFERENCES

1. A. Papoulis, " Probability, Random Variables, and Stochastic Processes," McGraw-Hill, New York, 1965.
2. W. B. Davenport, Jr. and W. L. Root, "Random Signals and Noise," McGraw-Hill, New York, 1958.
3. B. P. Lathi, "An Introduction to Random Signals and Communication Theory," International Textbook, Scranton, 1968.
4. G. R. Cooper and C. D. McGillem, " Probabilistic Methods of Signal and System Analysis," Holt, New York, 1971.

PROBLEMS

10.1 A random variable x has a probability-density function

$$p_x(x) = \begin{cases} Ke^{-x} & \text{for } 1 \leq x \leq \infty \\ 0 & \text{otherwise} \end{cases}$$

(a) Find K.
(b) Find $\Pr[0 \leq x \leq 2]$.

10.2 A random variable x has a probability-density function

$$p_x(x) = \begin{cases} \dfrac{1}{q} & 0 \leq x \leq q \\ 0 & \text{otherwise} \end{cases}$$

Find its mean, mean square, and variance.

10.3 Find the mean, mean square, and variance for the random variable of Prob. 10.1.

10.4 Demonstrate the validity of Eq. (10.7).

10.5 A gaussian random variable $\mathbf{x}$ has a mean η and a variance σ^2. Show that

$$P_x(x_1 - \eta) = 1 - P_x(\eta - x_1)$$

where $P_x(x)$ is the probability distribution function of a gaussian random variable with zero mean.

10.6 A gaussian random variable $\mathbf{x}$ has $\eta = 0$ and $\sigma = 2$.

    (a) Find Pr $[\mathbf{x} \geq 2]$.

    (b) Find Pr $[|\mathbf{x}| \geq 2]$.

    (c) Find x_1 if Pr $[|\mathbf{x}| \leq x_1] = 0.95$.

10.7 The random variable of Prob. 10.5 satisfies the relations

$$\text{Pr } [\mathbf{x} \leq 60] = 0.2 \qquad \text{Pr } [\mathbf{x} \geq 90] = 0.1$$

Find η and σ^2.

10.8 A random variable $\mathbf{x}$ has a Rayleigh probability-density function given by

$$p_x(x) = \begin{cases} \dfrac{xe^{-x^2/2\alpha^2}}{\alpha^2} & \text{for } 0 \leq x \leq \infty \\ 0 & \text{otherwise} \end{cases}$$

Show that

    (a) $E\{\mathbf{x}\} = \alpha\sqrt{\dfrac{\pi}{2}}$

    (b) $E\{\mathbf{x}^2\} = 2\alpha^2$

    (c) $\sigma_{\mathbf{x}}^2 = \left(2 - \dfrac{\pi}{2}\right)\alpha^2$

10.9 A random process is given by

$$\mathbf{x}(t) = \mathbf{y}e^{-t}u(t - \mathbf{z})$$

where $\mathbf{y}$ and $\mathbf{z}$ are random variables uniformly distributed in the range $(-1, 1)$. Sketch five sample functions.

10.10 A random process is given by

$$\mathbf{x}(t) = 2 + \frac{\mathbf{y}t}{\sqrt{2}}$$

where $\mathbf{y}$ is a random variable with a probability-density function

$$p_y(y) = \frac{1}{\sqrt{2\pi}}e^{-y^2/2} \qquad -\infty \leq y \leq \infty$$

Find the first-order probability-density function of $\mathbf{x}(t)$.

10.11 A random process is given by

$$\mathbf{x}(t) = \mathbf{z} \cos (\omega_0 t + \mathbf{y})$$

Find the first-order probability-density function of $\mathbf{x}(t)$.

    (a) If $\mathbf{z}$ is a random variable distributed uniformly in the range $(-1, 1)$ and $\mathbf{y}$ is a constant.

    (b) If $\mathbf{y}$ is a random variable distributed uniformly in the range $(-\pi, \pi)$ and $\mathbf{z}$ is a constant.

10.12 Find the mean, mean square, and autocorrelation for the process in Prob. 10.10. Is the process stationary?

10.13 Repeat Prob. 10.12 for the processes in Prob. 10.11.

10.14 A stationary discrete-time random process is given by

$$x(nT) = E\{x(nT)\} + x_0(nT)$$

where $x_0(nT)$ is a zero-mean process. Show that
  (a) $r_x(0) = E\{x^2(nT)\}$
  (b) $r_x(-kT) = r_x(kT)$
  (c) $r_x(0) \geq |r_x(kT)|$
  (d) $r_x(kT) = E\{x^2(nT)\} + r_{x_0}(kT)$

10.15 Explain the physical significance of
  (a) $E\{x(nT)\}$
  (b) $E^2\{x(nT)\}$
  (c) $E\{x^2(nT)\}$
  (d) $\sigma_x^2 = E\{x^2(nT)\} - E^2\{x(nT)\}$

10.16 A discrete-time random process is given by

$$x(nT) = 3 + 4nTy$$

where y is a random variable with a probability-density function

$$p_y(y) = \frac{1}{2\sqrt{2\pi}} e^{-(y-4)^2/8} \qquad -\infty \leq y \leq \infty$$

Find its mean, mean square, and autocorrelation.

10.17 A discrete-time random process is given by

$$x(nT) = z \cos\left(\omega_0 nT + \frac{\pi}{8}\right)$$

where z is a random variable distributed uniformly in the range $(0, 1)$. Find the mean, mean square, and autocorrelation of $x(nT)$. Is the process stationary?

10.18 A discrete-time random process is given by

$$x(nT) = \sqrt{2} \cos(\omega_0 nT + y)$$

where y is a random variable uniformly distributed in the range $(-\pi, \pi)$.
  (a) Find the mean, mean square, and autocorrelation of $x(nT)$.
  (b) Show that the process is wide-sense-stationary.
  (c) Find the PSD of $x(nT)$.

10.19 The random process of Prob. 10.18 is passed through a digital filter characterized by

$$H(e^{j\omega T}) = \begin{cases} 1 & \text{for } |\omega| \leq \omega_c \\ 0 & \text{otherwise} \end{cases}$$

Sketch the input and output power-density spectra if $\omega_0 < \omega_c$.

10.20 A random process $x(nT)$ with a probability density function

$$p_x(x; nT) = \begin{cases} 1 & -\frac{1}{2} \leq x \leq \frac{1}{2} \\ 0 & \text{otherwise} \end{cases}$$

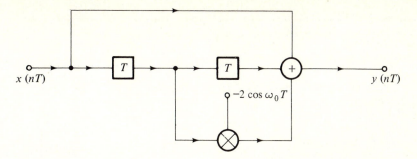

Figure P10.20

is applied at the input of the filter depicted in Fig. P10.20. Find the output PSD if $x(nT)$ and $x(kT)$ $(n \neq k)$ are statistically independent.

ELEVEN

EFFECTS OF FINITE WORD LENGTH IN DIGITAL FILTERS

11.1 INTRODUCTION

In software as well as hardware digital-filter implementations numbers are ultimately stored in finite-length registers. Consequently, coefficients and signal values must be quantized by rounding or truncation before they can be stored.

Number quantization gives rise to three types of errors:

1. Coefficient-quantization errors
2. Product-quantization errors
3. Input-quantization errors

The transfer-function coefficients are normally evaluated to a high degree of accuracy during the approximation step. If they are quantized, the frequency response of the resulting filter may differ appreciably from the desired response, and if the quantization step is coarse, the filter may actually fail to meet the desired specifications.

Product-quantization errors arise at the outputs of multipliers. Each time a signal represented by b_1 digits is multiplied by a coefficient represented by b_2 digits, a product having as many as $b_1 + b_2$ digits is generated. Since a uniform register length must, in practice, be used throughout the filter, each multiplier output must be rounded or truncated before processing can continue.

Input-quantization errors arise in applications where digital filters are used to process continuous-time signals. These are the errors inherent in the analog-to-digital conversion (see Sec. 6.7).

This chapter begins with a review of the various number systems and types of arithmetic that can be used in digital-filter implementations. It then proceeds to describe various methods of analysis which can be applied to study the effects of coefficient and product quantizations. The chapter concludes with a discussion of a nonlinear effect known as the *deadband effect*.

11.2 NUMBER REPRESENTATION

Binary System

The hardware implementation of digital filters, like the implementation of other digital hardware, is based on the binary-number representation.

In general, any number N can be expressed as

$$N = \sum_{i=-m}^{n} b_i r^i \tag{11.1}$$

where $$0 \le b_i \le r - 1$$

If distinct symbols are assigned to the permissible values of b_i, the number N can be represented by the notation

$$N = (b_n b_{n-1} \cdots b_0 . b_{-1} \cdots b_{-m})_r \tag{11.2}$$

The parameter r is said to be the *radix* of the representation, and the point separating N into two parts is called the *radix point*.

If $r = 10$, Eq. (11.2) becomes the decimal representation of N and the radix point is the decimal point. Similarly, if $r = 2$ Eq. (11.2) becomes the binary representation of N and the radix point is referred to as the *binary point*. The common symbols used to represent the two permissible values of b_i are 0 and 1. These are called *bits*.

A mixed decimal number can be converted into a binary number through the following steps:

1. Divide the integer part by 2 repeatedly and arrange the resulting remainders in the reverse order.
2. Multiply the fraction part by 2 and remove the resulting integer part; repeat as many times as necessary, and then arrange the integers obtained in the forward order.

A binary number can be converted into a decimal number by using Eq. (11.1).

Example 11.1 (*a*) Form the binary representation of $N = 18.375_{10}$. (*b*) Form the decimal representation of $N = 11.101_2$.

SOLUTION (*a*)

$$
\begin{array}{r|l}
2 & 18 \\
2 & 9 \rightarrow 0 \\
2 & 4 \rightarrow 1 \\
2 & 2 \rightarrow 0 \\
2 & 1 \rightarrow 0 \\
 & 0 \rightarrow 1
\end{array}
$$

$$0 \leftarrow 2 \times 0.375 = 0.75$$
$$1 \leftarrow 2 \times 0.75 \ = 1.5$$
$$1 \leftarrow 2 \times 0.5 \ \ = 1.0$$
$$0 \leftarrow 2 \times 0 \ \ \ \ = 0$$

Hence
$$18.375_{10} = 10010.011_2$$

(*b*) From Eq. (11.1)

$$11.101_2 = 1(2^1) + 1(2^0) + 1(2^{-1}) + 0(2^{-2}) + 1(2^{-3}) = 3.625_{10}$$

The most basic electronic memory device is the *flip-flop*, which can be either in a low or a high state. By assigning a 0 to the low state and a 1 to the high state, a single-bit binary number can be stored. By arranging *n* flip-flops in juxtaposition, as in Fig. 11.1*a*, a register can be formed which will store an *n*-bit number.

A rudimentary 4-bit filter implementation is shown in Fig. 11.1*b*. Registers R_y and R_b are used to store the the past output $y(n-1)$ and the multiplier coefficient *b*, respectively. The output of the multiplier at steady state is $by(n-1)$. Once a new input sample is received, the adder goes into action to form the new output $y(n)$, which is then used to update register R_y. Subsequently, the multiplier is triggered into operation and the product $by(n-1)$ is formed. The cycle is repeated when a new input sample is received.

A filter implementation like that in Fig. 11.1*b* can assume many forms, depending on the type of machine arithmetic used. The arithmetic can be of the fixed-point or floating-point type and in each case various conventions can be used for the representation of negative numbers. In a fixed-point arithmetic, as the name implies, the true binary point occupies a specific physical position in the register. In a floating-point arithmetic, on the other hand, no specific physical position of the register is assigned to the true binary point.

Fixed-Point Arithmetic

In fixed-point arithmetic, the numbers are usually assumed to be proper fractions. Mixed numbers or integers are avoided because (1) mixed numbers are more difficult to multiply and (2) the number of bits representing an integer cannot be reduced by rounding or truncation. Hence the binary point is usually set between the first and second bit positions of the register, as depicted in Fig. 11.2*a*. The first position is reserved for the sign of the number.

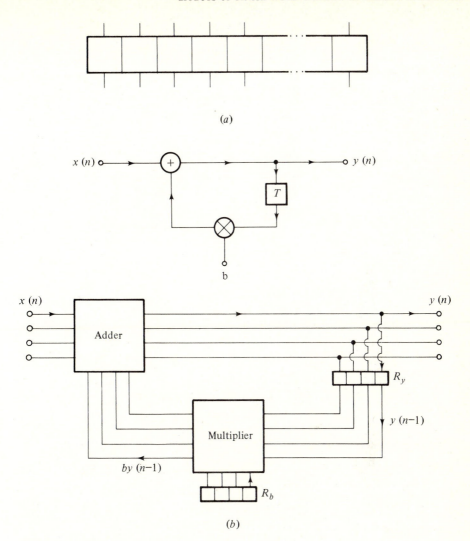

Figure 11.1 (*a*) Register, (*b*) rudimentary filter implementation.

Depending on the representation of negative numbers, fixed-point arithmetic can assume three forms:

1. Signed magnitude
2. One's complement
3. Two's complement

In the signed-magnitude arithmetic a fractional number

$$N = \pm 0.b_{-1}b_{-2} \cdots b_{-m}$$

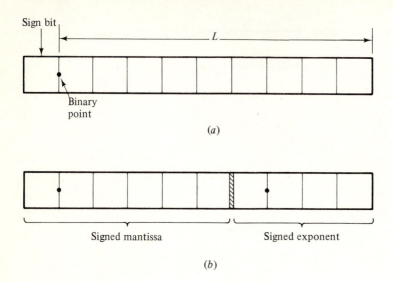

Figure 11.2 Storage of (a) fixed-point numbers and (b) floating-point numbers.

is represented as

$$N_{sm} = \begin{cases} 0.b_{-1}b_{-2}\cdots b_{-m} & \text{for } N \geq 0 \\ 1.b_{-1}b_{-2}\cdots b_{-m} & \text{for } N \leq 0 \end{cases}$$

The most significant bit is said to be the sign bit; e.g., if $N = +0.1101$ or -0.1001, $N_{sm} = 0.1101$ or 1.1001.

The one's-complement representation of a number N is defined as

$$N_1 = \begin{cases} N & \text{for } N \geq 0 \\ 2 - 2^{-L} - |N| & \text{for } N \leq 0 \end{cases} \tag{11.3}$$

where L, referred to as the *word length*, is the number of bit locations in the register to the right of the binary point. The binary form of $2 - 2^{-L}$ is a string of 1s filling the $L + 1$ locations of the register. Thus, the one's complement of a negative number can be deduced by representing the number by $L + 1$ bits, including zeros if necessary, and then complementing (changing 0s into 1s and 1s into 0s) all bits; e.g., if $N = -0.11010$, then $N_1 = 1.00101$ for $L = 5$ and $N_1 = 1.00101111$ for $L = 8$.

The two's-complement representation is similar. We now have

$$N_2 = \begin{cases} N & \text{for } N \geq 0 \\ 2 - |N| & \text{for } N < 0 \end{cases}$$

The two's complement of a negative number can be formed by adding 1 at the least significant position of the one's complement. Similarly, a negative number can be recovered from its two's complement by complementing and then adding 1 at the least significant position.

The possible numbers that can be stored in a 4-bit register together with their decimal equivalents are listed in Table 11.1. Some peculiarities of the three systems are evident. The signed-magnitude and the one's-complement systems have two representations for zero whereas the two's-complement system has only one. On the other hand, -1 is represented in the two's-complement system but not in the other two.

The merits and demerits of the three types of arithmetic can be envisaged by examining how arithmetic operations are performed in each case.

One's-complement addition of any two numbers is carried out by simply adding their one's complements bit by bit. A carry bit at the most significant position, if one is generated, is added at the least significant position (*end-around carry*). Two's-complement addition is exactly the same except that a carry bit at the most significant position is ignored. Signed-magnitude addition, on the other hand, is much more complicated as it can involve sign checks as well as complementing and end-around carry [1].

In the one's- or two's-complement arithmetic, direct multiplication of the complements does not always yield the product, and as a consequence special algorithms must be employed (see Sec. 14.5). By contrast, signed-magnitude multiplication is accomplished by simply multiplying the magnitudes of the two numbers bit by bit and then adjusting the sign bit of the product.

Table 11.1 Decimal equivalents of numbers 0.000 to 1.111

Binary number	Decimal equivalent (eighths)		
	Signed magnitude	One's complement	Two's complement
0.000	0	0	0
0.001	1	1	1
0.010	2	2	2
0.011	3	3	3
0.100	4	4	4
0.101	5	5	5
0.110	6	6	6
0.111	7	7	7
1.000	-0	-7	-8
1.001	-1	-6	-7
1.010	-2	-5	-6
1.011	-3	-4	-5
1.100	-4	-3	-4
1.101	-5	-2	-3
1.110	-6	-1	-2
1.111	-7	-0	-1

Example 11.2 Form the sum $0.53125 + (-0.40625)$ using the one's- and two's-complement additions assuming a word length of 5 bits.

SOLUTION

$$0.53125_{10} = 0.10001_2$$

$$0.40625_{10} = 0.01101_2$$

	One's complement	Two's complement
0.53125	0.10001	0.10001
−0.40625	1.10010	1.10011
0.12500	0.00011	$1 \leftarrow 0.00100$
	$\searrow 1$	
	0.00100	

An important feature of the one's- or two's-complement addition is that a machine-representable sum $S = n_1 + n_2 + \cdots + n_i + \cdots$ will always be evaluated correctly, even if overflow does occur in the evaluation of partial sums.

Example 11.3 Form the sum $\frac{7}{8} + \frac{4}{8} + (-\frac{6}{8})$ using the two's-complement addition. Assume $L = 3$.

SOLUTION From Table 11.1

	7/8	0.111	
+	4/8	0.100	
	11/8	1.011	incorrect partial sum
−	6/8	+ 1.010	
	5/8	0.101	correct sum

Floating-Point Arithmetic

There are two basic disadvantages in a fixed-point arithmetic: (1) The range of numbers that can be handled is small; e.g., in the two's-complement representation the smallest number is -1 and the largest is $1 - 2^{-L}$. (2) The percentage error produced by truncation or rounding tends to increase as the magnitude of the number is decreased. For example, if numbers 0.11011010 and 0.000110101 are both truncated such that only 4 bits are retained to the right of the binary point, the respective errors will be 4.59 and 39.6 percent.

These problems can be alleviated to a large extent by using a floating-point arithmetic. In this type of arithmetic, a number N is expressed as

$$N = M \times 2^e \qquad (11.4)$$

where e is an integer and

$$\tfrac{1}{2} \le M < 1$$

M and e are referred to as the *mantissa* and *exponent*, respectively. For example, numbers 0.00110101 and 1001.11 are represented by 0.110101×2^{-2} and 0.100111×2^4, respectively. Negative numbers are handled in the same way as in fixed-point arithmetic.

Floating-point numbers are stored in registers, as depicted in Fig. 11.2b. The register is subdivided into two segments, one for the signed mantissa and one for the signed exponent.

Floating-point addition is carried out by shifting the mantissa of the smaller number to the right and increasing the exponent until the exponents of the two numbers are equal. The mantissas are then added to form the sum, which is subsequently put back into the normalized representation of Eq. (11.4). Multiplication is accomplished by multiplying mantissas, adding exponents, and then readjusting the product.

Floating-point arithmetic, as implied above, leads to increased dynamic range and improved accuracy of processing. Unfortunately, it also leads to increased cost of hardware and to reduced speed of processing. The reason is that hardware is in a sense duplicated since both the mantissa and exponent have to be manipulated. For software non-real-time implementations on general purpose digital computers, floating-point arithmetic is always preferred since neither the cost of hardware nor the speed of processing is a significant factor.

Number Quantization

Once the register length in a fixed-point implementation is assigned, the set of machine representable numbers is fixed. If the word length is L bits (excluding the sign bit), the smallest number variation that can be represented is a 1 at the least significant register position, which corresponds to 2^{-L}. Therefore, any number consisting of b bits (excluding the sign bit), where $b > L$, must be quantized. This can be accomplished (1) by truncating all bits that cannot be accommodated in the register, and (2) by rounding the number to the nearest machine-representable number.

Obviously, if a number x is quantized, an error ε will be introduced given by

$$\varepsilon = x - Q[x] \qquad (11.5)$$

where $Q[x]$ denotes the quantized value of x. The range of ε tends to depend on the type of number representation and also on the type of quantization. Let us examine the various possibilities, starting with truncation.

As can be seen in Table 11.1, the representation of positive numbers is identical in all three fixed-point representations. Since truncation can only reduce a positive number, ε is positive. Its maximum value occurs when all disregarded bits are 1s, in which case

$$0 \leq \varepsilon_T \leq 2^{-L} - 2^{-b} \qquad \text{for } x \geq 0$$

For negative numbers the three representations must be considered individually. For the signed-magnitude representation, truncation will decrease the magnitude of the number or increase its signed value, and hence $Q[x] > x$ or

$$-(2^{-L} - 2^{-b}) \leq \varepsilon_T \leq 0 \qquad \text{for } x < 0$$

The one's-complement representation of a negative number

$$x = -\sum_{i=1}^{b} b_{-i} 2^{-i} \tag{11.6}$$

(where $b_{-i} = 0$ or 1) is obtained from Eq. (11.3) as

$$x_1 = 2 - 2^{-L} - \sum_{i=1}^{b} b_{-i} 2^{-i}$$

If all the disregarded bits are 0s, obviously $\varepsilon = 0$. At the other extreme if all the disregarded bits are 1s, we have

$$Q[x_1] = 2 - 2^{-L} - \sum_{i=1}^{b} b_{-i} 2^{-i} - (2^{-L} - 2^{-b})$$

Consequently, the decimal equivalent of $Q[x_1]$ is

$$Q[x] = -\left[\sum_{i=1}^{b} b_{-i} 2^{-i} + (2^{-L} - 2^{-b}) \right] \tag{11.7}$$

and, therefore, from Eqs. (11.5) to (11.7)

$$0 \leq \varepsilon_T \leq 2^{-L} - 2^{-b} \qquad \text{for } x < 0$$

The same inequality holds for two's-complement numbers, as can easily be shown. In summary, for signed-magnitude numbers

$$-q < \varepsilon_T < q$$

where $q = 2^{-L}$ is the quantization step, whereas for one's- or two's-complement numbers

$$0 \leq \varepsilon_T < q$$

Evidently, quantization errors can be kept as low as desired by using a sufficiently large value of L.

For rounding, the quantization error can be positive as well as negative by definition, and its maximum value is $q/2$. If numbers lying halfway between quantization levels are rounded up, we have

$$\frac{-q}{2} \le \varepsilon_R < \frac{q}{2} \qquad (11.8)$$

Rounding can be effected, in practice, by adding 1 at position $L + 1$ and then truncating the number to L bits.

A convenient way of visualizing the process of quantization is to imagine a quantizer with input x and output $Q[x]$. Depending on the type of quantization, the transfer characteristic of the device can assume one of the forms illustrated in Fig. 11.3.

The range of quantization error in floating-point arithmetic can be evaluated by a similar approach.

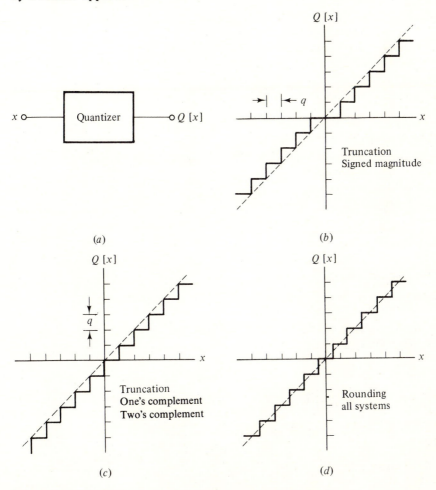

Figure 11.3 Number quantization: (a) quantizer, (b) to (d) $Q[x]$ versus x.

11.3 COEFFICIENT QUANTIZATION

Coefficient-quantization errors introduce perturbations in the zeros and poles of the transfer function, which in turn manifest themselves as errors in the frequency response. Product-quantization errors, on the other hand, can be regarded as noise sources which give rise to output noise. Since the two types of errors are different, it is frequently advantageous to use different word lengths for coefficients and signal values. The coefficient word length can be chosen to satisfy the frequency-response specifications, whereas the signal word length can be chosen to satisfy the signal-to-noise-ratio specification.

Consider a digital filter characterized by $H(z)$ and let

$M(\omega)$ = amplitude response without coefficient quantization
$\quad\quad [= |H(e^{j\omega T})|]$
$M_Q(\omega)$ = amplitude response with quantization
$M_I(\omega)$ = idealized amplitude response
$\delta_p(\delta_a)$ = passband (stopband) tolerance on amplitude response

These quantities are illustrated in Fig. 11.4.

The effect of coefficient quantization is to introduce an error ΔM in $M(\omega)$ given by

$$\Delta M = M(\omega) - M_Q(\omega)$$

The maximum permissible value of $|\Delta M|$, denoted by $\Delta M_{max}(\omega)$, can be deduced from Fig. 11.4 as

$$\Delta M_{max}(\omega) = \begin{cases} \delta_p - |M(\omega) - M_I(\omega)| & \text{for } \omega \le \omega_p \\ \delta_a - |M(\omega) - M_I(\omega)| & \text{for } \omega \ge \omega_a \end{cases}$$

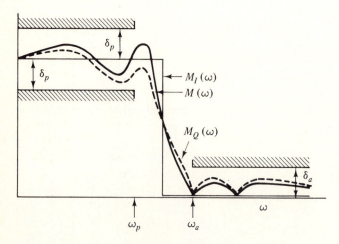

Figure 11.4 Coefficient quantization.

and if

$$|\Delta M| \leq \Delta M_{max}(\omega) \tag{11.9}$$

the desired specification will be met. The optimum word length can thus be determined exactly by evaluating $|\Delta M|$ as a function of frequency for successively larger values of the word length until Eq. (11.9) is satisfied. Evidently, this is a trial-and-error approach and may entail considerable computation.

An alternative approach is to employ a statistical method proposed by Avenhaus [2] and later modified by Crochiere [3]. This method yields a fairly accurate estimate of the required word length and is, in general, more efficient than the exact method outlined above. Its details follow.

Consider a fixed-point implementation and assume that quantization is carried out by rounding. From Eq. (11.8) the error in coefficient c_i ($i = 1, 2, \ldots, m$), denoted by Δc_i, can assume any value in the range $-q/2$ to $+q/2$; that is, Δc_i is a random variable. If the probability density of Δc_i is assumed to be uniform, i.e.,

$$p(\Delta c_i) = \begin{cases} \dfrac{1}{q} & \text{for } \dfrac{-q}{2} \leq \Delta c_i \leq \dfrac{q}{2} \\ 0 & \text{otherwise} \end{cases}$$

then from Eqs. (10.8) and (10.10)

$$E\{\Delta c_i\} = 0 \tag{11.10}$$

$$\sigma^2_{\Delta c_i} = \frac{q^2}{12} \tag{11.11}$$

The variation ΔM in $M(\omega)$ is also a random variable. By virtue of Taylor's theorem we can write

$$\Delta M = \sum_{i=1}^{m} \Delta c_i\, S_{c_i}$$

where

$$S_{c_i} = \frac{\partial M(\omega)}{\partial c_i}$$

is the sensitivity of $M(\omega)$ with respect to variations in c_i. Evidently,

$$E\{\Delta M\} = \sum_{i=1}^{m} S_{c_i} E\{\Delta c_i\} = 0$$

according to Eq. (11.10). If Δc_i and Δc_j ($i \neq j$) are assumed to be statistically independent, then from Eq. (10.7)

$$\sigma^2_{\Delta M} = \sum_{i=1}^{m} \sigma^2_{\Delta c_i}\, S^2_{c_i}$$

and therefore, from Eq. (11.11),

$$\sigma_{\Delta M}^2 = \frac{q^2 S_T^2}{12} \tag{11.12}$$

where

$$S_T^2 = \sum_{i=1}^{m} S_{c_i}^2 \tag{11.13}$$

For a large value of m, ΔM is approximately gaussian by virtue of the central-limit theorem [4], and since $E\{\Delta M\} = 0$, Eq. (10.1) gives

$$p(\Delta M) = \frac{1}{\sigma_{\Delta M} \sqrt{2\pi}} e^{-\Delta M^2/2\sigma_{\Delta M}^2} \qquad -\infty \leq \Delta M \leq \infty$$

Consequently, ΔM will be in some range $-\Delta M_1 \leq \Delta M \leq \Delta M_1$ with a probability y given by

$$y = \Pr\left[|\Delta M| \leq \Delta M_1\right] = \frac{2}{\sigma_{\Delta M} \sqrt{2\pi}} \int_0^{\Delta M_1} e^{-\Delta M^2/2\sigma_{\Delta M}^2} \, d(\Delta M) \tag{11.14}$$

With the variable transformation

$$\Delta M = x\sigma_{\Delta M} \qquad \Delta M_1 = x_1 \sigma_{\Delta M} \tag{11.15}$$

Equation (11.4) can be put in the standard form

$$y = \frac{2}{\sqrt{2\pi}} \int_0^{x_1} e^{-x^2/2} \, dx$$

Once an acceptable confidence factor y is selected, the corresponding value of x_1 can be obtained from published tables or by using a numerical method. The quantity ΔM_1 is essentially a statistical bound on ΔM, and if the word length is chosen such that

$$\Delta M_1 \leq \Delta M_{\max}(\omega) \tag{11.16}$$

the desired specifications will be satisfied to within a confidence factor y. The resulting word length can be referred to as the *statistical word length*. A statistical bound on the quantization step can be deduced from Eqs. (11.12), (11.15), and (11.16) as

$$q \leq \frac{\sqrt{12} \, \Delta M_{\max}(\omega)}{x_1 S_T} \tag{11.17}$$

The register length should be sufficiently large to accommodate the quantized value of the largest coefficient; so let

$$Q[\max c_i] = \sum_{i=-K}^{J} b_i 2^i$$

where b_J and $b_{-K} \neq 0$. The required word length must be

$$L = 1 + J + K \tag{11.18}$$

and since $q = 2^{-K}$ or

$$K = \log_2 \frac{1}{q} \tag{11.19}$$

Eqs. (11.17) to (11.19) give the desired result as

$$L \geq L(\omega) = 1 + J + \log_2 \frac{x_1 S_T}{\sqrt{12}\,\Delta M_{max}(\omega)}$$

A reasonable agreement between the statistical and exact word lengths is achieved [3, 5] by using $x_1 = 2$. This value of x_1 corresponds to a confidence factor of 0.95.

The sensitivities S_{c_i} in Eq. (11.13) can be computed efficiently by using the transpose or adjoint approach described in Sec. 4.7. This approach gives

$$S_{c_i}^H(e^{j\omega T}) = \frac{\partial H(e^{j\omega T})}{\partial c_i} = \text{Re}\,[S_{c_i}^H(e^{j\omega T})] + j\,\text{Im}\,[S_{c_i}^H(e^{j\omega T})]$$

and if

$$H(e^{j\omega T}) = M(\omega)e^{j\theta(\omega)}$$

we can show that

$$\text{Re}\,[S_{c_i}^H(e^{j\omega T})] = \cos\,[\theta(\omega)]\frac{\partial M(\omega)}{\partial c_i} - M(\omega)\sin\,[\theta(\omega)]\frac{\partial\theta(\omega)}{\partial c_i}$$

$$\text{Im}\,[S_{c_i}^H(e^{j\omega T})] = \sin\,[\theta(\omega)]\frac{\partial M(\omega)}{\partial c_i} + M(\omega)\cos\,[\theta(\omega)]\frac{\partial\theta(\omega)}{\partial c_i}$$

Therefore

$$S_{c_i} = \frac{\partial M(\omega)}{\partial c_i} = \cos\,[\theta(\omega)]\,\text{Re}\,[S_{c_i}^H(e^{j\omega T})] + \sin\,[\theta(\omega)]\,\text{Im}\,[S_{c_i}^H(e^{j\omega T})]$$

The statistical word length is a convenient figure of merit of a specific filter structure. It can serve as a sensitivity measure in studies where a general comparison of various structures is desired. It can also be used as an objective function in word-length optimization algorithms [3].

A different approach for the study of quantization effects was proposed by Jenkins and Leon [6]. In this a computer-aided analysis scheme is used to generate confidence-interval error bounds on the time-domain response of the filter. The method can be used to study the effects of coefficient or product quantization in fixed-point or floating-point implementations. Furthermore, the quantization can be by rounding or truncation.

11.4 PRODUCT QUANTIZATION

The output of a finite-word-length multiplier can be expressed as

$$Q[c_i x(n)] = c_i x(n) + e(n)$$

where $c_i x(n)$ and $e(n)$ are the exact product and quantization error, respectively. A machine multiplier can thus be represented by the model depicted in Fig. 11.5a, where $e(n)$ is a noise source.

Consider the filter structure of Fig. 11.5b and assume a fixed-point implementation. Each multiplier can be replaced by the model of Fig. 11.5a, as in Fig. 11.5c. If product quantization is carried out by rounding, each noise signal $e_i(n)$ can be regarded as a random process with uniform probability density, i.e.,

$$p(e_i; n) = \begin{cases} \dfrac{1}{q} & \text{for } -\dfrac{q}{2} \le e_i(n) \le \dfrac{q}{2} \\ 0 & \text{otherwise} \end{cases}$$

Hence from Sec. 10.8 and Eqs. (10.8) and (10.9)

$$E\{e_i(n)\} = 0 \tag{11.20}$$

$$E\{e_i^2(n)\} = \frac{q^2}{12} \tag{11.21}$$

$$r_{e_i}(k) = E\{e_i(n)e_i(n+k)\} \tag{11.22}$$

If the signal levels throughout the filter are much larger than q, the following reasonable assumptions can be made: (1) $e_i(n)$ and $e_i(n+k)$ are statistically independent for any value of n ($k \ne 0$), and (2) $e_i(n)$ and $e_j(n+k)$ are statistically independent for any value of n or k ($i \ne j$). Let us examine the implications of these assumptions starting with the first assumption. From Eqs. (11.20) to (11.22) and (10.4)

$$r_{e_i}(0) = E\{e_i^2(n)\} = \frac{q^2}{12}$$

and $$r_{e_i}(k)\Big|_{k \ne 0} = E\{e_i(n)\}E\{e_i(n+k)\} = 0$$

that is, $$r_{e_i}(k) = \frac{q^2}{12}\delta(k)$$

where $\delta(k)$ is the impulse function. Therefore, the power spectral density (PSD) of $e_i(n)$ is

$$S_{e_i}(z) = \mathscr{Z}r_{e_i}(k) = \frac{q^2}{12} \tag{11.23}$$

that is, $e_i(n)$ is a white-noise process.

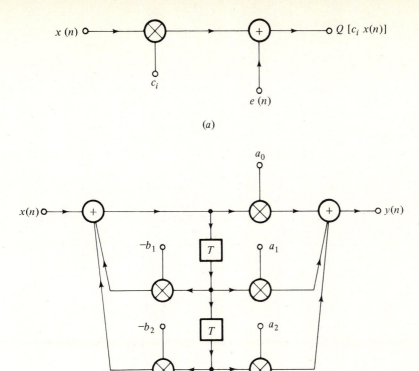

(a)

(b)

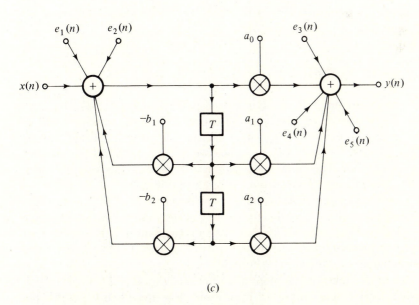

(c)

Figure 11.5 Product quantization: (*a*) noise model for a multiplier, (*b*) second-order canonic section, (*c*) noise model for a second-order canonic section.

Let us now consider the implications of the second assumption. The autocorrelation of sum $e_i(n) + e_j(n)$ is

$$r_{e_i+e_j}(k) = E\{[e_i(n) + e_j(n)][e_i(n + k) + e_j(n + k)]\}$$
$$= E\{e_i(n)e_i(n + k)\} + E\{e_i(n)\}E\{e_j(n + k)\}$$
$$+ E\{e_j(n)\}E\{e_i(n + k)\} + E\{e_j(n)e_j(n + k)\}$$

or $\quad r_{e_i+e_j}(k) = r_{e_i}(k) + r_{e_j}(k)$

Therefore

$$S_{e_i+e_j}(z) = \mathscr{L}[r_{e_i}(k) + r_{e_j}(k)] = S_{e_i}(z) + S_{e_j}(z)$$

i.e., the PSD of a sum of two processes is equal to the sum of their respective PSDs. In effect, superposition can be employed.

Now from Fig. 11.5c and Eq. (10.22)

$$S_y(z) = H(z)H(z^{-1}) \sum_{i=1}^{2} S_{e_i}(z) + \sum_{i=3}^{5} S_{e_i}(z)$$

where $H(z)$ is the transfer function of the filter, and hence from Eq. (11.23) the output PSD is given by

$$S_y(z) = \frac{q^2}{6} H(z)H(z^{-1}) + \frac{q^2}{4}$$

The above approach is applicable to any filter structure. Furthermore, it can be used to study the effects of input quantization.

11.5 SIGNAL SCALING

If the amplitude of any internal signal in a fixed-point implementation is allowed to exceed the dynamic range, overflow will occur and the output signal will be severely distorted. On the other hand, if all the signal amplitudes throughout the filter are unduly low, the filter will be operating inefficiently and the signal-to-noise ratio will be poor. Therefore, for optimum filter performance suitable signal scaling must be employed to adjust the various signal levels.

A scaling technique applicable to one's- or two's-complement implementations was proposed by Jackson [7]. In this technique a scaling multiplier is used at the input of a filter section, as in Fig. 11.6, with its constant λ chosen such that amplitudes of multiplier inputs are bounded by M if $|x(n)| \leq M$. Under these circumstances, adder outputs are also bounded by M and cannot overflow. This is due to the fact that a machine-representable sum is always evaluated correctly in one's- or two's-complement arithmetic, even if overflow does occur in one of the partial sums (see Example 11.3). There are two methods for the determination of λ, as follows.

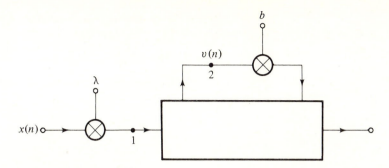

Figure 11.6 Signal scaling.

Consider the filter section of Fig. 11.6, where $v(n)$ is a multiplier input. The transfer function between nodes 1 and 2 can be denoted by $F(z)$. From the convolution summation

$$v(n) = \sum_{k=0}^{\infty} \lambda f(k) x(n-k) \tag{11.24}$$

where

$$f(n) = \mathscr{L}^{-1} F(z)$$

Evidently

$$|v(n)| \leq \sum_{k=0}^{\infty} |\lambda f(k)| \cdot |x(n-k)|$$

and if

$$|x(n)| \leq M$$

then

$$|v(n)| \leq M \sum_{k=0}^{\infty} |\lambda f(k)|$$

Thus a sufficient condition for $|v(n)| \leq M$ is

$$\sum_{k=0}^{\infty} |\lambda f(k)| \leq 1$$

or

$$\lambda \leq \frac{1}{\displaystyle\sum_{k=0}^{\infty} |f(k)|} \tag{11.25}$$

Now consider the specific signal

$$x(n-k) = \begin{cases} M & \text{for } \lambda f(k) > 0 \\ -M & \text{for } \lambda f(k) < 0 \end{cases}$$

where $M > 0$. From Eq. (11.24)

$$v(n) = M \sum_{k=0}^{\infty} |\lambda f(k)|$$

and therefore $|v(n)| \leq M$ if and only if Eq. (11.25) holds.

The second method uses L_p-norm notation. The L_p norm of an arbitrary periodic function $A(\omega)$ with period ω_s is defined as

$$\|A\|_p = \left(\frac{1}{\omega_s} \int_0^{\omega_s} |A(\omega)|^p \, d\omega \right)^{1/p}$$

where $p \geq 1$. This exists if

$$\int_0^{\omega_s} |A(\omega)|^p < \infty$$

and if $A(\omega)$ is continuous, the following limit exists:

$$\lim_{p \to \infty} \|A\|_p = \|A\|_\infty = \max_{0 \leq \omega \leq \omega_s} |A(\omega)| \tag{11.26}$$

Now let

$$X(z) = \sum_{n=-\infty}^{\infty} x(n)z^{-n} \qquad a < |z| < b$$

$$F(z) = \sum_{n=-\infty}^{\infty} f(n)z^{-n} \qquad c < |z| < b$$

where $c < 1$ for a stable filter and $b > 1$. From Eq. (11.24)

$$V(z) = \lambda F(z)X(z) \qquad d < |z| < b$$

where $d = \max (a, c)$. The inverse z transform of $V(z)$ is

$$v(n) = \frac{1}{2\pi j} \oint_\Gamma \lambda F(z)X(z)z^{n-1} \, dz \tag{11.27}$$

where Γ is a contour in the annulus of convergence. If $a < 1$, Γ can be taken to be the unit circle $|z| = 1$. With $z = e^{j\omega T}$ Eq. (11.27) becomes

$$v(n) = \frac{1}{\omega_s} \int_0^{\omega_s} \lambda F(e^{j\omega T})X(e^{j\omega T})e^{jn\omega T} \, d\omega$$

We can thus write

$$|v(n)| \leq \left[\max_{0 \leq \omega \leq \omega_s} |X(e^{j\omega T})| \right] \frac{1}{\omega_s} \int_0^{\omega_s} |\lambda F(e^{j\omega T})| \, d\omega \tag{11.28}$$

or

$$|v(n)| \leq \left[\max_{0 \leq \omega \leq \omega_s} |\lambda F(e^{j\omega T})| \right] \frac{1}{\omega_s} \int_0^{\omega_s} |X(e^{j\omega T})| \, d\omega \tag{11.29}$$

and by virtue of the Schwarz inequality [7]

$$|v(n)| \leq \left[\frac{1}{\omega_s} \int_0^{\omega_s} |\lambda F(e^{j\omega T})|^2 \, d\omega \right]^{1/2} \left[\frac{1}{\omega_s} \int_0^{\omega_s} |X(e^{j\Omega T})|^2 \, d\Omega \right]^{1/2} \tag{11.30}$$

When L_p-norm notation is used, Eqs. (11.28) to (11.30) can be put in the compact form

$$|v(n)| \leq \|X\|_\infty \|\lambda F\|_1 \qquad |v(n)| \leq \|X\|_1 \|\lambda F\|_\infty \qquad |v(n)| \leq \|X\|_2 \|\lambda F\|_2$$

In fact these inequalities are particular cases of the general inequality [7, 8]

$$|v(n)| \leq \|X\|_q \|\lambda F\|_p \tag{11.31}$$

where the relation

$$p = \frac{q}{q - 1} \tag{11.32}$$

must hold.

Equation (11.31) is valid for any transfer function $\lambda F(z)$ including $\lambda F(z) = 1$, in which case $v(n) = x(n)$ and $\|1\|_p = 1$ for all $p \geq 1$. Consequently, from Eq. (11.31)

$$|x(n)| \leq \|X\|_q \qquad \text{for all } q \geq 1$$

Now if

$$|x(n)| \leq \|X\|_q \leq M$$

Eq. (11.31) gives

$$|v(n)| \leq M \|\lambda F\|_p$$

Therefore

$$|v(n)| \leq M$$

provided that

$$\|\lambda F\|_p \leq 1$$

or

$$\lambda \leq \frac{1}{\|F\|_p} \qquad \text{for } \|X\|_q \leq M \tag{11.33}$$

where Eq. (11.32) must hold.

From an engineering viewpoint, the significant values of p and q are 1, 2, and ∞. The case $p = \infty$ and $q = 1$ represents the most stringent condition since

$$\|F\|_\infty \geq \|F\|_p$$

according to Eq. (11.26).

If there are m multipliers in the filter of Fig. 11.6, then $|v_i(n)| \leq M$ provided that

$$\lambda_i \leq \frac{1}{\|F_i\|_p}$$

for $i = 1, 2, \ldots, m$. Therefore, in order to ensure that all multiplier inputs are bounded by M we must assign

$$\lambda = \min (\lambda_1, \lambda_2, \ldots, \lambda_m)$$

or

$$\lambda = \frac{1}{\max (\|F_1\|_p, \|F_2\|_p, \ldots, \|F_m\|_p)} \tag{11.34}$$

In the case of parallel or cascade realizations, efficient scaling can be accomplished by using one scaling multiplier per section.

The preceding analysis is based on the assumption that $X(z)$ converges on the unit circle $|z| = 1$, which is not always valid; e.g., if $x(n) = A \cos \omega_0 nT$. The problem can be resolved, however, by using the interrelation between the Fourier and z transforms. For the cosine input, Eq. (6.14) gives

$$X(e^{j\omega T}) = X^*(j\omega) = \frac{1}{T} \sum_{k=-\infty}^{\infty} X_A(j\omega + jk\omega_s)$$

where

$$X_A(j\omega) = \mathscr{F}(A \cos \omega_0 t) = \pi A[\delta(\omega - \omega_0) + \delta(\omega + \omega_0)]$$

Thus

$$X(e^{j\omega T}) = \frac{\pi A}{T} \sum_{k=-\infty}^{\infty} [\delta(\omega + k\omega_s - \omega_0) + \delta(\omega + k\omega_s + \omega_0)]$$

and

$$\|X\|_1 = \frac{1}{\omega_s} \int_0^{\omega_s} |X(e^{j\omega T})| \, d\omega = A \tag{11.35}$$

Therefore, Eq. (11.31) holds with $p = \infty$ and $q = 1$, in which case $M = A$, as may be expected.

Example 11.4 Deduce the scaling formulation for the cascade filter of Fig. 11.7a assuming $p = \infty$ and $q = 1$.

SOLUTION The only critical signals are $y'_j(n)$ and $y_j(n)$ since the inputs of the feedback multipliers are delayed versions of $y'_j(n)$. The filter can be represented by the flow graph of Fig. 11.7b, where

$$F'_j(z) = \frac{z^2}{z^2 + b_{1j}z + b_{2j}} \qquad F_j(z) = \frac{(z + 1)^2}{z^2 + b_{1j}z + b_{2j}}$$

By using Eq. (11.34) we obtain

$$\lambda_0 = \frac{1}{\max (\|F'_1\|_\infty, \|F_1\|_\infty)} \qquad \lambda_1 = \frac{1}{\lambda_0 \max (\|F_1 F'_2\|_\infty, \|F_1 F_2\|_\infty)}$$

$$\lambda_2 = \frac{1}{\lambda_0 \lambda_1 \max (\|F_1 F_2 F'_3\|_\infty, \|F_1 F_2 F_3\|_\infty)}$$

The scaling constants can be evaluated by noting that

$$\|\Pi F_i\|_\infty = \max_{0 \le \omega \le \omega_s} |\Pi F_i(e^{j\omega T})|$$

according to Eq. (11.26).

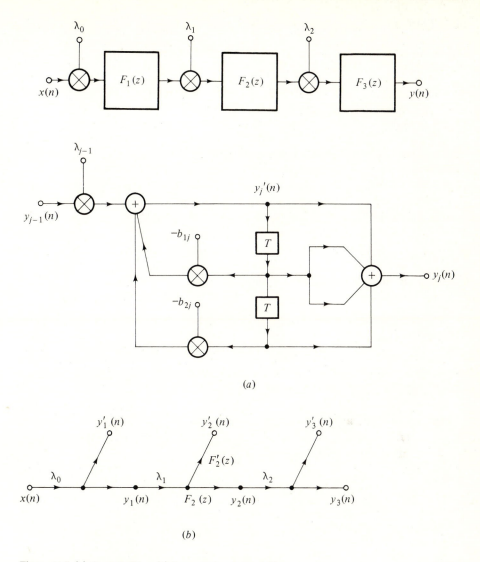

(a)

(b)

Figure 11.7 (a) Cascade filter, (b) flow-graph representation.

The scaling constants are usually chosen to be the nearest powers of 2 satisfying the overflow constraints. In this way scaling multiplications can be reduced to simple data shifts.

In cascade filters, the ordering of sections has an influence on scaling, which in turn has an influence on the output noise. Analytical techniques for determining the optimum sequential ordering have not yet been devised. Nevertheless, some guidelines suggested by Jackson [9] lead to a good ordering.

11.6 THE DEADBAND EFFECT

The effect of product quantization in fixed-point implementations has been studied in Sec. 11.4. As may be recalled, the fundamental assumption was made that signal levels throughout the filter are much larger than the quantization step. This allowed us to assume statistically independent noise signals from sample to sample and from source to source. On many occasions, signal levels can become very low, at least for short periods of time, e.g., during pauses in speech and music signals. Under such circumstances, quantization errors tend to become highly correlated and can actually cause a filter to lock in an unstable mode whereby a steady output oscillation is generated. This phenomenon is known as the *deadband effect*. The oscillation generated is called a *limit cycle*. The deadband effect can be studied by using a technique developed by Jackson [10].

First-Order Filter

Consider the filter of Fig. 11.8*a*, in which

$$H(z) = \frac{H_0 z}{z - b}$$

and
$$y(n) = H_0 x(n) + by(n - 1) \tag{11.36}$$

The impulse response of the filter is

$$h(n) = H_0(b)^n$$

If $b = 1$ or -1, the filter is unstable and has an impulse response

$$h(n) = \begin{cases} H_0 & \text{for } b = 1 \\ H_0(-1)^n & \text{for } b = -1 \end{cases}$$

With $H_0 = 10$ and $b = 0.9$, the exact impulse response given in the second column of Table 11.2 can be obtained.

Table 11.2 Impulse response of first-order filter

n	$h(n)$	$h_q(n)$
0	10.0	10.0
1	9.0	9.0
2	8.1	8.0
3	7.29	7.0
4	6.561	6.0
5	5.9049	5.0
6	5.31441	5.0
. . . .		
100	2.65614×10^{-4}	5.0

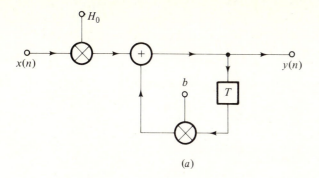

(a)

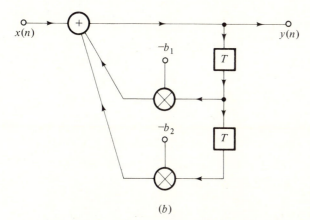

(b)

Figure 11.8 (a) First-order filter, (b) second-order filter.

Now assume that the filter is implemented using fixed-point decimal arithmetic, where each product $by(n-1)$ is rounded to the nearest integer according to the rule

$$Q[|by(n-1)|] = \text{Int}\,[|by(n-1)| + 0.5] \tag{11.37}$$

With $H_0 = 10$ and $b = 0.9$, the response in the third column of Table 11.2 is obtained. As can be seen, for $n \geq 5$ the response is constant and, in a sense, quantization has rendered the filter unstable.

If Eq. (11.36) is assumed to hold during the unstable mode, the effective value of b must be 1 for $b > 0$ or -1 for $b < 0$. If this is the case

$$Q[|by(n-1)|] = |y(n-1)|$$

and from Eq. (11.37)

$$\text{Int}\,[|b| \cdot |y(n-1)| + 0.5] = |y(n-1)|$$

or $\text{Int}\,[|y(n-1)| - (1 - |b|)|y(n-1)| + 0.5] = |y(n-1)|$

This equation can be satisfied if

$$0 \leq -(1 - |b|)|y(n-1)| + 0.5 < 1$$

Therefore, instability will occur if

$$|y(n-1)| \leq \frac{0.5}{1-|b|} = k \tag{11.38}$$

Since $y(n-1)$ is an integer, instability cannot arise if $|b| < 0.5$. On the other hand, if $|b| \geq 0.5$, the response will tend to decay to zero once the input is removed, and eventually $|y(n-1)|$ will assume values in the so-called *deadband range* $[-k, k]$. When this happens, the filter will become unstable. Any tendency of $|y(n-1)|$ to exceed k will restore stability, but in the absence of an input signal the response will again decay to a value within the deadband. Thus the filter will lock into a limit cycle of amplitude equal to or less than k. Since the effective value of b is $+1$ for $0.5 \leq b < 1$ or -1 for $-1 < b \leq -0.5$, the frequency of the limit cycle will be 0 or $\omega_s/2$.

Second-Order Filter

For the filter of Fig. 11.8b

$$H(z) = \frac{z^2}{z^2 + b_1 z + b_2} \quad \text{and} \quad y(n) = x(n) - b_1 y(n-1) - b_2 y(n-2)$$

If the poles are complex,

$$h(n) = \frac{r^n}{\sin \theta} \sin [(n+1)\theta]$$

where

$$r = \sqrt{b_2} \tag{11.39}$$

$$\theta = \cos^{-1} \frac{-b_1}{2\sqrt{b_2}} \tag{11.40}$$

For $b_2 = 1$, the impulse response is a sinusoid with constant amplitude and frequency

$$\omega = \frac{1}{T} \cos^{-1} \left(\frac{-b_1}{2} \right) \tag{11.41}$$

Product quantization will cause instability if the effective value of b_2 is 1, in which case

$$Q[|b_2 y(n-2)|] = |y(n-2)|$$

Hence, as before, a condition for instability can be deduced as

$$|y(n-2)| \leq \frac{0.5}{1-|b_2|} = k \tag{11.42}$$

With k an integer, values of b_2 in the ranges

$$0.5 \le |b_2| < 0.75$$

$$0.75 \le |b_2| < 0.833$$

$$\frac{2k - 1}{2k} \le |b_2| < \frac{2k + 1}{2(k + 1)}$$

---

will yield deadbands $[-1, 1]$, $[-2, 2]$, ..., $[-k, k]$, ...

If the poles are close to the unit circle, the limit cycle is approximately sinusoidal with a frequency close to the value given by Eq. (11.41).

For signed-magnitude binary arithmetic Eq. (11.42) becomes

$$|y(n - 2)| \le \frac{q}{2(1 - |b_2|)}$$

where q is the quantization step.

Limit-Cycle Bounds

Jackson's approach to the limit-cycle problem attempts to linearize a highly non-linear problem, and for this reason the results are sometimes inaccurate. In recent times, a number of alternative approaches to the problem have been proposed by Parker and Hess [11], Sandberg and Kaiser [12], and Long and Trick [13]. In all these methods bounds on the limit-cycle amplitude are developed, which are often very useful in determining the signal word length. The technique of Long and Trick gives reasonably tight bounds and is relatively simple to apply. It will now be described.

Consider a direct realization characterized by

$$y(n) = \sum_{i=0}^{N} a_i x(n - i) - \sum_{i=1}^{N} b_i y(n - i)$$

For a fixed-point implementation with the quantization carried out by rounding, the filter can be represented by the network of Fig. 11.9a, where $e_{ai}(n)$ and $e_{bi}(n)$ are noise sources due to feedforward and feedback multipliers, respectively. Hence

$$y(n) = \sum_{i=0}^{N} [a_i x(n - i) + e_{ai}(n)] - \sum_{i=1}^{N} [b_i y(n - i) - e_{bi}(n)]$$

and with zero input, $e_{ai}(n) = 0$ and

$$y(n) = e(n) - \sum_{i=1}^{N} b_i y(n - i)$$

where

$$e(n) = \sum_{i=1}^{N} e_{bi}(n) \tag{11.43}$$

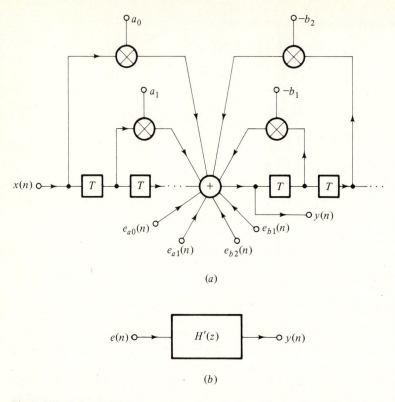

(a)

(b)

Figure 11.9 Noise model for general direct realization: (a) network, (b) block diagram.

The filter can thus be represented by the block diagram of Fig. 11.9b, where

$$H'(z) = \frac{1}{1 + \sum\limits_{i=1}^{N} b_i z^{-i}}$$

The technique is based on the assumption that the limit cycle is a periodic function of nT; so let the period be MT. Since $H'(z)$ represents a linear filter, $e(n)$ must also be periodic with the same period, i.e.,

$$e(n) = e(n + M)$$

From the convolution summation

$$y(n) = \sum_{k=-\infty}^{n} h'(n-k)e(k) \qquad (11.44)$$

where

$$h'(n) = \mathscr{L}^{-1} H'(z)$$

and hence we can write

$$y(n) = \sum_{k=n-M+1}^{n} h'(n-k)e(k) + \sum_{k=n-2M+1}^{n-M} h'(n-k)e(k) + \cdots$$

$$+ \sum_{k=n-(i+1)M+1}^{n-iM} h'(n-k)e(k) + \cdots$$

In this way Eq. (11.44) can be expressed as

$$y(n) = \sum_{i=0}^{\infty} \sum_{k=n-(i+1)M+1}^{n-iM} h'(n-k)e(k)$$

and if $k = n - iM - p$, we have

$$y(n) = \sum_{i=0}^{\infty} \sum_{p=0}^{M-1} h'(p+iM)e(n-iM-p)$$

By noting that

$$e(n-iM-p) = e(n-p)$$

and then changing the order of summation we deduce

$$y(n) = \sum_{p=0}^{M-1} e(n-p) \sum_{i=0}^{\infty} h'(p+iM) \tag{11.45}$$

Now from Eq. (11.43)

$$e(n-p) = \sum_{i=1}^{N} e_{bi}(n-p)$$

and since

$$|e_{bi}(n-p)| \le \frac{q}{2}$$

we can write

$$|e(n-p)| \le \sum_{i=1}^{N} |e_{bi}(n-p)| \le \frac{Nq}{2}$$

Therefore, Eq. (11.45) yields a bound on the limit-cycle amplitude as

$$|y(n)| \le \frac{Nq}{2} \sum_{p=0}^{M-1} \left| \sum_{i=0}^{\infty} h'(p+iM) \right| \tag{11.46}$$

Evidently, this depends on the limit-cycle period, which is unknown. The problem can be overcome, however, by computing the bound for values of M corresponding to frequencies in the range 0 to $\omega_s/2$. The maximum value so obtained can then be used as a conservative estimate of the bound.

An absolute bound independent of M can be deduced from Eq. (11.46) by writing

$$|y(n)| \leq \frac{Nq}{2} \sum_{p=0}^{M-1} \sum_{i=0}^{\infty} |h'(p + iM)|$$

and then letting $p = k - iM$, in which case

$$|y(n)| \leq \frac{Nq}{2} \sum_{k=iM}^{(i+1)M-1} \sum_{i=0}^{\infty} |h'(k)|$$

$$\leq \frac{Nq}{2} \sum_{i=0}^{\infty} \sum_{k=iM}^{(i+1)M-1} |h'(k)|$$

$$\leq \frac{Nq}{2} \left[\sum_{k=0}^{M-1} |h'(k)| + \sum_{k=M}^{2M-1} |h'(k)| + \cdots \right]$$

$$\leq \frac{Nq}{2} \sum_{k=0}^{\infty} |h'(k)| \qquad (11.47)$$

Bounds for Second-Order Filters

For second-order filters

$$H'(z) = \frac{z^2}{z^2 + b_1 z + b_2}$$

and

$$h'(n) = \frac{1}{2\pi j} \oint_{\Gamma} \frac{z^{n+1}}{(z - p_1)(z - p_2)} dz$$

where the poles are given by

$$p_1, p_2 = -\frac{b_1}{2} \pm \frac{1}{2} \sqrt{b_1^2 - 4b_2}$$

Hence

$$\sum_{i=0}^{\infty} h'(p + iM) = \frac{1}{2\pi j} \sum_{i=0}^{\infty} \oint_{\Gamma} \frac{z^{p+iM+1}}{(z - p_1)(z - p_2)} dz \qquad (11.48)$$

For distinct real poles ($b_1^2 > 4b_2$), Eqs. (11.46) and (11.48) yield

$$|y(n)| \leq q \sum_{p=0}^{M-1} \left| \frac{1}{D_1} \left(\frac{p_1^{p+1}}{1 - p_1^M} - \frac{p_2^{p+1}}{1 - p_2^M} \right) \right|$$

where

$$D_1 = 2 \sqrt{\left(\frac{b_1}{2} \right)^2 - b_2}$$

The limit-cycle frequency is usually 0 ($M = 1$) or $\omega_s/2$ ($M = 2$), and thus

$$|y(n)| \leq \begin{cases} \left|\dfrac{q}{1 + b_1 + b_2}\right| & \text{for } M = 1 \\[3ex] \dfrac{q}{1 - |b_1| + b_2} & \text{for } M = 2 \end{cases}$$

The corresponding absolute bound is obtained from Eq. (11.47) as

$$|y(n)| \leq \frac{q}{1 - |b_1| + b_2}$$

For repeated poles ($b_1^2 = 4b_2$), the tight bound of Eq. (11.46) is given by

$$|y(n)| \leq \sum_{p=0}^{M-1} \left| p_1^p \left[\frac{p}{1 - p_1^M} + \frac{1 + (M - 1)p_1^M}{(1 - p_1^M)^2} \right] \right| \qquad \text{where } p_1 = -\frac{b_1}{2}$$

The absolute bound in this case is

$$|y(n)| \leq \frac{q}{(1 - \sqrt{b_2})^2}$$

For complex poles ($4b_2 > b_1^2$), Eqs. (11.46) and (11.47) yield complicated expressions. However, a simple expression exists for a conservative estimate of the absolute bound. The impulse response is

$$h'(n) = \frac{r^n}{\sin \theta} \sin [(n + 1)\theta]$$

where r and θ are given by Eqs. (11.39) and (11.40). Hence from Eq. (11.47)

$$|y(n)| \leq q \sum_{k=0}^{\infty} \left| \frac{r^k}{\sin \theta} \sin [(k + 1)\theta] \right| \leq q \sum_{k=0}^{\infty} \left| \frac{r^k}{\sin \theta} \right|$$

and after some manipulation

$$|y(n)| \leq \frac{q}{(1 - \sqrt{b_2})\sqrt{1 - b_1^2/4b_2}}$$

A closer approximation for the absolute bound, due to Chang [14], is

$$|y(n)| \leq q \frac{\sin \theta + 4K \sin (\theta/2)}{(1 - |b_1| + b_2) \sin \theta}$$

where $\quad K = \dfrac{r^m}{1 - r^m} \qquad m = \text{Int } \dfrac{\pi}{\theta} \qquad \theta = \cos^{-1} \dfrac{|b_1|}{2\sqrt{b_2}} \qquad r = \sqrt{b_2}$

REFERENCES

1. I. Flores, "The Logic of Computer Arithmetic," Prentice Hall, Englewood Cliffs, N.J., 1963.
2. E. Avenhaus, On the Design of Digital Filters with Coefficients of Limited Word Length, *IEEE Trans. Audio Electroacoust.*, vol. AU-20, pp. 206–212, August 1972.
3. R. E. Crochiere, A New Statistical Approach to the Coefficient Word Length Problem for Digital Filters, *IEEE Trans. Circuits Syst.*, vol. CAS-22, pp. 190–196, March 1975.
4. A. Papoulis, "Probability, Random Variables, and Stochastic Processes," McGraw-Hill, New York, 1965.
5. R. E. Crochiere and A. V. Oppenheim, Analysis of Linear Digital Networks, *Proc. IEEE*, vol. 63, pp. 581–595, April 1975.
6. W. K. Jenkins and B. J. Leon, An Analysis of Quantization Error in Digital Filters Based on Interval Algebras, *IEEE Trans. Circuits Syst.*, vol. CAS-22, pp. 223–232, March 1975.
7. L. B. Jackson, On the Interaction of Roundoff Noise and Dynamic Range in Digital Filters, *Bell Syst. Tech. J.*, vol. 49, pp. 159–184, February 1970.
8. G. Bachman and L. Naria, "Functional Analysis," Academic, New York, 1966.
9. L. B. Jackson, Roundoff-Noise Analysis for Fixed-Point Digital Filters Realized in Cascade or Parallel Form, *IEEE Trans. Audio Electroacoust.*, vol. AU-18, pp. 107–122, June 1970.
10. L. B. Jackson, An Analysis of Limit Cycles Due to Multiplication Rounding in Recursive Digital Filters, *Proc. 7th Annu. Allerton Conf. Circuit Syst. Theory*, pp. 69–78, 1969.
11. S. R. Parker and S. F. Hess, Limit-Cycle Oscillations in Digital Filters, *IEEE Trans. Circuit Theory*, vol. CT-18, pp. 687–697, November 1971.
12. I. W. Sandberg and J. F. Kaiser, A Bound on Limit Cycles in Fixed-Point Implementations of Digital Filters, *IEEE Trans. Audio Electroacoust.*, vol. AU-20, pp. 110–114, June 1972.
13. J. L. Long and T. N. Trick, An Absolute Bound on Limit Cycles Due to Roundoff Errors in Digital Filters, *IEEE Trans. Audio Electroacoust.*, vol. AU-21, pp. 27–30, February 1973.
14. T. L. Chang, A Note on Upper Bounds on Limit Cycles in Digital Filters, *IEEE Trans. Acoust., Speech, Signal Process.*, vol. ASSP-24, pp. 99–100, February 1976.

ADDITIONAL REFERENCES

Avenhaus, E.: A Proposal to Find Suitable Canonical Structures for the Implementation of Digital Filters with Small Coefficient Wordlength, *Nachrichtentech. Z.*, vol. 25, pp. 377–382, August 1972.

——— and W. Schüssler: On the Approximation Problem in the Design of Digital Filters with Limited Wordlength, *Arch. Elektron. Uebertrag.*, vol. 24, pp. 571–572, 1970.

Brglez, F.: Digital Filter Design with Coefficients of Reduced Word Length, *Proc. 1977 IEEE Int. Symp. Circuits Syst.*, pp. 52–55.

Chan, D. S. K., and L. R. Rabiner: Analysis of Quantization Errors in the Direct Form of Finite Impulse Response Digital Filters, *IEEE Trans. Audio Electroacoust.*, vol. AU-21, pp. 354–366, August 1973.

Charalambous, C., and M. J. Best: Optimization of Recursive Digital Filters with Finite Word Lengths, *IEEE Trans. Acoust., Speech, Signal Process.*, vol. ASSP-22, pp. 424–431, December 1974.

Claasen, T. A. C. M., and L. O. G. Kristiansson: Necessary and Sufficient Conditions for the Absence of Overflow Phenomena in a Second-Order Recursive Digital Filter, *IEEE Trans. Acoust., Speech, Signal Process.*, vol. ASSP-23, pp. 509–515, December 1975.

———, W. F. G. Mecklenbräuker, and J. B. H. Peek: Some Remarks on the Classification of Limit Cycles in Digital Filters, *Philips Res. Rep.*, vol. 28, pp. 297–305, August 1973.

———, ———, and ———: Quantization Noise Analysis for Fixed-Point Digital Filters Using Magnitude Truncation for Quantization, *IEEE Trans. Circuits Syst.*, vol. CAS-22, pp. 887–895, November 1975.

Ebert, P. M., J. E. Mazo, and M. G. Taylor: Overflow Oscillations in Digital Filters, *Bell Syst. Tech. J.*, vol. 48, pp. 2999–3020, November 1969.

Fettweis, A.: Roundoff Noise and Attenuation Sensitivity in Digital Filters with Fixed-Point Arithmetic, *IEEE Trans. Circuit Theory*, vol. CT-20, pp. 174–175, March 1973.

Herrmann, O., and W. Schüssler: On the Accuracy Problem in the Design of Nonrecursive Digital Filters, *Arch. Elekron. Uebertrag.*, vol. 24, pp. 525–526, 1970.

Huang, T. C., and D. C. Huey: Errors Caused by the Quantization of the Results of Finite Length Arithmetic Operations for Different Number Representations, *Proc. 1975 IEEE Int. Symp. Circuits Syst.*, pp. 60–63.

Jackson, L. B.: Roundoff Noise Bounds Derived from Coefficient Sensitivities for Digital Filters, *IEEE Trans. Circuits Syst.*, vol. CAS-23, pp. 481–485, August 1976.

Kaiser, J. F.: Some Practical Considerations in the Realization of Linear Digital Filters, *Proc. 3rd Annu. Allerton Conf. Circuit Theory*, pp. 621–633, 1965.

Kan, E. P. F., and J. K. Aggarwal: Error Analysis of Digital Filter Employing Floating-Point Arithmetic, *IEEE Trans. Circuit Theory*, vol. CT-18, pp. 678–686, November 1971.

—— and ——: Minimum-Deadband Design of Digital Filters, *IEEE Trans. Audio Electroacoust.*, vol. AU-19, pp. 292–296, December 1971.

Kaneko, T.: Limit-Cycle Oscillations in Floating-Point Digital-Filters, *IEEE Trans. Audio Electroacoust.*, vol. AU-21, pp. 100–106, April 1973.

Kieburtz, R. B.: An Experimental Study of Roundoff Effects in a Tenth-Order Recursive Digital Filter, *IEEE Trans. Commun.*, vol. COM-21, pp. 757–763, June 1973.

Knowles, J. B., and R. Edwards: Effect of a Finite-Word-Length Computer in a Sampled-Data Feedback System, *Proc. IEE*, vol. 112, pp. 1197–1207, June 1965.

—— and E. M. Olcayto: Coefficient Accuracy and Digital Filter Response, *IEEE Trans. Circuit Theory*, vol. CT-15, pp. 31–41, March 1968.

Liu, B.: Effect of Finite Word Length on the Accuracy of Digital Filters: A Review, *IEEE Trans. Circuit Theory*, vol. CT-18, pp. 670–677, November 1971.

Maria, G. A., and M. M. Fahmy: Limit Cycle Oscillations in a Cascade of First- and Second-Order Digital Sections, *IEEE Trans. Circuits Syst.*, vol. CAS-22, pp. 131–134, February 1975.

Mitra, S. K., K. Hirano, and H. Sakaguchi: A Simple Method of Computing the Input Quantization and Multiplication Roundoff Errors in Digital Filters, *IEEE Trans. Acoust., Speech, Signal Process.*, vol. ASSP-22, pp. 326–329, October 1974.

Rahman, M. H., and M. M. Fahmy: A Roundoff-Noise Minimization Technique for Cascade Realization of Digital Filters, *Proc. 1977 Int. Symp. Circuits Syst.*, pp. 45–48.

Steiglitz, K.: Designing Short-Word Recursive Digital Filters, *Proc. 9th Annu. Allerton Conf. Circuit Syst. Theory, October 1971*, pp. 778–788.

Weinstein, C., and A. V. Oppenheim: A Comparison of Roundoff Noise in Floating Point and Fixed Point Digital Filter Realizations, *Proc. IEEE*, vol. 57, pp. 1181–1183, June 1969.

PROBLEMS

11.1 (*a*) Convert the decimal numbers

$$730.796875 \quad \text{and} \quad -3521.8828125$$

into binary representation.

(*b*) Convert the binary numbers

$$11011101.011101 \quad \text{and} \quad -100011100.1001101$$

into decimal representation.

11.2 Deduce the signed-magnitude, one's-complement, and two's-complement representations of (*a*) 0.810546875 and (*b*) −0.9462890625. Assume a word length $L = 10$.

11.3 The two's complement of a number x can be designated as

$$\tilde{x} = x_0 . x_1 x_2 \cdots x_L$$

(a) Show that

$$x = -x_0 + \sum_{i=1}^{L} x_i 2^{-i}$$

(b) Find x if $\tilde{x} = 0.1110001011$.

(c) Find x if $\tilde{x} = 1.1001110010$.

11.4 Perform the following operations by using the one's- and two's-complement additions.

(a) $0.6015625 - 0.4218750$

(b) $-0.359375 + (-0.218750)$

Assume that $L = 7$.

11.5 The two's complement of x is given by

$$\tilde{x} = x_0 . x_1 x_2 \cdots x_L$$

(a) Show that

$$\text{Two's complement } (2^{-1}x) = \begin{cases} 2^{-1}\tilde{x} & \text{if } x_0 = 0 \\ 1 + 2^{-1}\tilde{x} & \text{if } x_0 = 1 \end{cases}$$

(b) Find the two's complement of $2^{-4}x$ if $\tilde{x} = 1.00110$.

11.6 (a) The register length in a fixed-point digital-filter implementation is 9 bits (including the sign bit), and the arithmetic is of the two's-complement type. Find the largest and smallest machine-representable decimal numbers.

(b) Show that the addition $0.8125 + 0.65625$ will cause overflow.

(c) Show that the addition $0.8125 + 0.65625 + (-0.890625)$ will be evaluated correctly despite of the overflow in the first partial sum.

11.7 The mantissa and exponent register segments in a floating-point implementation are 8 and 4 bits long, respectively.

(a) Deduce the register contents for -0.0234375, -5.0, 0.359375, and 11.5.

(b) Determine the dynamic range of the implementation.

Both mantissa and exponent are stored in signed-magnitude form.

11.8 A floating-point number

$$x = M \times 2^e \qquad \text{where} \qquad M = \sum_{i=1}^{b} b_{-i} 2^{-i}$$

is to be stored in a register whose mantissa and exponent segments comprise $L + 1$ and $e + 1$ bits, respectively. Assuming signed-magnitude representation and quantization by rounding, find the range of the quantization error.

11.9 A filter section is characterized by the transfer function

$$H(z) = \frac{H(z + 1)^2}{z^2 + b_1 z + b_2}$$

where $\qquad H = -0.01903425 \qquad b_1 = -0.5596596 \qquad b_2 = 0.8638557$

(a) Find the quantization error for each coefficient if signed-magnitude fixed-point arithmetic is to be used. Assume quantization by truncation and a word length $L = 6$ bits.

(b) Repeat part (a) if the quantization is to be by rounding.

11.10 (*a*) Realize the transfer function of Prob. 11.9 by using a canonic structure.

(*b*) The filter obtained in part (*a*) is implemented by using the arithmetic described in Prob. 11.9a. Plot the amplitude-response error versus frequency for $10 \leq \omega \leq 30$ rad/s. The sampling frequency is 100 rad/s.

(*c*) Repeat part (*b*), assuming quantization by rounding.

(*d*) Compare the results obtained in parts (*b*) and (*c*).

11.11 (*a*) The transfer function

$$H(z) = \frac{z^2 + 2z + 1}{z^2 + b_1 z + b_2} \quad \text{where} \quad \begin{aligned} b_1 &= -r\sqrt{2} \\ b_2 &= r^2 \end{aligned}$$

is to be realized by using the canonic structure of Fig. 11.7a. Find the sensitivities $S_{b_1}^H(z)$ and $S_{b_2}^H(z)$.

(*b*) The section is to be implemented by using fixed-point arithmetic, and the coefficient quantization is to be by rounding. Compute the statistical word length $L(\omega)$ for $0.7 \leq r \leq 0.95$ in steps of 0.05. Assume that $\Delta M_{\max}(\omega) = 0.02$, $x_1 = 2$ (see Sec. 11.3).

(*c*) Plot the statistical word length versus r and discuss the results achieved.

11.12 The response of an A/D converter to a signal $x(t)$ is given by

$$y(n) = x(n) + e(n)$$

where $x(n)$ and $e(n)$ are random variables uniformly distributed in the ranges $-1 \leq x(n) \leq 1$ and $-2^{-(L+1)} \leq e(n) \leq 2^{-(L+1)}$, respectively.

(*a*) Find the signal-to-noise ratio. This is defined as

$$\text{SNR} = 10 \log \frac{\text{average signal power}}{\text{average noise power}}$$

(*b*) Find the PSD of $y(n)$ if $x(n)$, $e(n)$, $x(k)$, and $e(k)$ are statistically independent.

11.13 The transfer function

$$H(z) = \prod_{i=1}^{3} \frac{a_i(z + 1)^2}{z^2 + b_{1i} z + b_{2i}}$$

where a_i, b_{1i}, and b_{2i} are given in Table P11.13, represents a lowpass Butterworth filter.

Table P11.13

i	a_i	b_{1i}	b_{2i}
1	0.165765	-1.404385	0.735915
2	0.134910	-1.142981	0.412801
3	0.121819	-1.032070	0.275708

(*a*) Realize the transfer function using three canonic sections in cascade.

(*b*) The filter is to be implemented by employing a fixed-point arithmetic, and product quantization is to be by rounding. Plot the relative, output-noise PSD versus frequency. This is defined as

$$\text{RPSD} = 10 \log \frac{S_y(e^{j\omega T})}{S_e(e^{j\omega T})}$$

where $S_y(e^{j\omega T})$ is the PSD of output noise and $S_e(e^{j\omega T})$ is the PSD of a single noise source. The sampling frequency is 10^4 rad/s.

11.14 The filter in Prob. 11.13 is to be scaled according to the scheme of Fig. 11.7. Find the scaling constants λ_0, λ_1, and λ_2 if $p = \infty$ and $q = 1$.

11.15 Repeat Prob. 11.13, assuming the scaling obtained in Prob. 11.14.

11.16 The transfer function

$$H(z) = \prod_{i=1}^{3} \frac{a_{0i}z^2 + a_{1i}z + a_{0i}}{z^2 + b_{1i}z + b_{2i}}$$

where a_{0i}, a_{1i}, b_{1i}, and b_{2i} are given in Table P11.16 represents a bandstop elliptic filter.

Table P11.16

i	a_{0i}	a_{1i}	b_{1i}	b_{2i}
1	4.623281×10^{-1}	7.859900×10^{-9}	7.859900×10^{-9}	-7.534381×10^{-2}
2	4.879171×10^{-1}	5.904108×10^{-2}	8.883641×10^{-1}	8.051571×10^{-1}
3	1.269926	-1.536691×10^{-1}	-8.883640×10^{-1}	8.051571×10^{-1}

(a) Realize the transfer function using three canonic sections in cascade.

(b) Determine the scaling constants for optimum signal-to-noise ratio. Assume the section ordering implied by the transfer function.

(c) Plot the relative output-noise PSD versus frequency. The sampling frequency is 18 rad/s.

11.17 The transfer function

$$H(z) = \prod_{i=1}^{3} \frac{a_{0i}z^2 + a_{1i}z + 1}{z^2 + a_{1i}z + a_{0i}}$$

where a_{0i} and a_{1i} are given in Table P11.17, represents a digital equalizer. Repeat parts (a) to (c) of Prob. 11.16. The sampling frequency is 2.4π rad/s.

Table P11.17

i	a_{0i}	a_{1i}
1	0.973061	-1.323711
2	0.979157	-1.316309
3	0.981551	-1.345605

11.18 The transfer function

$$H(z) = \frac{1}{z^2 + b_1 z + b_2} \quad \text{where} \quad \begin{matrix} b_1 = -1.343503 \\ b_2 = 0.9025 \end{matrix}$$

is to be implemented by using signed-magnitude decimal arithmetic. Quantization is to be performed by rounding each product to the nearest integer, and $\omega_s = 2\pi$ rad/s.

(a) Estimate the peak-to-peak amplitude and frequency of the limit cycle by using Jackson's approach.

(b) Calculate the absolute bound of the limit cycle by using the formula due to Long and Trick.

(c) Repeat part (b) by using Chang's formula.

(d) Determine the actual amplitude and frequency of the limit cycle by simulation.

(e) Compare the results obtained.

11.19 Repeat Prob. 11.18 if $b_1 = -1.8$ and $b_2 = 0.99$.

11.20 Design a sinusoidal oscillator by using a digital filter in cascade with a bandpass filter. The frequency of oscillation is required to be $\omega_s/10$.

TWELVE

WAVE DIGITAL FILTERS

12.1 INTRODUCTION

The effects of coefficient quantization in digital filters can be kept small by using low-sensitivity structures. Such structures can be obtained by realizing the transfer function directly as a cascade or parallel connection of second-order filter sections of the type described in Chap. 4.

Alternative low-sensitivity structures can be obtained by using a synthesis advanced by Fettweis [1, 2] and developed by Sedlmeyer and Fettweis [3, 4]. In this approach an equally terminated LC filter satisfying prescribed specifications is first designed. Then by replacing analog elements by appropriate digital realizations, the LC filter is transformed into a digital filter. The synthesis is based on the wave network characterization, and for this reason the resulting structures are referred to collectively as *wave digital filters*. The low sensitivity comes about because equally terminated LC filters are inherently low-sensitivity structures.

This chapter begins with a qualitative justification of the low-sensitivity attribute of equally terminated LC filters. It then proceeds to the design and analysis details of structures of the Sedlmeyer-Fettweis type. Later, in Sec. 12.10, an alternative to the cascade synthesis of Sec. 4.4 is developed by using the concept of the generalized-immittance converter. This approach tends to yield filters with improved in-band signal-to-noise ratio. The chapter concludes with a list of guidelines that can be used in the choice of a digital-filter structure.

12.2 SENSITIVITY CONSIDERATIONS

An equally terminated LC filter like that in Fig. 12.1a can be characterized in terms of its insertion loss, defined as

$$L(\omega) = 10 \log \frac{P_m(\omega)}{P(\omega)}$$

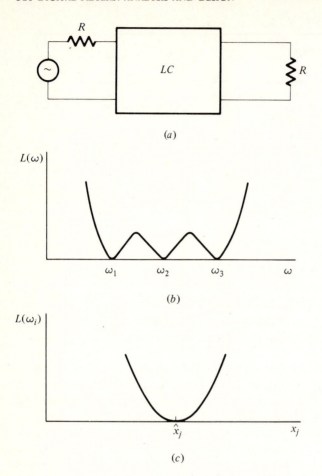

Figure 12.1 (a) Equally terminated, LC filter, (b) equiripple loss characteristic, (c) $L(\omega_i)$ versus x_j.

$P(\omega)$ is the actual output power, and $P_m(\omega)$ is the maximum output power under perfect matching conditions. Since the LC 2-port is a passive lossless network, $P(\omega) \leq P_m(\omega)$ and thus $L(\omega) \geq 0$. Now let us assume that $L(\omega_i) = 0$ for $i = 1$, 2, ..., as depicted in Fig. 12.1b, as in the case of an elliptic characteristic. At frequency ω_i the filter delivers the maximum available power, and if any lossless element x_j is increased above or decreased below its nominal value $\hat{x}_j$, $L(\omega_i)$ must necessarily increase above zero as illustrated in Fig. 12.1c. Clearly

$$\lim_{x_j \to \hat{x}_j} \frac{\Delta L(\omega_i)}{\Delta x_j} = \frac{dL(\omega_i)}{dx_j} = 0$$

for $i = 1, 2, \ldots$ and $j = 1, 2, \ldots$ independently of the order of the filter [5]. Consequently, the sensitivity of the passband loss to element variations in equally terminated LC filters is inherently low. Therefore, by simulating filters of this type digitally, low-sensitivity digital-filter structures can be obtained.

12.3 WAVE NETWORK CHARACTERIZATION

An analog n-port network like that in Fig. 12.2a can be represented by the set of equations

$$\left. \begin{array}{l} A_k = V_k + I_k R_k \\ B_k = V_k - I_k R_k \end{array} \right| \quad k = 1, 2, \ldots, n \qquad (12.1)$$

The parameters A_k and B_k are referred to as the *incident* and *reflected wave quantities*, respectively, and R_k is the *port resistance*. The representation can be either in the time domain or in the frequency domain.

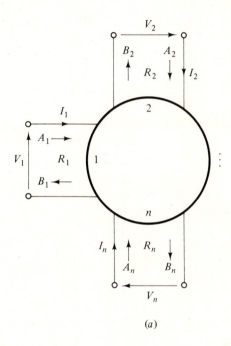

(a)

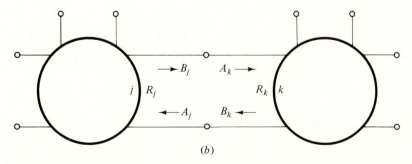

(b)

Figure 12.2 (a) Analog n-port network, (b) interconnected n-ports.

If two n-ports are cascaded as in Fig. 12.2b, it is necessary to assign

$$R_j = R_k$$

so that $\qquad\qquad A_k = B_j \qquad A_j = B_k$

i.e., a common resistance must be assigned to two interconnected ports to maintain continuity in the wave flow. Otherwise, R_k can be assigned on an arbitrary basis.

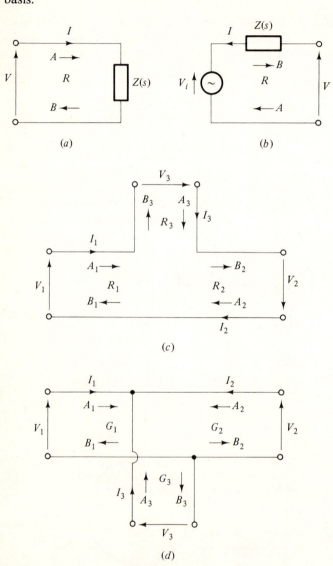

Figure 12.3 (a) Impedance, (b) voltage source, (c) series wire interconnection, (d) parallel wire interconnection.

An *LC* filter can be regarded as a conglomerate of a number of impedances (*R*, *sL*, or 1/*sC*), a source (voltage or current), and a number of 3-port series and parallel wire interconnections like those depicted in Fig. 12.3*c* and *d*. By realizing these elements digitally and subsequently replacing analog elements in *LC* filters by their realizations wave digital filters can be synthetized.

12.4 ELEMENT REALIZATIONS

Digital realizations for analog elements can be derived by using the following procedure:

1. Represent the element in terms of the wave characterization.
2. Eliminate variables V_k, I_k, and s, using the loop and node equations and the bilinear transformation

$$s = \frac{2}{T} \frac{z - 1}{z + 1}$$

3. Express the reflected wave quantities as functions of the incident wave quantities.
4. Realize the resulting set of equations using unit delays, adders, inverters, and multipliers.

Impedances

Consider an impedance

$$Z(s) = s^\lambda R_x \tag{12.2}$$

where R_x is a positive constant and $\lambda = -1$ for a capacitance, $\lambda = 0$ for a resistance, and $\lambda = 1$ for an inductance. From Eq. (12.1) and Fig. 12.3*a*

$$A = V + IR \qquad B = V - IR$$

where
$$V = IZ(s) \tag{12.3}$$

and if *s* in *A*, *B*, *V*, and *I* is eliminated, that is,

$$Q\Big|_{s = \frac{2}{T}\frac{z-1}{z+1}} \to Q \tag{12.4}$$

for $Q = A$, *B*, *V*, or *I*, we obtain

$$B = f(z)A \tag{12.5}$$

where
$$f(z) = \frac{Z(s) - R}{Z(s) + R}\bigg|_{s = \frac{2}{T}\frac{z-1}{z+1}} \tag{12.6}$$

Element	R	Realization	Symbol

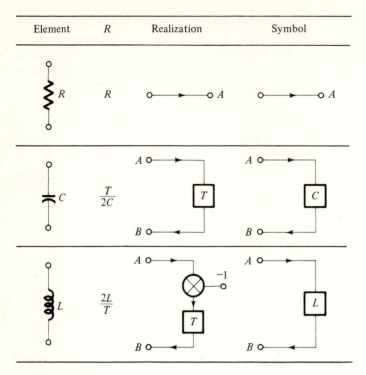

Figure 12.4 Digital realization of impedances.

Now on choosing

$$R = \left(\frac{2}{T}\right)^{\lambda} R_x \tag{12.7}$$

and then using Eqs. (12.2) and (12.6) we have

$$f(z) = \begin{cases} z^{-1} & \text{for } \lambda = -1 \\ 0 & \text{for } \lambda = 0 \\ -z^{-1} & \text{for } \lambda = 1 \end{cases}$$

Hence Eq. (12.5) results in the element realizations of Fig. 12.4; that is, a resistance translates into a digital sink, a capacitance into a unit delay, and an inductance into a unit delay in cascade with an inverter.

Voltage Sources

For the voltage source of Fig. 12.3b, where

$$Z(s) = s^{\lambda} R_x$$

we can write

$$A = V + IR \qquad B = V - IR \qquad V = IZ(s) + V_i$$

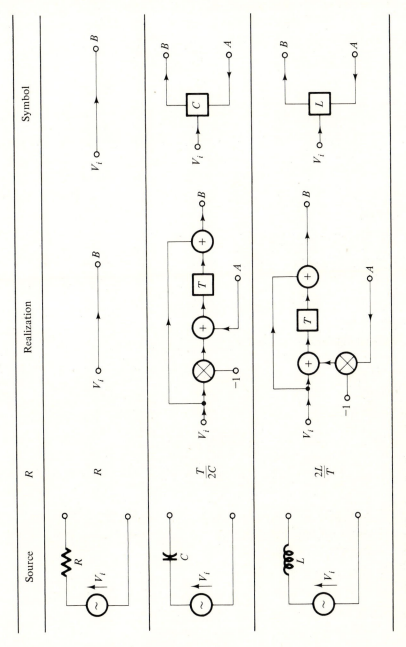

Figure 12.5 Digital realization of voltage sources.

and on eliminating V, I, and s we deduce

$$B = f_1(z)V_i + f_2(z)A \qquad (12.8)$$

where $\qquad f_1(z) = \dfrac{2R}{R + Z(s)}\bigg|_{s = \frac{2}{T}\frac{z-1}{z+1}} \qquad f_2(z) = \dfrac{Z(s) - R}{Z(s) + R}\bigg|_{s = \frac{2}{T}\frac{z-1}{z+1}}$

With

$$R = \left(\frac{2}{T}\right)^{\lambda} R_x$$

$f_1(z)$ and $f_2(z)$ simplify to

$$f_1(z) = \begin{cases} 1 - z^{-1} & \text{for } \lambda = -1 \\ 1 & \text{for } \lambda = 0 \\ 1 + z^{-1} & \text{for } \lambda = 1 \end{cases} \quad \text{and} \quad f_2(z) = \begin{cases} z^{-1} & \text{for } \lambda = -1 \\ 0 & \text{for } \lambda = 0 \\ -z^{-1} & \text{for } \lambda = 1 \end{cases}$$

Hence Eq. (12.8) yields realizations for capacitive, resistive, and inductive sources, as depicted in Fig. 12.5.

Series Wire Interconnection

The preceding approach can be readily extended to wire interconnections. For the series interconnection of Fig. 12.3c

$$I_1 = I_2 = I_3 \qquad V_1 + V_2 + V_3 = 0$$

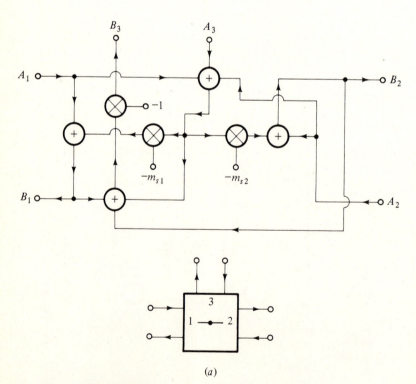

(a)

Figure 12.6 (a) Type S2 adaptor;

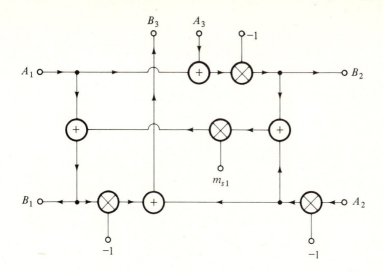

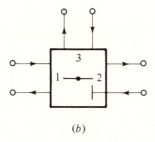

(b)

Figure 12.6 (b) type S1 adaptor.

and on eliminating voltages and currents in Eq. (12.1) we can show that

$$B = (I - M_s)A \tag{12.9}$$

where I is the 3×3 unit matrix, A and B are column vectors

$$M_s = \begin{bmatrix} m_{s1} & m_{s1} & m_{s1} \\ m_{s2} & m_{s2} & m_{s2} \\ m_{s3} & m_{s3} & m_{s3} \end{bmatrix} \qquad m_{s3} = 2 - m_{s1} - m_{s2}$$

and

$$m_{sk} = \frac{2R_k}{R_1 + R_2 + R_3} \tag{12.10}$$

A realization of Eq. (12.9) is shown in Fig. 12.6a. This can be referred to as type S2 adaptor, i.e., series 2-multiplier adaptor.

With R_2 unspecified, one can choose

$$R_2 = R_1 + R_3$$

so that

$$m_{s1} = \frac{R_1}{R_2} \qquad m_{s2} = 1$$

according to Eq. (12.10). As a consequence, the above adaptor can be simplified to the series 1-multiplier adaptor (type S1) of Fig. 12.6b.

Parallel Wire Interconnection

Similarly, for the parallel wire interconnection of Fig. 12.3d

$$V_1 = V_2 = V_3 \qquad I_1 + I_2 + I_3 = 0$$

and from Eq. (12.1)

$$\mathbf{B} = (\mathbf{M}_p - \mathbf{I})\mathbf{A} \tag{12.11}$$

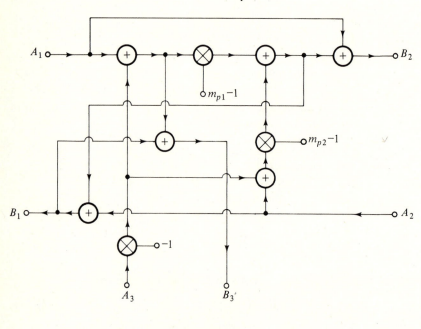

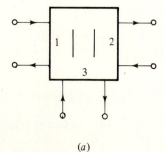

(a)

Figure 12.7 (a) Type P2 adaptor;

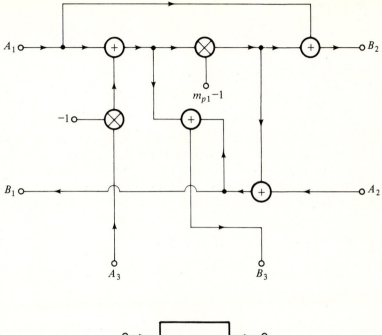

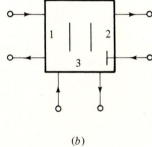

(b)

Figure 12.7 (b) type P1 adaptor.

where
$$\mathbf{M}_p = \begin{bmatrix} m_{p1} & m_{p2} & m_{p3} \\ m_{p1} & m_{p2} & m_{p3} \\ m_{p1} & m_{p2} & m_{p3} \end{bmatrix} \qquad m_{p3} = 2 - m_{p1} - m_{p2}$$

and
$$m_{pk} = \frac{2G_k}{G_1 + G_2 + G_3} \tag{12.12}$$

G_k is the port conductance. A realization of Eq. (12.11), referred to here as type P2 adaptor, is shown in Fig. 12.7a. The corresponding 1-multiplier realization (type P1 adaptor) shown in Fig. 12.7b is obtained by choosing the conductance at port 2 as

$$G_2 = G_1 + G_3$$

so that

$$m_{p1} = \frac{G_1}{G_2} \qquad m_{p2} = 1$$

Alternative adaptor configurations can be found in Ref. 6.

Realizability Constraint

Digital networks containing delay-free loops are said to be unrealizable because certain node signals in such networks cannot be computed. The networks derived so far do not contain delay-free loops. However, such can arise if adaptor ports with direct paths are interconnected. The only adaptor port without direct paths is port 2 in adaptors S1 and P1, as can be seen in Figs. 12.6 and 12.7. Therefore, for the sake of realizability, every direct connection between adaptor ports must necessarily involve port 2 of either an S1 or a P1 adaptor.

12.5 DIGITAL-FILTER REALIZATION

With digital realizations available for the various analog elements, given an *LC* filter, one can readily derive a corresponding wave digital filter by using the following procedure:

1. Identify the various series and parallel wire interconnections in the *LC* filter and number the ports such that every direct connection between wire-interconnection ports involves a port 2.
2. Assign port resistances to the wire-interconnection ports. For a port terminated by an impedance $s^\lambda R_x$ or by a voltage source with an internal impedance $s^\lambda R_x$ assign a port resistance $(2/T)^\lambda R_x$. Then choose the unspecified port resistances to give as far as possible type S1 and P1 adaptors, ensuring that a common resistance is assigned to any two interconnected ports.
3. Calculate the multiplier constants for the various adaptors.
4. Replace each analog element in the *LC* filter by its digital realization.

   Example 12.1 Figure 12.8*a* represents an elliptic lowpass filter with the following specifications:

   Passband ripple $= 1$ dB   Minimum stopband loss $= 34.5$ dB

   Passband edge $= \sqrt{0.5}$ rad/s   Stopband edge $= 1/\sqrt{0.5}$ rad/s

   The element values of the filter are

   $C_1 = C_3 = 2.6189$ F $C_2 = 0.31946$ F $L_2 = 1.2149$ H $R = 1\,\Omega$

   Derive a corresponding wave digital filter using a sampling frequency of 10 rad/s.

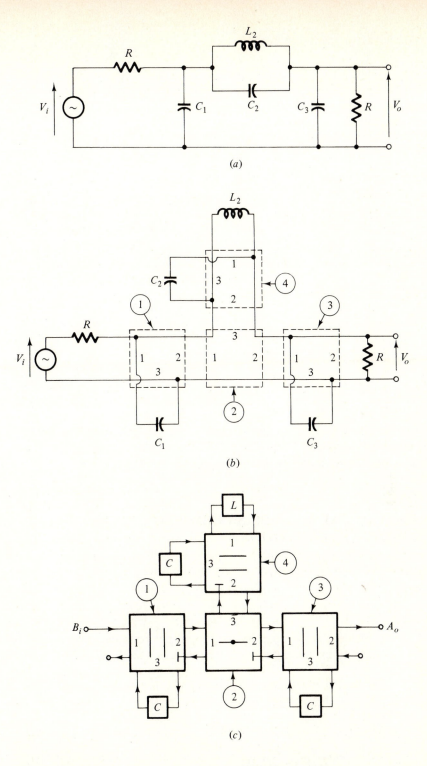

Figure 12.8 (*a*) Elliptic lowpass filter (Example 12.1); (*b*) identification of wire interconnections; (*c*) wave digital filter.

SOLUTION The wire interconnections can be identified as illustrated in Fig. 12.8b. Let G_{jk} (R_{jk}) represent the port conductance (resistance) assigned to the jth port of the kth wire interconnection. From step 2 of the above procedure the following assignments can be made:

Interconnection 1:

$$G_{11} = \frac{1}{R} \qquad G_{31} = \frac{2C_1}{T} \qquad G_{21} = G_{11} + G_{31} \qquad\qquad m_{p1} = 0.107110$$

Interconnection 4:

$$G_{14} = \frac{T}{2L_2} \qquad G_{34} = \frac{2C_2}{T} \qquad G_{24} = G_{14} + G_{34} \qquad\qquad m_{p1} = 0.202741$$

Interconnection 2:

$$R_{12} = \frac{1}{G_{21}} \qquad R_{32} = \frac{1}{G_{24}} \qquad R_{22} = R_{12} + R_{32} \qquad\qquad m_{s1} = 0.120194$$

Interconnection 3:

$$G_{13} = \frac{1}{R_{22}} \qquad G_{23} = \frac{1}{R} \qquad G_{33} = \frac{2C_3}{T} \qquad\qquad \begin{array}{l} m_{p1} = 0.214595 \\ m_{p2} = 0.191234 \end{array}$$

Interconnections 1, 2, 3, and 4 result in P1, S1, P2, and P1 adaptors, respectively, as depicted in Fig. 12.8c. The multiplier coefficients can be computed as shown above by using Eqs. (12.10) and (12.12).

12.6 WAVE DIGITAL FILTERS SATISFYING PRESCRIBED SPECIFICATIONS

The transfer function of the filter in Fig. 12.8c is given by

$$H_D(z) = \frac{A_o}{B_i} \qquad\qquad (12.13)$$

where A_o is the incident wave quantity for the output resistance and B_i is the reflected wave quantity for the input source. From Eqs. (12.3), (12.4), and (12.8)

$$A_o = 2V_o \Big|_{s = \frac{2}{T}\frac{z-1}{z+1}} \qquad \text{and} \qquad B_i = V_i \Big|_{s = \frac{2}{T}\frac{z-1}{z+1}}$$

Thus from Eq. (12.13)

$$H_D(z) = 2H_A(s) \Big|_{s = \frac{2}{T}\frac{z-1}{z+1}}$$

where $H_A(s)$ is the transfer function of the analog filter. Clearly, wave digital filters, like other digital filters based on the bilinear transformation, are subject to the warping effect discussed in Sec. 7.6.

Wave digital filters satisfying prescribed specifications can be designed by using the prewarping techniques of Chap. 8. A detailed design procedure is as follows:

1. Using the specifications, derive an appropriate normalized lowpass transfer function according to steps 1 to 3 in Sec. 8.4.
2. Realize the transfer function derived in step 1 as an equally terminated *LC* filter.
3. Transform the lowpass filter realized in step 2 using the appropriate formula in Table 8.1.
4. Form the desired digital filter using the procedure in Sec. 12.5.

Nowadays step 2 can be carried out by using filter-design tables like those found in Refs. 7 to 10.

Example 12.2 Design a wave bandpass digital filter satisfying the following specifications:

> Passband ripple = 1 dB   Minimum stopband loss $\geq$ 35 dB
> Lower and upper passband edges = 2.0, 3.0 rad/s
> Lower and upper stopband edges = 1.5, 3.5 rad/s
> Sampling frequency = 10 rad/s.

SOLUTION On choosing an elliptic approximation and then using the procedure in Sec. 8.4 (Program B.6), we obtain

$$n = 3 \quad k = 0.4472136 \quad \omega_0 = 3.183099 \quad B = 3.093133$$

where n and k are the order and selectivity factor of the normalized lowpass filter, respectively, and ω_0 and B are the parameters in the transformation

$$s = \frac{1}{B}\left(\bar{s} + \frac{\omega_0^2}{\bar{s}}\right)$$

A normalized lowpass *LC* filter with $n = 3$ and $k = 0.45$ can be obtained from Ref. 7 as depicted in Fig. 12.8a, where we now have

$$C_1 = C_3 = 2.8130 \text{ F} \quad C_2 = 0.26242 \text{ F} \quad L_2 = 1.3217 \text{ H} \quad R = 1 \ \Omega$$

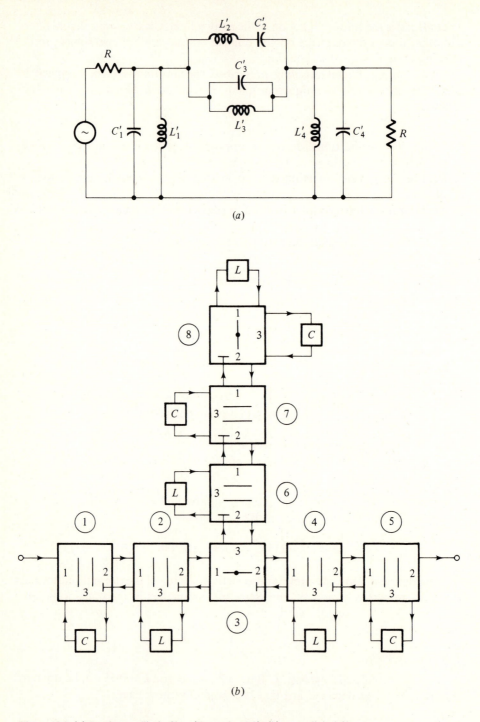

Figure 12.9 (a) Bandpass, elliptic filter (Example 12.2); (b) wave digital filter.

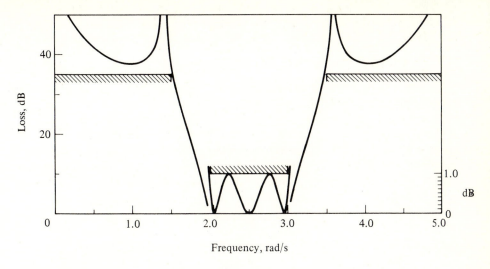

Figure 12.10 Loss characteristic of bandpass filter (Example 12.2).

On applying the above lowpass-to-bandpass transformation, the bandpass filter of Fig. 12.9*a* can be formed, where

$$L'_1 = L'_4 = 0.108525 \text{ H} \qquad C'_1 = C'_4 = 0.909434 \text{ F}$$

$$L'_2 = 0.427301 \text{ H} \qquad C'_2 = 0.230975 \text{ F}$$

$$L'_3 = 1.16333 \text{ H} \qquad C'_3 = 0.0848396 \text{ F}$$

$$R = 1 \text{ } \Omega$$

Subsequently, on using the procedure of Sec. 12.5, the wave digital filter of Fig. 12.9*b* can be derived. The resulting multiplier constants are given in Table 12.1. The loss characteristic achieved is plotted in Fig. 12.10.

Table 12.1 Multiplier constants (Example 12.2)

Adaptor	Type	k	m_{pk} or m_{sk}
1	P1	1	0.256751
2	P1	1	0.573642
3	S1	1	0.117926
4	P1	1	0.216662
5	P2	1	0.973738
		2	0.263494
6	P1	1	0.702492
7	P1	1	0.576495
8	S1	1	0.500000

12.7 REDUCTION IN THE NUMBER OF DIGITAL ELEMENTS

The cost and speed of a digital-filter implementation are closely related to the total number of multipliers, adders, and unit delays employed. The cost tends to increase and the speed tends to decrease as the number of elements increases. Hence the element count must be kept to a minimum in practice.

The element count in wave digital filters can sometimes be reduced [11] by employing a pair of impedance transformations first used by Bruton [12] in the domain of active filters. These transformations give rise to s^2-*impedance* and s^2-*admittance* elements, known collectively as *frequency-dependent negative resistances* (FDNRs). The terminal relation for an s^2-impendance elements is

$$V = s^2 EI$$

and, similarly, for an s^2-admittance element

$$I = s^2 DV$$

Element	R	Realization	Symbol

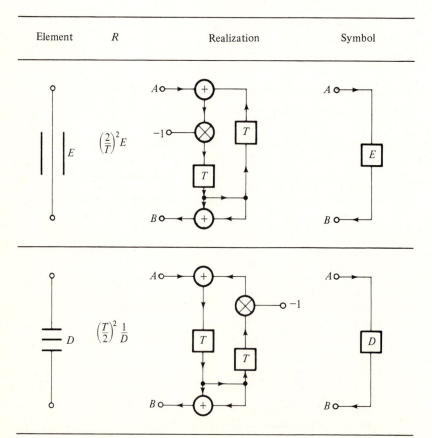

Figure 12.11 FDNR elements and their digital realizations.

where E and D are positive constants. The two FDNRs are usually represented by the symbols in Fig. 12.11. Digital realizations for these elements can readily be derived as depicted in Fig. 12.11 by letting $\lambda = -2$ or $+2$ in Eqs. (12.2) to (12.7).

If each impedance $Z(s)$ in an LC filter is modified according to the transformation

$$Z(s) \rightarrow sZ(s) \qquad (12.14)$$

impedances R_x, R_x/s, and sR_x will transform into sR_x, R_x, and s^2R_x, respectively; i.e., resistances will translate into inductances, capacitances into resistances, and inductances into s^2-impedance elements. Similarly, if

$$Z(s) \rightarrow \frac{Z(s)}{s} \qquad (12.15)$$

resistances will translate into capacitances, capacitances into s^2-admittance elements, and inductances into resistances.

Impedance transformations in general preserve the transfer function and sensitivity properties of the original network. Consequently, by transforming the LC filter before the digital realization using Eq. (12.14) or (12.15) a pair of alternative low-sensitivity digital structures can be derived.

Example 12.3 Derive an alternative digital structure from the LC filter of Example 12.1 (Fig. 12.8a) using the transformation in Eq. (12.14).

SOLUTION On multiplying impedances in Fig. 12.8a by s, the FDNR network of Fig. 12.12a is obtained. Then on using the procedure in Sec. 12.5 the wave digital filter of Fig. 12.12b can be formed. The multiplier coefficients can be shown to be the same as in Example 12.1.

The effects of the above transformations on the element count can be summarized as follows:

1. Translating capacitances (inductances) into resistances will decrease the element count because (1) the realization of a resistance is just a sink and (2) adaptors with resistive terminations can be simplified by eliminating redundant adders.
2. Translating inductances (capacitances) into s^2-impedance (s^2-admittance) elements and resistances into inductances (capacitances) will increase the number of elements, as can be seen by comparing the various element realizations.

In LC filters with a large number of capacitances and a small number of inductances or vice versa, the element economies in item 1 tend to exceed the losses in item 2. Under these circumstances, the use of either Eq. (12.14) or (12.15) will lead to a more economical design.

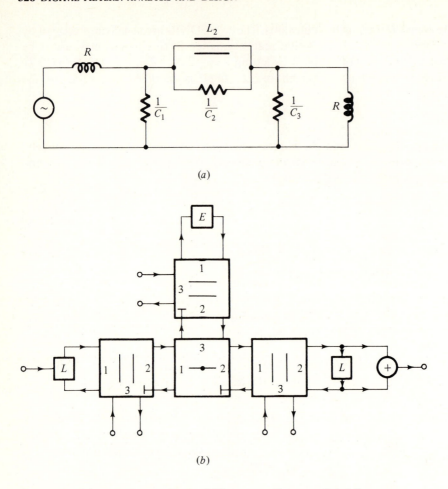

(a)

(b)

Figure 12.12 (a) FDNR lowpass filter and (b) corresponding wave digital filter.

The element counts for the LC and FDNR digital realizations of the lowpass filter shown in Fig. 12.13 are compared in Table 12.2. As can be seen, the FDNR realization is more economical for odd $n > 3$ and for even $n > 6$. Similar economies can be achieved in highpass filters.

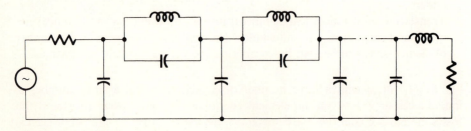

Figure 12.13 Nth-order lowpass filter.

Table 12.2 Number of digital elements in lowpass realization

Element type	LC realization		FDNR realization	
	Odd n	Even n	Odd n	Even n
Inverters	$3n - 2$	$3n - 3$	$2n$	$2n$
Adders	$6n - 2$	$6n - 4$	$5n + 1$	$5n + 3$
Unit delays	$\dfrac{3n - 1}{2}$	$\dfrac{3n - 2}{2}$	$n + 1$	$n + 2$
Multipliers	$\dfrac{3n + 1}{2}$	$\dfrac{3n}{2}$	$\dfrac{3n + 1}{2}$	$\dfrac{3n}{2}$

12.8 FREQUENCY-DOMAIN ANALYSIS

Once a wave digital filter is designed, a frequency-domain analysis is often necessary to study quantization effects or simply to verify the design. Such an analysis will now be described.

Consider the network in Fig. 12.14a, where adaptor q is terminated by subnetworks N_p, N_r, and N_s. Adaptor q can be characterized by

$$H_q(z) = \frac{B_{2q}}{A_{1q}} \qquad F_{1q} = \frac{B_{1q}}{A_{1q}} \qquad F_{2q} = \frac{B_{2q}}{A_{2q}}$$

$H_q(z)$ is the transfer function of the terminated adaptor, and F_{1q} and F_{2q} are its *input functions* at ports 1 and 2, respectively. Similarly, subnetworks N_p, N_r, and N_s can be characterized by the input functions

$$F_p = \frac{B_p}{A_p} = \frac{A_{1q}}{B_{1q}} \qquad F_r = \frac{B_r}{A_r} = \frac{A_{2q}}{B_{2q}} \qquad F_s = \frac{B_s}{A_s} = \frac{A_{3q}}{B_{3q}}$$

Expressions for $H_q(z)$, F_{1q}, and F_{2q} in terms of F_p, F_r, and F_s for series and parallel adaptors can be derived from Eqs. (12.9) and (12.11). For the S2 adaptor we have

$$H_q(z) = \frac{m_{s2}(F_s - 1)}{D_1} \qquad F_{1q} = \frac{-C_1 + C_3 F_r - C_2 F_s - F_r F_s}{D_1}$$

$$F_{2q} = \frac{-C_2 + C_3 F_p - C_1 F_s - F_p F_s}{D_2}$$

where

$$D_1 = 1 + C_2 F_r - C_3 F_s + C_1 F_r F_s \qquad D_2 = 1 + C_1 F_p - C_3 F_s + C_2 F_p F_s$$

$$C_1 = m_{s1} - 1 \qquad C_2 = m_{s2} - 1 \qquad C_3 = m_{s1} + m_{s2} - 1$$

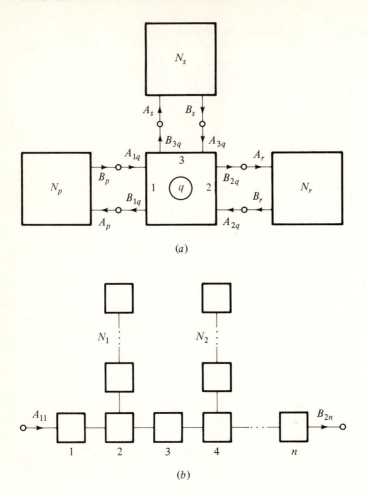

(a)

(b)

Figure 12.14 Analysis of wave digital filters: (a) embedded adaptor, (b) block diagram.

Similarly, for the P2 adaptor

$$H_q(z) = \frac{m_{p1}(1 + F_s)}{D_1} \qquad F_{1q} = \frac{C_1 + C_3 F_r - C_2 F_s + F_r F_s}{D_1}$$

$$F_{2q} = \frac{C_2 + C_3 F_p - C_1 F_s + F_p F_s}{D_2}$$

where

$$D_1 = 1 - C_2 F_r + C_3 F_s + C_1 F_r F_s \qquad D_2 = 1 - C_1 F_p + C_3 F_s + C_2 F_p F_s$$

$$C_1 = m_{p1} - 1 \qquad C_2 = m_{p2} - 1 \qquad C_3 = m_{p1} + m_{p2} - 1$$

These relations apply to S1 and P1 adaptors except that $m_{s2} = 1$ and $m_{p2} = 1$.

Now consider the filter of Fig. 12.14b and assume that the adaptors of the main path are numbered consecutively from input to output. The overall transfer function of the filter is

$$H(z) = \frac{B_{2n}}{A_{11}}$$

Since the reflected and incident wave quantities at the output of adaptor q become the incident and reflected wave quantities at the input of adaptor $q + 1$, respectively, we can write

$$H(z) = \frac{B_{21}}{A_{11}} \frac{B_{22}}{B_{21}} \cdots \frac{B_{2n}}{B_{2(n-1)}}$$

$$= \frac{B_{21}}{A_{11}} \frac{B_{22}}{A_{12}} \cdots \frac{B_{2n}}{A_{1n}}$$

Therefore

$$H(z) = \prod_{q=1}^{n} H_q(z) \tag{12.16}$$

For the connection of Fig. 12.14a, $H_q(z)$ and F_{1q} depend on F_s and F_r, as was shown earlier. If $N_s(N_r)$ comprises a cascade of adaptors, $F_s(F_r)$ will depend on the input function of the second adaptor in the cascade, which will in turn depend on the input function of the third adaptor, and so on. Consequently, for the filter in Fig. 12.14b, the input functions of branches $N_1, N_2, \ldots$ must be evaluated first, starting with the last adaptor and proceeding to the branch input in each case. Subsequently, the main-path adaptors should be analyzed, starting with the output adaptor and proceeding to the filter input. With the frequency responses of the individual main-path adaptors known, the overall response of the filter can be evaluated by using Eq. (12.16).

12.9 ALTERNATIVE APPROACH TO THE SYNTHESIS OF WAVE DIGITAL FILTERS

An alternative approach to the synthesis of wave digital filters, which obviates the need for adaptors, was proposed independently by Constantinides [13] and Swamy and Thyagarajan [14]. In this approach, the prototype LC filter is viewed as a cascade connection of elemental 2-ports, as depicted in Fig. 12.15. The rth 2-port can be characterized by

$$\begin{bmatrix} V_{1r} \\ I_{1r} \end{bmatrix} = \begin{bmatrix} \alpha_r & \beta_r \\ \gamma_r & \delta_r \end{bmatrix} \begin{bmatrix} V_{2r} \\ I_{2r} \end{bmatrix} \tag{12.17}$$

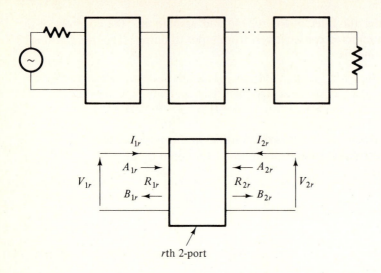

Figure 12.15 Alternative approach to the synthesis of wave digital filters.

Hence on assigning wave quantities and port resistances, as in Sec. 12.3, and then using Eqs. (12.1) and (12.17) we can show that

$$\begin{bmatrix} A_{1r} \\ B_{1r} \end{bmatrix} = \begin{bmatrix} a_r & b_r \\ c_r & d_r \end{bmatrix} \begin{bmatrix} A_{2r} \\ B_{2r} \end{bmatrix}$$

or

$$\begin{bmatrix} B_{1r} \\ B_{2r} \end{bmatrix} = \begin{bmatrix} \dfrac{d_r}{b_r} & c_r - \dfrac{a_r d_r}{b_r} \\ \dfrac{1}{b_r} & -\dfrac{a_r}{b_r} \end{bmatrix} \begin{bmatrix} A_{1r} \\ A_{2r} \end{bmatrix}$$ (12.18)

where

$$a_r = \frac{1}{2}\left(\alpha_r + \gamma_r R_{1r} + \frac{\beta_r + \delta_r R_{1r}}{R_{2r}} \right) \qquad b_r = \frac{1}{2}\left(\alpha_r + \gamma_r R_{1r} - \frac{\beta_r + \delta_r R_{1r}}{R_{2r}} \right)$$

$$c_r = \frac{1}{2}\left(\alpha_r - \gamma_r R_{1r} + \frac{\beta_r - \delta_r R_{1r}}{R_{2r}} \right) \qquad d_r = \frac{1}{2}\left(\alpha_r - \gamma_r R_{1r} - \frac{\beta_r - \delta_r R_{1r}}{R_{2r}} \right)$$

By applying the bilinear transformation to Eq. (12.18) and then realizing the resulting equation by means of digital elements a digital structure for the rth 2-port can be derived. Realizations for the common types of 2-ports as well as further details about the synthesis can be found in Refs. 15 to 18.

12.10 A CASCADE SYNTHESIS BASED ON THE WAVE CHARACTERIZATION

The wave characterization along with the concept of the generalized-immittance converter (GIC) [19] can be used to develop an alternative to the cascade synthesis of Sec. 4.4 [20]. The details of this approach are as follows.

Generalized-Immittance Converters

A GIC is a 2-port whose input admittance Y_i is related to the load admittance Y_L by

$$Y_i = h(s)Y_L$$

where $h(s)$ is the *admittance conversion function* of the device. Two specific types of GIC can be identified, namely voltage- and current-conversion GICs. The current-conversion GIC (CGIC) is characterized by the terminal relations

$$V_1 = V_2 \qquad I_1 = -h(s)I_2 \tag{12.19}$$

This is usually represented by the symbol of Fig. 12.16a.

Analog G-CGIC Configuration

By interconnecting three conductances and two CGICs, we can construct the G-CGIC configuration of Fig. 12.17a [21]. If each CGIC is assumed to have a conversion function $h(s) = s$, straightforward analysis yields

$$\frac{V_o}{V_i} = \frac{k_0 G_0 + k_1 G_1 s + k_2 G_2 s^2}{G_0 + G_1 s + G_2 s^2}$$

and if $G_r = b_r$ and $k_r = a_r/b_r$ for $r = 0, 1, 2$, the network realizes the transfer function

$$H(s) = \frac{a_0 + a_1 s + a_2 s^2}{b_0 + b_1 s + b_2 s^2} \tag{12.20}$$

By cascading a number of sections like the above any stable continuous-time transfer function can be realized.

Digital G-CGIC Configuration

Like an *LC* network, the G-CGIC network of Fig. 12.17a can readily be simulated by digital elements. We need only develop a digital realization for the CGIC by using the procedure outlined in Sec. 12.4.

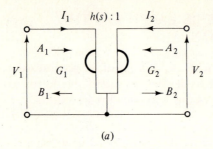

$$(a)$$

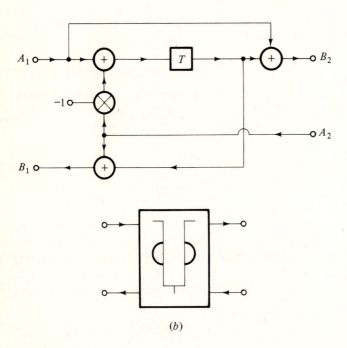

$$(b)$$

Figure 12.16 (a) Current-conversion generalized-immittance converter; (b) digital realization.

On assigning wave quantities and conductances to the CGIC ports, as illustrated in Fig. 12.16a, and then using Eqs. (12.1), (12.4), and (12.19), we can show that

$$B_1 = A_2 + (A_1 - A_2)F(z) \qquad B_2 = A_1 + (A_1 - A_2)F(z)$$

where

$$F(z) = \frac{G_1 - G_2 h(z)}{G_1 + G_2 h(z)} \qquad (12.21)$$

$$h(z) = h(s) \Big|_{s \to \frac{2}{T}\frac{z-1}{z+1}}$$

Hence with $h(s) = s$ and $G_1 = 2G_2/T$ Eq. (12.21) reduces to

$$F(z) = z^{-1}$$

Therefore, a digital realization for the CGIC can be obtained, as depicted in Fig. 12.16b.

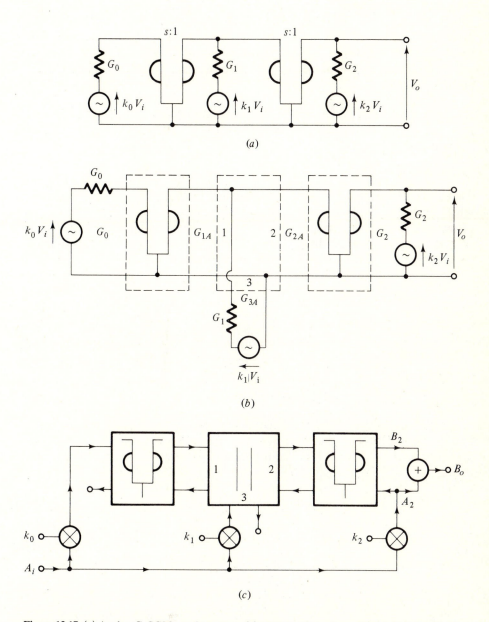

(a)

(b)

(c)

Figure 12.17 (a) Analog G-CGIC configuration; (b) identification of n-ports; (c) digital realization.

The individual n-ports of the G-CGIC configuration can now be identified, as indicated in Fig. 12.17b. On assigning the port conductances

$$G_{1A} = \frac{TG_0}{2} \qquad G_{2A} = \frac{2G_2}{T} \qquad G_{3A} = G_1$$

the general second-order digital section of Fig. 12.17c can be derived. An output proportional to V_o can be formed by using an adder at the input or output of any one of the CGICs, as in Fig. 12.17c, or at port 3 of the adaptor. This is permissible by virtue of Eq. (12.19).

The transfer function of the derived structure can be obtained from Eqs. (12.1) and (12.4) as

$$H_D(z) = \frac{B_o}{A_i} = \frac{B_2 + A_2}{A_i} = \frac{2V_o}{V_i} = 2H(s)\Big|_{s \rightarrow \frac{2}{T}\frac{z-1}{z+1}}$$

Cascade Synthesis

Almost invariably recursive filters are designed by using Butterworth, Tschebyscheff, Bessel, or elliptic transfer functions which have zeros at the origin of the s plane, on the imaginary axis, or at infinity (see Chap 5). Hence the continuous-time transfer function can be realized as a cascade connection of second-order sections characterized by transfer functions of the type

$$H_A(s) = \frac{N_A(s)}{b_0 + b_1 s + s^2}$$

where $N_A(s)$ can take the form b_0, s^2, $b_1 s$, or $a_0 + s^2$ for a lowpass (LP), highpass (HP), bandpass (BP), or notch (N) section, respectively. On the other hand, delay equalizers are designed by using allpass (AP) sections, in which

$$H_A(s) = \frac{b_0 - b_1 s + s^2}{b_0 + b_1 s + s^2}$$

Evidently, the above transfer functions are special cases of the transfer function in Eq. (12.20), and therefore they can all be readily realized by using the digital structure of Fig. 12.17c. The resulting structures are shown in Fig. 12.18, where

$$k_0 = \frac{a_0}{b_0} \tag{12.22}$$

$$m_1 = \frac{b_0 - (2/T)b_1 - (2/T)^2}{b_0 + (2/T)b_1 + (2/T)^2} \tag{12.23}$$

$$m_2 = -\frac{b_0 + (2/T)b_1 - (2/T)^2}{b_0 + (2/T)b_1 + (2/T)^2} \tag{12.24}$$

in each case.

With a set of universal sections available, any Butterworth, Tschebyscheff, Bessel, or elliptic digital filter satisfying prescribed specifications can be designed by using the following procedure:

1. Using the specifications, derive the appropriate normalized lowpass transfer function according to steps 1 to 3 in Sec. 8.4.
2. Apply the transformation in Eq. (8.1).
3. Select suitable sections from Fig. 12.18.
4. Calculate the multiplier constants using Eqs. (12.22) to (12.24).
5. Connect the various sections in cascade.

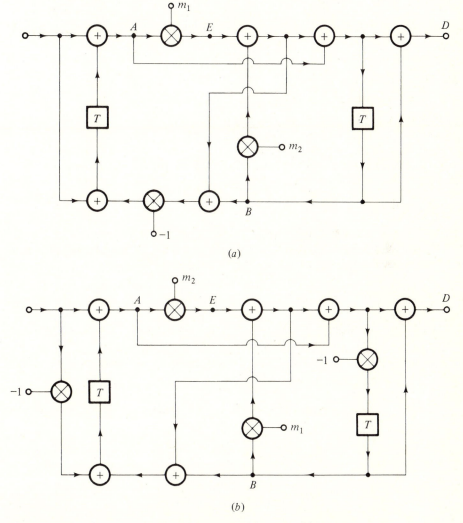

(a)

(b)

Figure 12.18 Universal second-order CGIC sections: (a) lowpass, (b) highpass;

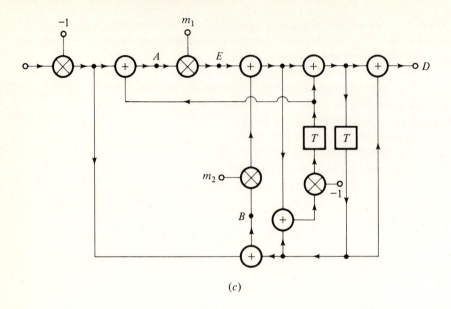

(c)

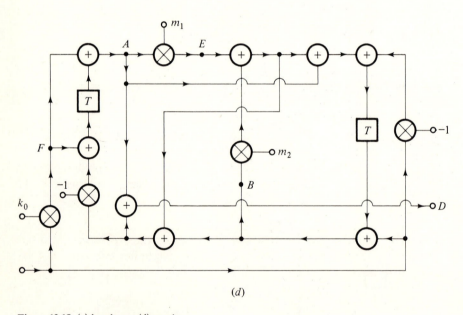

(d)

Figure 12.18 (c) bandpass, (d) notch;

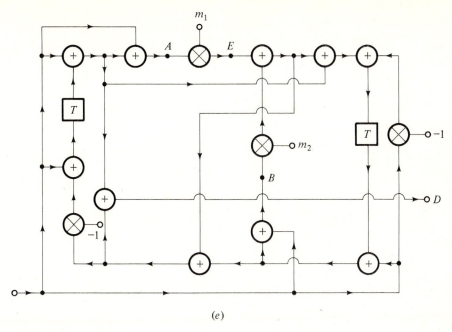

(e)

Figure 12.18 (e) allpass.

Example 12.4 A Butterworth lowpass filter is characterized by

$$H(s) = \prod_{j=1}^{3} \frac{b_{0j}}{b_{0j} + b_{1j}s + s^2}$$

where coefficients b_{ij} are given in Table 12.3. Design a corresponding digital filter by using the CGIC cascade synthesis. The sampling frequency is 10^4 rad/s.

SOLUTION The filter can be designed by cascading three LP sections of the type shown in Fig. 12.18a. The values of the multiplier constants can be readily evaluated as in Table 12.3.

Table 12.3 Lowpass-filter parameters (Example 12.4)

j	b_{0j}	b_{1j}	m_{1j}	m_{2j}
1	1.069676×10^6	5.353680×10^2	-8.342350×10^{-1}	5.701500×10^{-1}
2	1.069676×10^6	1.462653×10^3	-8.650900×10^{-1}	2.778910×10^{-1}
3	1.069676×10^6	1.998021×10^3	-8.781810×10^{-1}	1.538890×10^{-1}

Signal Scaling

Assuming a fixed-point implementation, the CGIC sections of Fig. 12.18 can be scaled by using Jackson's technique (see Sec. 11.5). For this purpose each of the five sections can be represented by the flow graph of Fig. 12.19a, where

$$H_A(z) = \frac{N_A(z)}{D(z)} \qquad H_B(z) = \frac{N_B(z)}{D(z)} \qquad \text{and} \qquad H_D(z) = \frac{N_D(z)}{D(z)}$$

are the transfer functions between section input and nodes A, B, and D respectively. The above polynomials are given in Table 12.4. The optimum value of λ, for maximum signal-to-noise ratio, is given by

$$\lambda = \frac{1}{\max\left[\left\|H_A(e^{j\omega T})\right\|_\infty, \left\|H_B(e^{j\omega T})\right\|_\infty, \left\|H_D(e^{j\omega T})\right\|_\infty\right]}$$

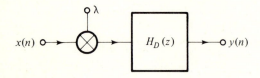

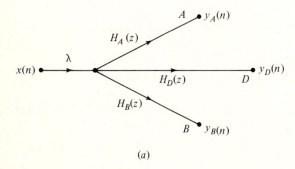

(a)

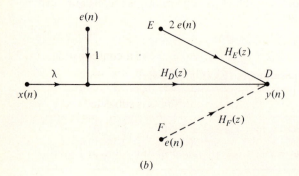

(b)

Figure 12.19 Universal CGIC sections: (a) scaling model; (b) noise model.

Table 12.4 Polynomials in CGIC sections

Type	$N_A(s)$	$N_D(z)$	$N_A(z)$	$N_B(z)$
LP	b_0	$(1 + m_1)(z + 1)^2$	$(z - m_2)(z + 1)$	$(1 + m_1)(z + 1)$
HP	s^2	$(1 + m_2)(z - 1)^2$	$(z + m_1)(z - 1)$	$-(1 + m_2)(z - 1)$
BP	$b_1 s$	$(m_1 + m_2)(1 - z^2)$	$-(z^2 - 2m_2 z + 1)$	$-(z^2 + 2m_1 z + 1)$
N	$a_0 + s^2$	$k_0(1 + m_1)(z + 1)^2$ $+ (1 + m_2)(z - 1)^2$	$k_0(z - m_2)(z + 1)$ $- (1 + m_2)(z - 1)$	$k_0(1 + m_1)(z + 1)$ $+ (z + m_1)(z - 1)$
AP	$b_0 - b_1 s + s^2$	$2[(1 + m_1 + m_2)z^2$ $+ (m_1 - m_2)z + 1]$	$2(z^2 - 2m_2 z + 1)$	$2(z^2 + 2m_1 z + 1)$

$D(z) = z^2 + (m_1 - m_2)z + (1 + m_1 + m_2)$

Output Noise

For the purpose of noise analysis, the five CGIC sections can be represented by the model of Fig. 12.19b, where $e(n)$ is the noise component generated by one multiplier and

$$H_E(z) = \frac{(z + 1)(z - 1)}{D(z)} \qquad H_F(z) = \frac{(1 + m_1)(z + 1)^2}{D(z)}$$

are the transfer functions between nodes E and F and section output. The dotted line in Fig. 12.19b applies to the N section only.

On using the approach of Sec. 11.4, the output-noise PSD can be deduced as

$$S_o(e^{j\omega T}) = [|H_D(e^{j\omega T})|^2 + 2|H_E(e^{j\omega T})|^2 + |H_F(e^{j\omega T})|^2]S_e(e^{j\omega T})$$

where $$S_e(e^{j\omega T}) = \frac{q^2}{12} \quad \text{and} \quad H_F(e^{j\omega T}) = 0$$

in all sections except for the N section in which

$$H_F(e^{j\omega T}) \neq 0.$$

A useful property of the CGIC sections can be identified at this point. $H_E(z)$ is a bandpass transfer function in each of the five sections. As a consequence, noise generated by multipliers m_1 and m_2 will be attenuated at low as well as high frequencies, becoming zero at $\omega = 0$ as well as at $\omega = \omega_s/2$. By contrast, in the conventional canonic sections noise due to the multipliers is subjected to the same transfer function as the signal; e.g., in a lowpass section, the quantization noise is subjected to a lowpass transfer function. Because of this property, the CGIC synthesis tends to yield lowpass, highpass, and bandstop filters and also equalizers with improved in-band signal-to-noise ratio [20].

12.11 CHOICE OF STRUCTURE

This chapter, like Chap. 4, has demonstrated that many distinct structures are possible for a given filter specification. Hence one of the initial tasks of the filter designer is to choose a structure. The principal factors in this task are the sensitivity of the structure to coefficient quantization, the level of output noise due to product quantization, and the computational efficiency of the structure. As may be expected, these factors tend to depend to a large extent on the desired specifications, on the type of filter, i.e., lowpass, bandpass, etc.; on the type of approximation used, i.e., Butterworth, elliptic, etc.; on the type of arithmetic, i.e., fixed-point or floating point; on the number system used, i.e., two's complement, signed magnitude, etc.; on the ordering of filter sections (in cascade structures), and so on. Consequently, categorical statements about one or the other type of structure are difficult if not impossible to make. Nevertheless, certain tendencies are beginning to emerge, as follows:

1. Direct and continued-fraction ladder structures tend to be very sensitive to coefficient quantization [22–24].
2. Cascade, parallel, and wave structures tend to have similar sensitivity properties for fixed-point arithmetic [22–25].
3. Wave structures tend to be less sensitive than cascade structures for floating-point arithmetic [26–27].
4. Direct and continued-fraction ladder structures tend to generate a high level of product-quantization noise [23–25].
5. Parallel structures tend to generate a lower level of product-quantization noise than cascade structures [23–24, 28–29].
6. CGIC cascade structures tend to yield an improved in-band signal-to-noise ratio relative to that in conventional cascade structures [20].
7. For filters with zeros on the unit circle of the z plane, cascade canonic structures involve the lowest number of arithmetic operations.
8. Wave structures entail an excessive number of additions (see Table 12.2).

It should be mentioned that the choice of structure involves many other issues besides the above, e.g., limit-cycle effects, dynamic range, the amenability of the structure to large-scale integration and to multiplexing (see Chap. 14), and the cost of hardware. Also in applications where very high sampling rates are employed, the degree of parallelism inherent in the various structures should be considered. In canonic structures, all multiplications can be performed simultaneously, and as a consequence the time taken to process one filter cycle can be as short as the time taken to perform one multiplication. In wave structures, on the other hand, multiplications must be performed in sequence according to a certain hierarchy, because of topological constraints, and hence the minimum time to process one filter cycle can be much longer [22].

REFERENCES

1. A. Fettweis, Digital Filter Structures Related to Classical Filter Networks, *Arch. Elektron. Uebertrag.*, vol. 25, pp. 79–89, 1971.
2. A. Fettweis, Some Principles of Designing Digital Filters Imitating Classical Filter Structures, *IEEE Trans. Circuit Theory*, vol. CT-18, pp. 314–316, March 1971.
3. A. Sedlmeyer and A. Fettweis, Digital Filters with True Ladder Configuration, *Int. J. Circuit Theory Appl.*, vol. 1, pp. 5–10, March 1973.
4. A. Sedlmeyer and A. Fettweis, Realization of Digital Filters with True Ladder Configuration, *Proc. 1973 IEEE Int. Symp. Circuit Theory*, pp. 149–152.
5. H. J. Orchard,. Inductorless Filters, *Electron. Lett.*, vol. 2, pp. 224–225, June 1966.
6. A. Fettweis and K. Meerkötter, On Adaptors for Wave Digital Filters, *IEEE Trans. Acoust., Speech, Signal Process.*, vol. ASSP-23, pp. 516–525, December 1975.
7. J. K. Skwirzynski, "Design Theory and Data for Electrical Filters," Van Nostrand, London, 1965.
8. R. Saal, "The Design of Filters Using the Catalogue of Normalized Low-Pass Filters," Telefunken AG, Backnang, 1966.
9. A. I. Zverev, "Handbook of Filter Synthesis," Wiley, New York, 1967.
10. L. Weinberg, "Network Analysis and Synthesis," McGraw-Hill, New York, 1962.
11. A. Fettweis, Wave Digital Filters with Reduced Number of Delays, *Int. J. Circuit Theory Appl.*, vol. 2, pp. 319–330, December 1974.
12. L. T. Bruton, Network Transfer Functions Using the Concept of Frequency-dependent Negative Resistance, *IEEE Trans. Circuit Theory*, vol. CT-16, pp. 406–408, August 1969.
13. A. G. Constantinides, Alternative Approach to Design of Wave Digital Filters, *Electron. Lett.*, vol. 10, pp. 59–60, March 1974.
14. M. N. S. Swamy and K. S. Thyagarajan, A New Wave Digital Filter, *Proc. 2d InterAm Conf. Syst. Inf. Mexico, November 1974*, pap. 265.
15. S. S. Lawson and A. G. Constantinides, A Method for Deriving Digital Filter Structures from Classical Filter Networks, *Proc. 1975 IEEE Int. Symp. Circuits Syst.*, pp. 170–173.
16. M. N. S. Swamy and K. S. Thyagarajan, A New Type of Wave Digital Filter, *Proc. 1975 IEEE Int. Symp. Circuits Syst.*, pp. 174–178.
17. M. N. S. Swamy and K. S. Thyagarajan, A New Type of Wave Digital Filter, *J. Franklin Inst.*, vol. 300, pp. 41–58, July 1975.
18. A. G. Constantinides, Design of Digital Filters from *LC* Ladder Networks, *Proc. IEE*, vol. 123, pp. 1307–1312, December 1976.
19. A. Antoniou, Realization of Gyrators Using Operational Amplifiers, and Their Use in *RC*-active-Network Synthesis, *Proc. IEE*, vol. 116, pp. 1838–1850, November 1969.
20. A. Antoniou and M. G. Rezk, Digital-Filter Synthesis Using Concept of Generalized-Immittance Convertor, *IEE J. Electron. Circuits Syst.*, vol. 1, pp. 207–216, November 1977.
21. A. Antoniou, Novel *RC*-Active-Network Synthesis Using Generalized-Immittance Converters, *IEEE Trans. Circuit Theory*, vol. CT-17, pp. 212–217, May 1970.
22. R. E. Crochiere and A. V. Oppenheim, Analysis of Linear Digital Networks, *Proc. IEEE*, vol. 63, pp. 581–595, April 1975.
23. W. K. Jenkins and B. J. Leon, Algebraic Techniques for the Analysis and Design of Digital Filters, *Purdue Univ. Sch. Elect. Eng. Rep.*, TR-EE 74–27, August 1974.
24. W. K. Jenkins and B. J. Leon, An Analysis of Quantization Error in Digital Filters Based on Interval Algebras, *IEEE Trans. Circuits Syst.*, vol. CAS-22, pp. 223–232, March 1975.
25. J. L. Long and T. N. Trick, Sensitivity and Noise Comparison of Some Fixed-Point Recursive Digital Filter Structures, *Proc. 1975 IEEE Int. Symp. Circuits Syst.*, pp. 56–59.
26. R. E. Crochiere, Digital Ladder Structures and Coefficient Sensitivity, *IEEE Trans. Audio Electroacoust.*, vol. AU-20, pp. 240–246, October 1972.
27. W. H. Ku and S. M. Ng, Floating-Point Coefficient Sensitivity and Roundoff Noise of Recursive Digital Filters Realized in Ladder Structures, *IEEE Trans. Circuits Syst.*, vol. CAS-22, pp. 927–936, December 1975.

28. L. B. Jackson, Roundoff-Noise Analysis for Fixed-Point Digital Filters Realized in Cascade or Parallel Form, *IEEE Trans. Audio Electroacoust.*, vol. AU-18, pp. 107–122, June 1970.
29. L. B. Jackson, Roundoff Noise Bounds Derived from Coefficient Sensitivities for Digital Filters, *IEEE Trans. Circuits Syst.*, vol. CAS-23, pp. 481–485, August 1976.

ADDITIONAL REFERENCES

Antoniou, A. and M. G. Rezk: Digital-Filter Synthesis Using Concept of Generalized-Immittance Convertor, *IEE J. Electron. Circuits Syst. (Corr.)*, vol. 2, p. 88, May 1978.
Bruton, L. T.: Low-Sensitivity Digital Ladder Filters, *IEEE Trans. Circuits Syst.*, vol. CAS-22, pp. 168–176, March 1975.
——— and D. A. Vaughan-Pope: Synthesis of Digital Ladder Filters from *LC* filters, *IEEE Trans. Circuits Syst.*, vol. CAS-23, pp. 395–402, June 1976.
Fettweis, A.: On the Connection between Multiplier Word Length Limitation and Roundoff Noise in Digital Filters, *IEEE Trans. Circuit Theory*, vol. CT-19, pp. 486–491, September 1972.
———: Pseudopassivity, Sensitivity, and Stability of Wave Digital Filters, *IEEE Trans. Circuit Theory*, vol. CT-19, pp. 668–673, November 1972.
———, G. J. Mandeville, and C. Y. Kao: Design of Wave Digital Filters for Communications Applications, *Proc. 1975 IEEE Int. Symp. Circuits Syst.*, pp. 162–165.
——— and K. Meerkötter: Suppression of Parasitic Oscillations in Wave Digital Filters, *IEEE Trans. Circuits Syst.*, vol. CAS-22, pp. 239–246, March 1975.
——— and ———: On Parasitic Oscillations in Digital Filters under Looped Conditions, *Proc. 1977 IEEE Int. Symp. Circuits Syst.*, pp. 187–190.
Jackson, L. B., A. G. Lindgren, and Y. Kim: Synthesis of State-Space Digital Filters with Low Roundoff Noise and Coefficient Sensitivity, *Proc. 1977 IEEE Int. Symp. Circuits Syst.*, pp. 41–44.
Lawson, S. S., and A. G. Constantinides: The Sensitivity of the Attenuation to Multiplier Variations in Wave Digital Filters, *DFG-Kolloq. Digitale Syst. Signalverarbeitung, Erlangen, Germany, March 1974*.
Renner, K., and S. C. Gupta: On the Design of Wave Digital Filters with Low Sensitivity Properties, *IEEE Trans. Circuit Theory*, vol. CT-20, pp. 555–567, September 1973.
——— and ———: Reduction of Roundoff Noise in Wave Digital Filters, *IEEE Trans. Circuits Syst.*, vol. CAS-21, pp. 305–310, March 1974.
Swamy, M. N. S., and K. S. Thyagarajan: Digital Bandpass and Bandstop Filters with Variable Center Frequency and Bandwidth, *Proc. IEEE*, vol. 64, pp. 1632–1634, November 1976.

PROBLEMS

12.1 Figure P12.1 represents an independent current source with an internal impedance $Z(s) = s^\lambda R_x$. Obtain corresponding digital realizations for $\lambda = -1, 0, 1$ if $R = (2/T)^\lambda R_x$.

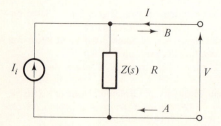

Figure P12.1

12.2 Derive a digital realization for the ideal transformer of Fig. P12.2.

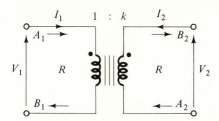

Figure P12.2

12.3 The 2-port of Fig. P12.3, where $V_1 = -RI_2$ and $V_2 = RI_1$, represents a gyrator circuit. Obtain a corresponding digital realization.

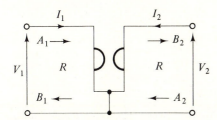

Figure P12.3

12.4 A 2-port in which the input impedance Z_i is related to the load impedance Z_L by $Z_i = -kZ_L$ is said to be a *negative-impedance converter* (NIC). The parameter k is referred to as the *impedance-conversion factor of* the device. Two types of NICs can be identified, namely voltage-conversion NICs, in which

$$V_1 = -kV_2 \qquad I_1 = -I_2$$

and current-conversion NICs, in which

$$V_1 = V_2 \qquad I_1 = kI_2$$

Derive digital realizations for each case if port resistances R_1 and R_2 are assigned to the input and output ports, respectively.

12.5 Analyze the series adaptors of Fig. 12.6a and b.

12.6 Analyze the parallel adaptors of Fig. 12.7a and b.

12.7 Figure P12.7 shows an elliptic lowpass filter. Obtain a corresponding wave structure, assuming a sampling frequency $\omega_s = 10$ rad/s.

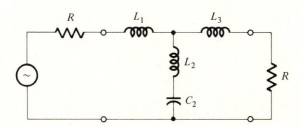

$L_1 = L_3 = 3.0316$ H, $L_2 = 0.21286$ H, $C_2 = 1.4396$ F, $R = 1$ Ω **Figure P12.7**

12.8 Figure P12.8 shows an elliptic highpass filter in which

$$A_p = 0.5 \text{ dB} \qquad A_a = 31.2 \text{ dB} \qquad \omega_p = 1/\sqrt{0.5} \text{ rad/s} \qquad \omega_a = \sqrt{0.5} \text{ rad/s}$$

(a) Obtain a corresponding wave digital filter, assuming that $\omega_s = 10$ rad/s.
(b) Determine the resulting passband and stopband edges.

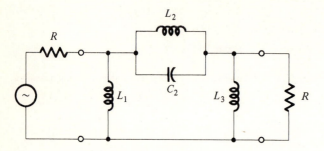

$L_1 = L_3 = 0.48948$ H, $L_2 = 3.4132$ H, $C_2 = 0.75489$ F, $R = 1$ Ω

Figure P12.8

12.9 An analog bandpass filter can be obtained by applying the lowpass-to-bandpass transformation

$$s = \frac{1}{10}\left(\bar{s} + \frac{625}{\bar{s}}\right)$$

to the lowpass filter of Fig. P12.7. Derive a corresponding wave digital filter if $\omega_s = 250$ rad/s.

12.10 By using the tables in Ref. 7 (or any other filter-design tables) design an elliptic lowpass wave digital filter satisfying the following specifications:

$$A_p = 1.0 \text{ dB} \qquad A_a \geq 60.0 \text{ dB} \qquad \tilde{\Omega}_p \approx 100 \text{ rad/s} \qquad \tilde{\Omega}_a \approx 200 \text{ rad/s} \qquad \omega_s = 1000 \text{ rad/s}$$

12.11 By applying the impedance transformation $Z(s) \to Z(s)/s$ to the filter of Fig. P12.7, derive a corresponding FDNR wave digital filter.

12.12 Repeat Prob. 12.11 using the highpass filter of Fig. P12.8 as a prototype.

12.13 The multiplier constants for the filter of Fig. 12.8c are given in Table P12.13. Compute the amplitude response of the filter if $\omega_s = 10$ rad/s.

Table P12.13

Adaptor	Multiplier constants
1	$m_{p1} = 1.341381 \times 10^{-1}$
	$m_{p2} = 1.0$
2	$m_{s1} = 9.615504 \times 10^{-2}$
	$m_{s2} = 1.0$
3	$m_{p1} = 1.720167 \times 10^{-1}$
	$m_{p2} = 2.399664 \times 10^{-1}$
4	$m_{p1} = 2.436145 \times 10^{-1}$
	$m_{p2} = 1.0$

12.14 Compute the amplitude response of the digital filter depicted in Fig. P12.14, assuming that $\omega_s = 10$ rad/s. The values of the multiplier constants are given in Table P12.14.

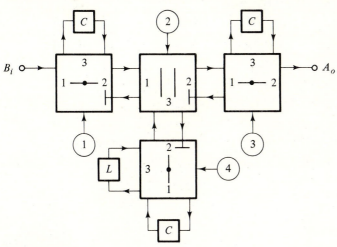

Figure P12.14

Table P12.14

Adaptor	Multiplier constants
1	$m_{s1} = 5.846557 \times 10^{-1}$ $m_{s2} = 1.0$
2	$m_{p1} = 4.307685 \times 10^{-1}$ $m_{p2} = 1.0$
3	$m_{s1} = 6.021498 \times 10^{-1}$ $m_{s2} = 8.172611 \times 10^{-1}$
4	$m_{s1} = 1.466151 \times 10^{-2}$ $m_{s2} = 1.0$

12.15 Derive Eq. (12.18).

12.16 (a) Obtain a digital realization for the 2-port of Fig. P12.16a, assuming that $R_1 = R_2 + L$ and $\omega_s = \pi$ rad/s.

(b) Repeat part (a) for the 2-port of Fig. P12.16b if $G_1 = G_2 + C$ and $\omega_s = \pi$ rad/s.

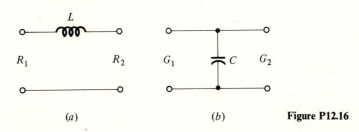

(a)　　　　　　　　　(b)　　　　　**Figure P12.16**

12.17 Analyze the configuration of Fig. 12.17a.

12.18 (a) Derive the lowpass section of Fig. 12.18a.

(b) Derive the highpass section of Fig. 12.18b.

12.19 An analog highpass filter is characterized by

$$H(s) = \prod_{j=1}^{3} \frac{s^2}{b_{0j} + b_{1j}s + s^2}$$

where

$$b_{01} = b_{02} = b_{03} = 31.15762 \qquad b_{11} = 10.78340$$

$$b_{12} = 7.8940 \qquad b_{13} = 2.889405$$

Obtain a corresponding digital filter by using the CGIC synthesis, assuming that $\omega_s = 10$ rad/s.

12.20 Design a CGIC digital lowpass filter satisfying the following specifications:

$$A_p = 0.5 \text{ dB} \qquad A_a \geq 65 \text{ dB} \qquad \tilde{\Omega}_p = 200 \text{ rad/s}$$

$$\tilde{\Omega}_a = 300 \text{ rad/s} \qquad \omega_s = 1000 \text{ rad/s}$$

Use an elliptic approximation.

THIRTEEN

THE DISCRETE FOURIER TRANSFORM

13.1 INTRODUCTION

An important mathematical tool in the software implementation of digital filters is the discrete Fourier transform (DFT). It is closely related to the z transform on the one hand and to the continuous Fourier transform (CFT) on the other. Its importance arises because it can be efficiently computed by using some very powerful algorithms known collectively as the *fast-Fourier-transform* (FFT) *method* [1–4].

We begin here by reviewing the principal properties of the DFT and then proceed to examine its relations with the other transforms. Later we describe two specific FFT algorithms and discuss their implementation and application.

13.2 DEFINITION

Given a finite-duration, real discrete-time signal $x(nT)$, a corresponding periodic signal $x_p(nT)$ with period NT can be formed as

$$x_p(nT) = \sum_{r=-\infty}^{\infty} x(nT + rNT) \tag{13.1}$$

(see Fig. 13.1). The DFT of $x_p(nT)$ is defined by

$$X_p(jk\Omega) = \sum_{n=0}^{N-1} x_p(nT)W^{-kn} = \mathscr{D}x_p(nT) \tag{13.2}$$

where

$$W = e^{j2\pi/N} \qquad \Omega = \frac{\omega_s}{N} \qquad \omega_s = \frac{2\pi}{T}$$

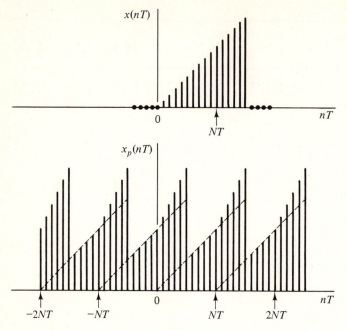

Figure 13.1 Signals $x(nT)$ and $x_p(nT)$.

In general, $X_p(jk\Omega)$ is complex and can be put in the form

$$X_p(jk\Omega) = A(k\Omega)e^{j\phi(k\Omega)}$$

where $\qquad A(k\Omega) = |X_p(jk\Omega)| \qquad$ and $\qquad \phi(k\Omega) = \arg X_p(jk\Omega)$

are discrete-frequency functions. The plots of $A(k\Omega)$ and $\phi(k\Omega)$ versus $k\Omega$ are referred to as the *amplitude spectrum* and *phase spectrum* of $x_p(nT)$, respectively. They are analogous to the corresponding spectra of the CFT.

13.3 INVERSE DFT

Definition The function $x_p(nT)$ is the *inverse DFT* (IDFT) of $X_p(jk\Omega)$ and is given by

$$x_p(nT) = \frac{1}{N}\sum_{k=0}^{N-1} X_p(jk\Omega)W^{kn} = \mathscr{D}^{-1}X_p(jk\Omega) \qquad (13.3)$$

Proof From the definition of the DFT

$$\frac{1}{N}\sum_{k=0}^{N-1} X_p(jk\Omega)W^{kn} = \frac{1}{N}\sum_{k=0}^{N-1}\left[\sum_{m=0}^{N-1} x_p(mT)W^{-km}\right]W^{kn}$$

$$= \frac{1}{N}\sum_{m=0}^{N-1} x_p(mT)\sum_{k=0}^{N-1} W^{k(n-m)}$$

where one can show that

$$\sum_{k=0}^{N-1} W^{k(n-m)} = \begin{cases} N & \text{for } m = n \\ 0 & \text{otherwise} \end{cases}$$

Therefore

$$\frac{1}{N} \sum_{k=0}^{N-1} X_p(jk\Omega)W^{kn} = x_p(nT)$$

13.4 PROPERTIES

Linearity

The DFT obeys the law of linearity; that is, for any two constants a and b,

$$\mathcal{D}[ax_p(nT) + by_p(nT)] = aX_p(jk\Omega) + bY_p(jk\Omega)$$

Periodicity

From Eq. (13.2)

$$X_p[j(k + rN)\Omega] = \sum_{n=0}^{N-1} x_p(nT)W^{-(k+rN)n} = \sum_{n=0}^{N-1} x_p(nT)W^{-kn}$$

$$= X_p(jk\Omega)$$

since $W^{-rnN} = 1$. In effect, $X_p(jk\Omega)$ is a periodic function of $k\Omega$ with period $N\Omega \ (= \omega_s)$.

Symmetry

The DFT has certain symmetry properties which are often useful. For example,

$$X_p[j(N - k)\Omega] = \sum_{n=0}^{N-1} x_p(nT)W^{-(N-k)n} = \sum_{n=0}^{N-1} x_p(nT)W^{kn}$$

$$= \left[\sum_{n=0}^{N-1} x_p(nT)W^{-kn} \right]^* = X_p^*(jk\Omega)$$

and as a result

$$\text{Re } X_p[j(N - k)\Omega] = \text{Re } X_p(jk\Omega) \qquad \text{Im } X_p[j(N - k)\Omega] = -\text{Im } X_p(jk\Omega)$$

$$A[(N - k)\Omega] = A(k\Omega) \qquad \phi[(N - k)\Omega] = -\phi(k\Omega) + 2\pi r$$

where r is any integer. If

$$x_p(nT) = \pm x_p[(N - n)T]$$

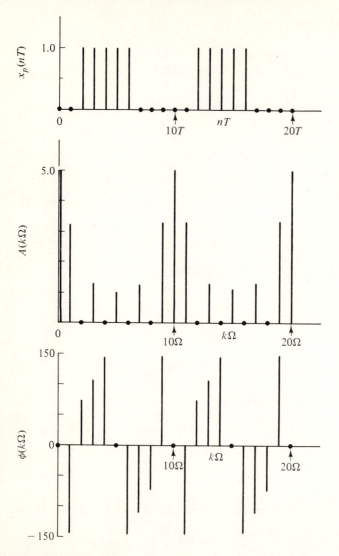

Figure 13.2 Amplitude and phase spectra of $x_p(nT)$ (Example 13.1).

we obtain

$$X_p(jk\Omega) = \pm \sum_{n=0}^{N-1} x_p[(N-n)T]W^{-kn} = \pm \sum_{m=1}^{N} x_p(mT)W^{-k(N-m)}$$

$$= \pm \left[\sum_{n=0}^{N-1} x_p(nT)W^{-kn}\right]^* = \pm X_p^*(jk\Omega)$$

Thus if

$$x_p(nT) = x_p[(N-n)T]$$

we have

$$\text{Im } X_p(jk\Omega) = 0$$

and if

$$x_p(nT) = -x_p[(N-n)T]$$

then

$$\text{Re } X_p(jk\Omega) = 0$$

Example 13.1 Find the DFT of $x_p(nT)$ if

$$x_p(nT) = \begin{cases} 1 & \text{for } 2 \le n \le 6 \\ 0 & \text{for } n = 0, 1, 7, 8, 9 \end{cases}$$

assuming that $N = 10$.

SOLUTION From Eq. (13.2)

$$X_p(jk\Omega) = \sum_{n=2}^{6} W^{-kn} = \frac{W^{-2k} - W^{-7k}}{1 - W^{-k}}$$

$$= e^{-j4\pi k/5} \frac{\sin(\pi k/2)}{\sin(\pi k/10)}$$

The amplitude and phase spectra of $x_p(nT)$ are shown in Fig. 13.2.

13.5 INTERRELATION BETWEEN THE DFT AND THE z TRANSFORM

The DFT of $x_p(nT)$ can be derived from the z transform of $x(nT)$ as we shall now show.

From Eqs. (13.1) and (13.2)

$$X_p(jk\Omega) = \sum_{n=0}^{N-1} \sum_{r=-\infty}^{\infty} x(nT + rNT)W^{-kn} = \sum_{r=-\infty}^{\infty} \sum_{n=0}^{N-1} x(nT + rNT)W^{-kn}$$

and by letting $n = m - rN$ we have

$$X_p(jk\Omega) = \sum_{r=-\infty}^{\infty} \sum_{m=rN}^{rN+N-1} x(mT)W^{-k(m-rN)}$$

$$= \cdots + \sum_{m=-N}^{-1} x(mT)W^{-km} + \sum_{m=0}^{N-1} x(mT)W^{-km} + \sum_{m=N}^{2N-1} x(mT)W^{-km} + \cdots$$

$$= \sum_{m=-\infty}^{\infty} x(mT)W^{-km}$$

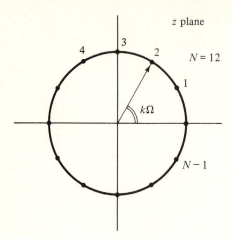

Figure 13.3 Relation between $X_p(jk\Omega)$ and $X_D(e^{jk\Omega T})$.

Alternatively, by replacing W by $e^{j2\pi/N}$ and m by n we have

$$X_p(jk\Omega) = \sum_{n=-\infty}^{\infty} x(nT)e^{-jk\Omega nT}$$

and therefore

$$X_p(jk\Omega) = X_D(e^{jk\Omega T}) \tag{13.4}$$

where

$$X_D(z) = \mathscr{Z}x(nT)$$

Hence the DFT of $x_p(nT)$ is numerically equal to the discrete-frequency function obtained by sampling the z transform of $x(nT)$ on the unit circle $|z| = 1$, as in Fig. 13.3.

Frequency-Domain Sampling Theorem

The application of the DFT in digital filtering is made possible by the frequency-domain sampling theorem, which is analogous to the time-domain sampling theorem considered in Chap. 6. It states that a z transform $X_D(z)$ for which

$$x(nT) = \mathscr{Z}^{-1}X_D(z) = 0 \tag{13.5}$$

for $n \geq N$ and $n < 0$ can be uniquely determined from its values $X_D(e^{jk\Omega T})$, where $\Omega = \omega_s/N$. Equivalently, subject to the above condition, $X_D(z)$ can be determined from the DFT of $x_p(nT)$ by virtue of Eq. (13.4).

The validity of this theorem can easily be demonstrated. With Eq. (13.5) satisfied, $x_p(nT)$ as given by Eq. (13.1) is a periodic continuation of $x(nT)$. Hence

$$x(nT) = [u(nT) - u(nT - NT)]x_p(nT) \tag{13.6}$$

as depicted in Fig. 13.4, and so

$$X_D(z) = \mathscr{Z}\{[u(nT) - u(nT - NT)]x_p(nT)\}$$

Now

$$x_p(nT) = \mathscr{D}^{-1}X_p(jk\Omega)$$

where

$$X_p(jk\Omega) = X_D(e^{jk\Omega T})$$

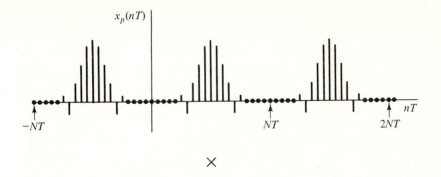

×

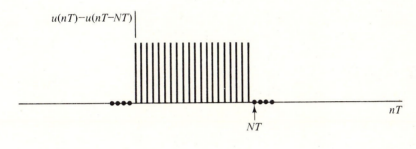

=

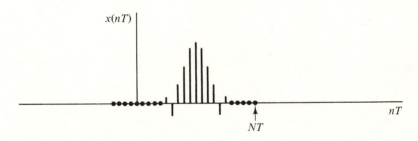

Figure 13.4 Derivation of $x(nT)$ from $x_p(nT)$.

and from Eq. (13.3)

$$X_D(z) = \mathscr{Z}\left\{[u(nT) - u(nT - NT)]\frac{1}{N}\sum_{k=0}^{N-1}X_p(jk\Omega)W^{kn}\right\}$$

$$= \frac{1}{N}\sum_{k=0}^{N-1}X_p(jk\Omega)\mathscr{Z}\{[u(nT) - u(nT - NT)]W^{kn}\}$$

Therefore, from Theorems 2.3 and 2.4

$$X_D(z) = \frac{1}{N}\sum_{k=0}^{N-1}X_p(jk\Omega)\frac{1 - z^{-N}}{1 - W^k z^{-1}} \tag{13.7}$$

since $W^{-kN} = 1$.

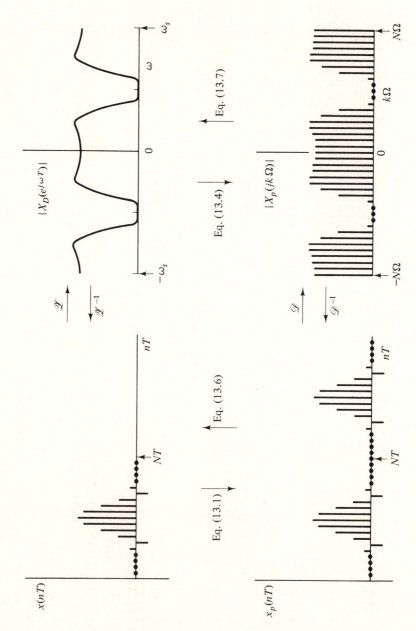

Figure 13.5 Interrelations between the DFT and the z transform.

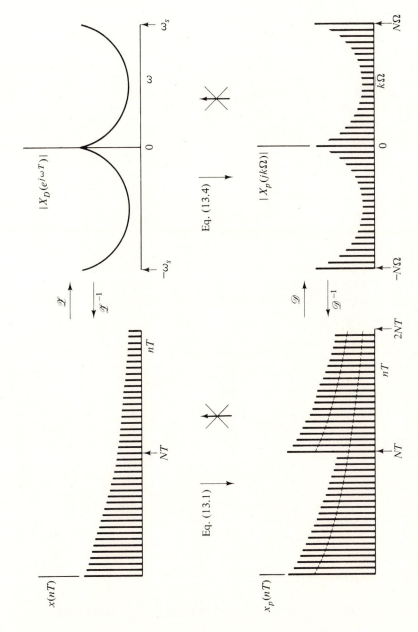

Figure 13.6 The effect of time-domain aliasing.

In summary, if $x(nT)$ is zero outside the range $0 \le nT \le (N-1)T$, $x_p(nT)$ and $X_p(jk\Omega)$ can be obtained from $x(nT)$ and $X_D(z)$ by using Eqs. (13.1) and (13.4), respectively. Conversely, $x(nT)$ and $X_D(z)$ can be obtained from $x_p(nT)$ and $X_p(jk\Omega)$ by using Eqs. (13.6) and (13.7), respectively, as illustrated in Fig. 13.5. Therefore, $x(nT)$ can be represented by the DFT of $x_p(nT)$. As a result, any finite-duration discrete-time signal can be processed by employing FFT algorithms provided that a sufficiently large value of N is chosen.

If

$$x(nT) \neq 0$$

for $n \ge N$ or $n < 0$, $x_p(nT)$ and $X_p(jk\Omega)$ can again be obtained from $x(nT)$ and $X_D(z)$ by using Eqs. (13.1) and (13.4), as depicted in Fig. 13.6. However, in this case $x(nT)$ cannot be recovered from $x_p(nT)$ by using Eq. (13.6), because of the inherent time-domain aliasing, and so Eq. (13.7) does not yield the z transform of $x(nT)$. Under these circumstances the DFT of $x_p(nT)$ is at best a distorted representation for $x(nT)$.

13.6 INTERRELATION BETWEEN THE DFT AND THE CFT

As is to be expected, a direct interrelation exists between the DFT and the CFT [5]. This can be readily established by using the results of Secs. 13.5 and 6.4.

Let $X(j\omega)$ and $X^*(j\omega)$ be the CFTs of $x(t)$ and $x^*(t)$, respectively, where $x^*(t)$ is the sampled (or impulse-modulated) version of $x(t)$. From Eqs. (6.11) and (13.4)

$$X_p(jk\Omega) = X_D(e^{jk\Omega T}) = X^*(jk\Omega)$$

and therefore, from Eq. (13.1) and (6.14),

$$\mathscr{D} \sum_{r=-\infty}^{\infty} x(nT + rNT) = \frac{1}{T} \sum_{r=-\infty}^{\infty} X(jk\Omega + jr\omega_s) \tag{13.8}$$

Now if

$$x(t) = 0 \qquad \text{for } t < 0 \text{ and } t \ge NT \tag{13.9}$$

and

$$X(j\omega) = 0 \qquad \text{for } |\omega| \ge \frac{\omega_s}{2} \tag{13.10}$$

the left- and right-hand summations in the above relation become periodic continuations of $x(nT)$ and $X(jk\Omega)$, respectively, and as a result

$$x_p(nT) = x(nT) \qquad \text{for } 0 \le nT \le (N-1)T$$

$$X_p(jk\Omega) = \frac{1}{T} X(jk\Omega) \qquad \text{for } |k\Omega| < \frac{\omega_s}{2}$$

Hence $x_p(nT)$ and $X_p(jk\Omega)$ can be obtained from $x(t)$ and $X(j\omega)$ and conversely, as depicted in Fig. 13.7. That is, $x(t)$ can be represented by a DFT, and accordingly it can be processed by using the FFT method.

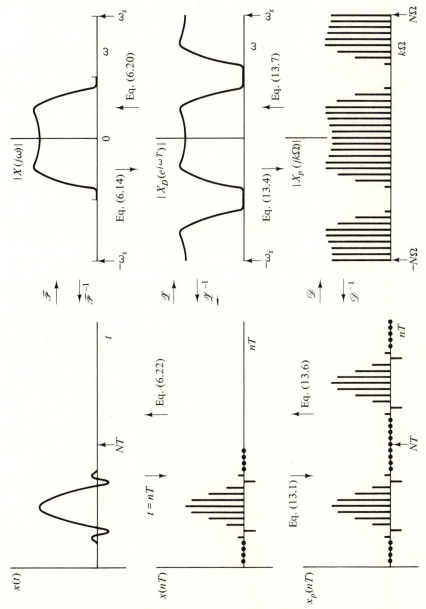

Figure 13.7 Interrelations between the DFT and the CFT.

359

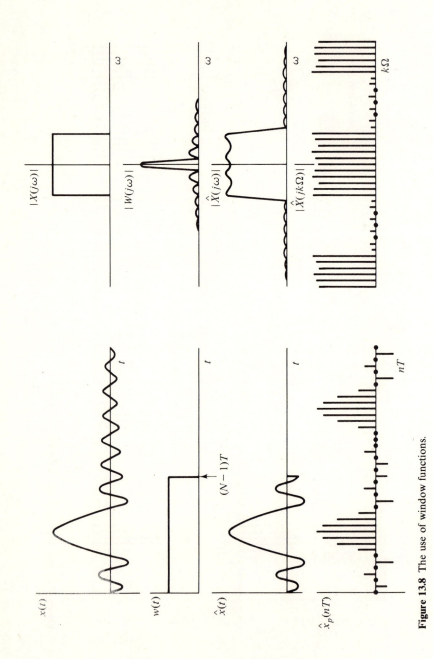

Figure 13.8 The use of window functions.

If Eq. (13.9) is violated, $x_p(nT)$ is no longer a periodic continuation of $x(nT)$ and the DFT of $x_p(nT)$ becomes an inaccurate representation for $x(t)$. However, for a bandlimited $x(t)$ this problem can often be overcome by using the window technique described in Sec. 9.4. We can form a truncated version of $x(t)$ [6] as

$$\hat{x}(t) = x(t)w(t)$$

where $w(t)$ is a window function such that

$$w(t) = 0$$

for $t < 0$ and $t \geq NT$. The Fourier transform of $\hat{x}(t)$ is given by Theorem 6.7b as

$$\hat{X}(j\omega) = \frac{1}{2\pi} \int_{-\infty}^{\infty} X(jv)W(j\omega - jv) \, dv$$

Now by repeating the arguments of Sec. 9.4 we can demonstrate that

$$\hat{X}(j\omega) \approx X(j\omega) \qquad \text{for } |\omega| < \frac{\omega_s}{2}$$

and since

$$\hat{x}(t) \approx x(t) \qquad \text{for } 0 < t \text{ and } t < NT$$

the DFT of $\hat{x}_p(nT)$ is an approximate representation for $x(t)$. This technique is illustrated in Fig. 13.8.

The type of window function and the value of N can be chosen as in Sec. 9.4. The value of T $(= 2\pi/\omega_s)$ must be chosen so that Eq. (13.10) is satisfied. However, as the spectrum of $x(t)$ is usually unknown at the outset, one may be forced to carry out the computations for progressively smaller values of T until two successive sets of computations yield approximately the same results.

The above technique can also be used to obtain approximate DFT representations for discrete-time signals in which

$$x(nT) \neq 0 \qquad \text{for } n < 0 \text{ or } n \geq N$$

13.7 INTERRELATION BETWEEN THE DFT AND THE FOURIER SERIES

The preceding results lead directly to a relationship between the DFT and the Fourier series [5].

A periodic signal $x_p(t)$ with a period T_0 can be expressed as

$$x_p(t) = \sum_{r=-\infty}^{\infty} x(t + rT_0) \tag{13.11}$$

where $x(t) = 0$ for $t < 0$ and $t \geq T_0$. Alternatively, by using the Fourier series

$$x_p(t) = \sum_{k=-\infty}^{\infty} A(k)e^{jk\omega_0 t}$$

where $\omega_0 = 2\pi/T_0$ and

$$A(k) = \frac{1}{T_0} \int_0^{T_0} x(t)e^{-jk\omega_0 t}\, dt$$

Now with $t = nT$ and $T_0 = NT$, Eq. (13.11) becomes

$$x_p(nT) = \sum_{r=-\infty}^{\infty} x(nT + rNT)$$

and consequently Eq. (13.8) yields

$$X_p(jk\Omega) = \frac{1}{T} \sum_{r=-\infty}^{\infty} X(jk\Omega + jr\omega_s) \tag{13.12}$$

where

$$X(jk\Omega) = \mathscr{F}x(t)\Big|_{\omega = k\Omega} = \int_0^{T_0} x(t)e^{-jk\Omega t}\, dt$$

or

$$X(jk\Omega) = \int_0^{T_0} x(t)e^{-jk\omega_0 t}\, dt$$

since $\Omega = \omega_s/N = 2\pi/NT = 2\pi/T_0 = \omega_0$. Evidently

$$X(jk\Omega) = T_0 A(k)$$

and since $T_0 = NT$, Eq. (13.12) can be put in the form

$$X_p(jk\Omega) = \frac{1}{T} \sum_{r=-\infty}^{\infty} X[j(k + rN)\Omega] = N \sum_{r=-\infty}^{\infty} A(k + rN) \tag{13.13}$$

In effect, the DFT of $x_p(nT)$ can be expressed in terms of the Fourier-series coefficients of $x_p(t)$.

Now with

$$A(k) \approx 0 \qquad \text{for } |k| \geq \frac{N}{2}$$

Eq. (13.13) gives

$$X_p(jk\Omega) \approx NA(k) \qquad \text{for } |k| < \frac{N}{2}$$

or

$$A(k) \approx \frac{1}{N} X_p(jk\Omega) \qquad \text{for } |k| < \frac{N}{2}$$

Thus the Fourier-series coefficients of $x_p(t)$ can be efficiently computed by using the FFT method.

13.8 NONRECURSIVE APPROXIMATIONS THROUGH THE USE OF THE DFT

The DFT can be used in the derivation of nonrecursive-filter approximations.

Section 13.5 has shown that a z transform $X_D(z)$ pertaining to a finite-duration signal can be deduced from its samples $X_D(e^{jk\Omega T})$ by using Eq. (13.7). Equivalently, if $H_p(jk\Omega)$ is a DFT, a corresponding transfer function can be formed as

$$H_D(z) = \frac{1}{N} \sum_{k=0}^{N-1} \frac{H_p(jk\Omega)(1 - z^{-N})}{1 - W^k z^{-1}}$$

After some manipulation, one can show that

$$H_D(e^{j\omega T}) = \frac{1}{N} e^{-j\omega(N-1)T/2} \sum_{k=0}^{N-1} \frac{H_p(jk\Omega)e^{-j\pi k/N} \sin(\omega NT/2)}{\sin(\omega T/2 - \pi k/N)}$$

and if

$$H_p(jk\Omega) = A(k\Omega)e^{j\theta(k\Omega)}$$

where $A(k\Omega) > 0$, then

$$|H_D(e^{jk\Omega T})| = A(k\Omega) \qquad \text{for } k = 0, 1, 2, \ldots, N-1$$

Furthermore, the above summation provides interpolation between samples, and as a consequence $H_D(z)$ constitutes a digital-filter approximation. These principles lead directly to the following approximation method:

1. Choose samples $A(k\Omega)$ and $\theta(k\Omega)$ according to the desired filter characteristic.
2. Find

$$h_p(nT) = \mathscr{D}^{-1} H_p(jk\Omega)$$

3. Form

$$h(nT) = [u(nT) - u(nT - NT)]h_p(nT)$$

4. Obtain the z transform

$$H_D(z) = \mathscr{Z} h(nT)$$

This approach is known as the *frequency-sampling approximation* method [7, 8] and is closely related to the Fourier-series method described in Sec. 9.3.

The samples of the phase response can be chosen such that a constant group delay may be achieved. According to Sec. 9.2, constant-group-delay nonrecursive filters are characterized either by a symmetrical or an antisymmetrical impulse response such that

$$h(nT) = h[(N - 1 - n)T] \qquad \text{or} \qquad h(nT) = -h[(N - 1 - n)T]$$

for $0 \leq n \leq N - 1$, and for each case N can be either odd or even. Hence there are four cases to consider.

Symmetrical Impulse Response

For a symmetrical and real impulse response, we can write

$$h_p(nT) = \frac{1}{N} \sum_{k=0}^{N-1} A(k\Omega) e^{j\theta(k\Omega)} W^{kn}$$

$$= h_p[(N-1-n)T] = h_p^*[(N-1-n)T]$$

$$= \frac{1}{N} \sum_{k=0}^{N-1} [A(k\Omega) e^{-j\theta(k\Omega)} W^{-(N-1)k}] W^{kn}$$

and hence a necessary condition for either odd or even N is obtained as

$$\theta(k\Omega) = \pi r - \frac{\pi k(N-1)}{N}$$

where r is some integer. By eliminating $\theta(k\Omega)$ in $h_p(nT)$, we can now determine the contraints on $A(k\Omega)$ and also the value (or values) of r that will guarantee a real impulse response.

For N odd, the constraints on $\theta(k\Omega)$ and $A(k\Omega)$ can be deduced as

$$\theta(k\Omega) = -\frac{\pi k(N-1)}{N} \qquad \text{for } 0 \leq k \leq N-1$$

$$A(k\Omega) = A[(N-k)\Omega] \qquad \text{for } 1 \leq k \leq N-1$$

in which case

$$h_p(nT) = \frac{1}{N}\left[A(0) + 2 \sum_{k=1}^{(N-1)/2} (-1)^k A(k\Omega) \cos \frac{\pi k(1+2n)}{N} \right]$$

Now from step 3 of the approximation method, the first formula in Table 13.1 follows.

Table 13.1 Formulas for frequency-sampling approximation method

Type	N	$h(nT)$ for $0 \leq n \leq N-1$	Restriction
Symmetrical	Odd	$\dfrac{1}{N}\left[A(0) + 2 \displaystyle\sum_{k=1}^{(N-1)/2} (-1)^k A(k\Omega) \cos k\Lambda \right]$	
	Even	$\dfrac{1}{N}\left[A(0) + 2 \displaystyle\sum_{k=1}^{N/2-1} (-1)^k A(k\Omega) \cos k\Lambda \right]$	$A\left(\dfrac{N\Omega}{2}\right) = 0$
Antisymmetrical	Odd	$\dfrac{2}{N} \displaystyle\sum_{k=1}^{(N-1)/2} (-1)^{k+1} A(k\Omega) \sin k\Lambda$	$A(0) = 0$
	Even	$\dfrac{1}{N}\left[(-1)^{N/2+n} A\left(\dfrac{N\Omega}{2}\right) + 2 \displaystyle\sum_{k=1}^{N/2-1} (-1)^k A(k\Omega) \sin k\Lambda \right]$	$A(0) = 0$
where $\Lambda = \dfrac{\pi(1+2n)}{N}$			

Similarly, for even N we can show that

$$\theta(k\Omega) = \begin{cases} -\dfrac{\pi k(N-1)}{N} & \text{for } 0 \le k \le \dfrac{N}{2} - 1 \\[3mm] \pi - \dfrac{\pi k(N-1)}{N} & \text{for } \dfrac{N}{2} + 1 \le k \le N - 1 \end{cases}$$

and

$$A(k\Omega) = A[(N-k)\Omega] \qquad \text{for } 1 \le k \le N - 1$$

$$A\left(\frac{N\Omega}{2}\right) = 0$$

The corresponding impulse response is given by the second formula in Table 13.1.

Antisymmetrical Impulse Response

For an antisymmetrical and real impulse response

$$h_p(nT) = -h_p[(N-1-n)T] = -h_p^*[(N-1-n)T]$$

$$= \frac{1}{N} \sum_{k=0}^{N-1} A(k\Omega)[-e^{-j\theta(k\Omega)} W^{-(N-1)k}] W^{kn}$$

and hence $\theta(k\Omega)$ must in this case satisfy the relation

$$\theta(k\Omega) = \frac{(1+2r)\pi}{2} - \frac{\pi k(N-1)}{N}$$

where r is some integer.

For odd N, $\theta(k\Omega)$ is given by

$$\theta(k\Omega) = \begin{cases} \dfrac{\pi}{2} - \dfrac{\pi k(N-1)}{N} & \text{for } 1 \le k \le \dfrac{N-1}{2} \\[3mm] -\dfrac{\pi}{2} - \dfrac{\pi k(N-1)}{N} & \dfrac{N+1}{2} \le k \le N - 1 \end{cases}$$

whereas for even N

$$\theta(k\Omega) = -\frac{\pi}{2} - \frac{\pi k(N-1)}{N} \qquad \text{for } 1 \le k \le N - 1$$

For both these cases, $A(k\Omega)$ must satisfy the relations

$$A(k\Omega) = A[(N-k)\Omega] \qquad A(0) = 0$$

The corresponding impulse responses are given by the third and fourth formulas in Table 13.1.

The frequency responses for the four types of filters can be calculated by using the formulas given in Table 9.1.

Example 13.2 Design a lowpass filter with a frequency response

$$H(e^{j\omega T}) \approx \begin{cases} 1 & \text{for } 0 \le |\omega| \le 16 \text{ rad/s} \\ 0 & \text{for } 17 \le |\omega| \le 32.5 \text{ rad/s} \end{cases}$$

Assume that $\omega_s = 65$ rad/s and $N = 65$.

SOLUTION Since $\Omega = \omega_s/N = 1$, we can assign

$$A(k\Omega) = \begin{cases} 1 & \text{for } 0 \le k \le 16 \\ 0 & \text{for } 17 \le k \le 32 \end{cases}$$

By using the first formula in Table 13.1 and the corresponding formula in Table 9.1, the amplitude response shown as curve A in Fig. 13.9 is obtained.

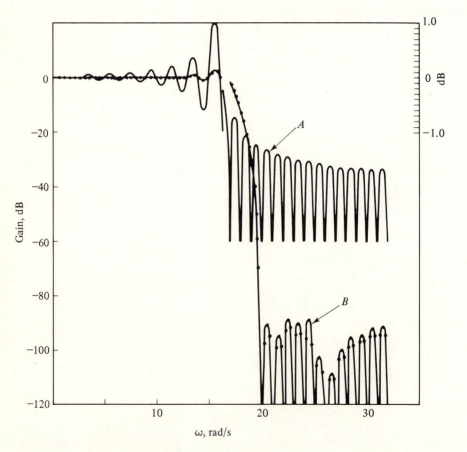

Figure 13.9 Amplitude response of lowpass filter (Example 13.2): curve A, without optimization; curve B, with optimization.

The direct application of the frequency-sampling method usually yields unsatisfactory results; e.g., compare Figs. 13.9 and 9.4. However, by leaving some values of $A(k\Omega)$ as parameters and then using optimization techniques some very good approximations can be obtained. This approach was used by Rabiner, Gold, and McGonegal [8] to develop extensive tables of optimized filters. According to these tables, an optimized version of the above lowpass filter can be obtained by choosing

$$A(17\Omega) = 0.71742143 \qquad A(18\Omega) = 0.24385557 \qquad A(19\Omega) = 0.02368774$$

The corresponding amplitude response is shown as curve B in Fig. 13.9.

13.9 SIMPLIFIED NOTATION

The preceding somewhat complicated notation for the DFT was adopted to eliminate possible confusion between the various transforms. As we shall be dealing exclusively with the DFT from now on, we can write

$$X(k) = \sum_{n=0}^{N-1} x(n)W^{-kn} \qquad x(n) = \frac{1}{N}\sum_{k=0}^{N-1} X(k)W^{kn}$$

where
$$x(n) \equiv x_p(nT) \qquad X(k) \equiv X_p(jk\Omega)$$

13.10 PERIODIC CONVOLUTIONS

The convolutions of the CFT, like those of the z transform, have been used extensively in previous chapters. Analogous and equally useful relations can be established for the DFT.

Time-Domain Convolution

The time-domain convolution of two periodic signals $x(n)$ and $h(n)$, each with period N, is defined as

$$y(n) = \sum_{m=0}^{N-1} x(m)h(n-m) = \sum_{m=0}^{N-1} x(n-m)h(m)$$

Like $x(n)$ or $h(n)$, $y(n)$ is a periodic function of n with period N.

Example 13.3 Find $y(5)$ if

$$x(n) = e^{-\alpha n} \qquad \text{for } 0 \leq n \leq 9$$

and
$$h(n) = \begin{cases} 1 & \text{for } 3 \leq n \leq 6 \\ 0 & \text{for } 0 \leq n \leq 2 \\ 0 & \text{for } 7 \leq n \leq 9 \end{cases}$$

Each signal has a period of 10.

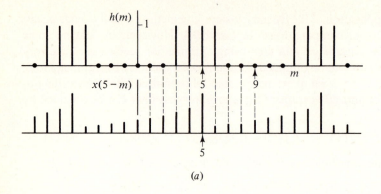

(a)

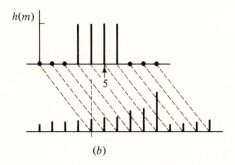

(b)

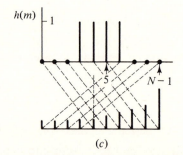

(c)

Figure 13.10 Time-domain, periodic convolution (Example 13.3).

SOLUTION From the construction of Fig. 13.10

$$y(5) = 1 + e^{-\alpha} + e^{-2\alpha} + e^{-9\alpha}$$

The DFT of $y(n)$ is

$$Y(k) = \sum_{n=0}^{N-1} \left[\sum_{m=0}^{N-1} h(m)x(n-m) \right] W^{-kn}$$

$$= \sum_{m=0}^{N-1} h(m)W^{-km} \sum_{n=0}^{N-1} x(n-m)W^{-k(n-m)}$$

and therefore

$$Y(k) = H(k)X(k) \qquad (13.14)$$

Frequency-Domain Convolution

The frequency-domain convolution is defined as

$$Y(k) = \frac{1}{N} \sum_{m=0}^{N-1} X(m)H(k-m) = \frac{1}{N} \sum_{m=0}^{N-1} X(k-m)H(m)$$

and, as above, we can show that

$$y(n) = h(n)x(n)$$

13.11 FAST-FOURIER-TRANSFORM ALGORITHMS

The direct evaluation of the DFT involves N complex multiplications and $N-1$ complex additions for each value of $X(k)$, and since there are N values to determine, N^2 multiplications and $N(N-1)$ additions are necessary. Consequently, for large values of N, say in excess of 1000, direct evaluation involves a considerable amount of computation.

The efficient and up-to-date approach for the evaluation of the DFT is through the use of FFT algorithms. We describe here two, the so-called *decimation-in-time* and *decimation-in-frequency algorithms*.

Decimation-in-Time Algorithm

Let the desired DFT be

$$X(k) = \sum_{n=0}^{N-1} x(n)W_N^{-kn} \qquad \text{where } W_N = e^{j2\pi/N}$$

and assume that

$$N = 2^r$$

where r is an integer. The above summation can be split into two parts as

$$X(k) = \sum_{\substack{n=0 \\ n \text{ even}}}^{N-1} x(n)W_N^{-kn} + \sum_{\substack{n=0 \\ n \text{ odd}}}^{N-1} x(n)W_N^{-kn}$$

Alternatively

$$X(k) = \sum_{n=0}^{N/2-1} x_{10}(n)W_N^{-2kn} + W_N^{-k} \sum_{n=0}^{N/2-1} x_{11}(n)W_N^{-2kn} \qquad (13.15)$$

where $$x_{10}(n) = x(2n) \qquad x_{11}(n) = x(2n+1) \qquad (13.16)$$

for $0 \le n \le N/2 - 1$. Since

$$W_N^{-2kn} = W_{N/2}^{-kn}$$

Eq. (13.15) can be expressed as

$$X(k) = \sum_{n=0}^{N/2-1} x_{10}(n)W_{N/2}^{-kn} + W_N^{-k} \sum_{n=0}^{N/2-1} x_{11}(n)W_{N/2}^{-kn}$$

Clearly

$$X(k) = X_{10}(k) + W_N^{-k}X_{11}(k) \tag{13.17}$$

and since $X_{10}(k)$ and $X_{11}(k)$ are periodic, each with period $N/2$, we have

$$X\left(k + \frac{N}{2}\right) = X_{10}\left(k + \frac{N}{2}\right) + W_N^{-(k+N/2)}X_{11}\left(k + \frac{N}{2}\right) = X_{10}(k) - W_N^{-k}X_{11}(k)$$
$$\tag{13.18}$$

Equations (13.17) and (13.18) can be represented by the "butterfly" flow graph of Fig. 13.11a, where the minus sign in $\pm W_N^{-k}$ is pertinent in the computation of $X(k + N/2)$. This flow graph can be represented by the simplified diagram of Fig. 13.11b for convenience.

What we have accomplished so far is to express the desired N-element DFT as a function of two $(N/2)$-element DFTs. Assuming that the values of $X_{10}(k)$ and $X_{11}(k)$ are available in corresponding arrays, the values of $X(k)$ can be readily computed as depicted in Fig. 13.12a.

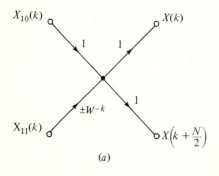

(a)

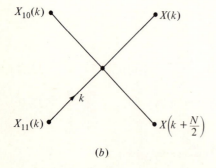

(b)

Figure 13.11 (a) Butterfly flow graph; (b) simplified diagram.

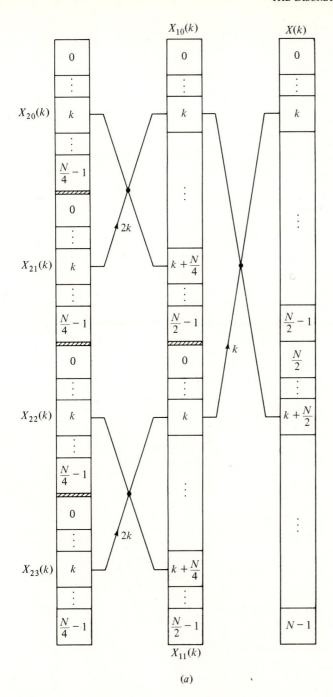

Figure 13.12 Decimation-in-time FFT algorithm: (*a*) first and second cycles;

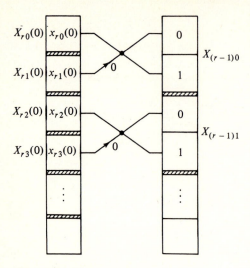

Figure 13.12 (*b*) *r*th cycle.

$X_{10}(k)$ and $X_{11}(k)$ can now be expressed in terms of $(N/4)$-element DFTs by repeating the above cycle. For $X_{10}(k)$, we can write

$$X_{10}(k) = \sum_{n=0}^{N/2-1} x_{10}(n)W_{N/2}^{-kn}$$

$$= \sum_{n=0}^{N/4-1} x_{10}(2n)W_{N/2}^{-2kn} + \sum_{n=0}^{N/4-1} x_{10}(2n+1)W_{N/2}^{-k(2n+1)}$$

$$= \sum_{n=0}^{N/4-1} x_{20}(n)W_{N/4}^{-kn} + W_N^{-2k} \sum_{n=0}^{N/4-1} x_{21}(n)W_{N/4}^{-kn} \qquad (13.19)$$

and, similarly, for $X_{11}(k)$

$$X_{11}(k) = \sum_{n=0}^{N/2-1} x_{11}(n)W_{N/2}^{-kn}$$

$$= \sum_{n=0}^{N/4-1} x_{22}(n)W_{N/4}^{-kn} + W_N^{-2k} \sum_{n=0}^{N/4-1} x_{23}(n)W_{N/4}^{-kn} \qquad (13.20)$$

where
$$\begin{array}{ll} x_{20}(n) = x_{10}(2n) & x_{21}(n) = x_{10}(2n+1) \\ x_{22}(n) = x_{11}(2n) & x_{23}(n) = x_{11}(2n+1) \end{array} \qquad (13.21)$$

for $0 \le n \le N/4 - 1$. Consequently, from Eqs. (13.19) and (13.20)

$$X_{10}(k) = X_{20}(k) + W_N^{-2k}X_{21}(k) \qquad X_{10}\left(k + \frac{N}{4}\right) = X_{20}(k) - W_N^{-2k}X_{21}(k)$$

$$X_{11}(k) = X_{22}(k) + W_N^{-2k}X_{23}(k) \qquad X_{11}\left(k + \frac{N}{4}\right) = X_{22}(k) - W_N^{-2k}X_{23}(k)$$

Thus if the values of $X_{20}(k)$, $X_{21}(k)$, $X_{22}(k)$, and $X_{23}(k)$ are available, those of $X_{10}(k)$ and $X_{11}(k)$ and in turn those of $X(k)$ can be computed, as illustrated in Fig. 13.12a.

In exactly the same way, the mth cycle of the procedure yields

$$X_{(m-1)0}(k) = X_{m0}(k) + W_N^{-2^{m-1}k}X_{m1}(k)$$

$$X_{(m-1)0}\left(k + \frac{N}{2^m}\right) = X_{m0}(k) - W_N^{-2^{m-1}k}X_{m1}(k)$$

$$X_{(m-1)1}(k) = X_{m2}(k) + W_N^{-2^{m-1}k}X_{m3}(k)$$

$$X_{(m-1)1}\left(k + \frac{N}{2^m}\right) = X_{m2}(k) - W_N^{-2^{m-1}k}X_{m3}(k)$$

$$\cdots\cdots\cdots\cdots\cdots\cdots\cdots\cdots\cdots$$

where

$$x_{m0}(n) = x_{(m-1)0}(2n)$$

$$x_{m1}(n) = x_{(m-1)0}(2n+1)$$

$$x_{m2}(n) = x_{(m-1)1}(2n) \tag{13.22}$$

$$x_{m3}(n) = x_{(m-1)1}(2n+1)$$

$$\cdots\cdots\cdots\cdots\cdots\cdots\cdots$$

for $0 \le n \le N/2^m - 1$. Clearly, the procedure terminates with the rth cycle $(N = 2^r)$ since $x_{r0}(n)$, $x_{r1}(n)$, ... reduce the one-element sequences, in which case

$$X_{ri}(0) = x_{ri}(0)$$

for $i = 0, 1, \ldots, N - 1$. The values of the penultimate DFTs can be obtained from the above equations as

$$X_{(r-1)0}(0) = x_{r0}(0) + W_N^0 x_{r1}(0)$$

$$X_{(r-1)0}(1) = x_{r0}(0) - W_N^0 x_{r1}(0)$$

$$X_{(r-1)1}(0) = x_{r2}(0) + W_N^0 x_{r3}(0)$$

$$X_{(r-1)1}(1) = x_{r2}(0) - W_N^0 x_{r3}(0)$$

$$\cdots\cdots\cdots\cdots\cdots\cdots\cdots$$

Assuming that the sequence $\{x_{r0}(0), x_{r1}(0), \ldots\}$ is available in an array, the values of $X_{(r-1)i}(k)$ for $i = 0, 1, \ldots$ can be computed as in Fig. 13.12b. Then the values of $X_{(r-2)i}(k)$, $X_{(r-3)i}(k)$, ... can be computed in sequence, and ultimately the values of $X(k)$ can be obtained.

The only remaining task at this point is to identify elements $x_{r0}(0), x_{r1}(0), \ldots$. Fortunately, this turns out to be easy. As can be shown, $x_{rp}(0)$ is given by

$$x_{rp}(0) = x(q)$$

where q is the r-bit binary representation of p reversed. For example, if $N = 16$, $r = 4$ and hence

$$x_{40}(0) = x(0)$$
$$x_{41}(0) = x(8)$$
$$x_{42}(0) = x(4)$$
$$\cdots\cdots\cdots\cdots$$
$$x_{4(15)}(0) = x(15)$$

In effect, sequence $\{x_{r0}(0), x_{r1}(0), \ldots\}$ is a reordered version of sequence $\{x(0), x(1), \ldots\}$.

In the above discussion, N has been assumed to be a power of 2. Nevertheless, the algorithm can be applied to any other finite-duration sequence by including a number of trailing zero elements in the given sequence.

Example 13.4 Construct the decimation-in-time algorithm for $N = 8$.

SOLUTION From Eq. (13.16)

$$\mathbf{x}_{10} = \{x(0), x(2), x(4), x(6)\}$$
$$\mathbf{x}_{11} = \{x(1), x(3), x(5), x(7)\}$$

and hence Eq. (13.21) gives

$$\mathbf{x}_{20} = \{x(0), x(4)\}$$
$$\mathbf{x}_{21} = \{x(2), x(6)\}$$
$$\mathbf{x}_{22} = \{x(1), x(5)\}$$
$$\mathbf{x}_{23} = \{x(3), x(7)\}$$

Finally, from Eq. (13.22)

$$\mathbf{x}_{30} = \{x(0)\}$$
$$\mathbf{x}_{31} = \{x(4)\}$$
$$\mathbf{x}_{32} = \{x(2)\}$$
$$\cdots\cdots\cdots\cdots$$
$$\mathbf{x}_{37} = \{x(7)\}$$

The complete algorithm is illustrated in Fig. 13.13.

The algorithm can be easily programmed as computations can be carried out in place in a single array. As can be observed in Fig. 13.13, once the outputs of each input butterfly are computed, the input elements are no longer needed for further processing and can be replaced by the corresponding outputs. When we proceed in the same way from left to right, at the end of computation the input

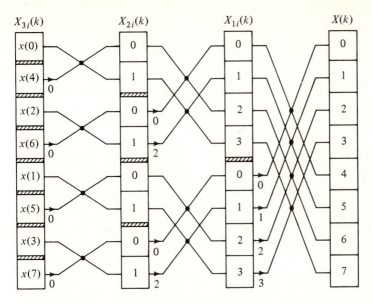

Figure 13.13 Decimation-in-time FFT algorithm for $N = 8$ (Example 13.4).

array will contain the elements of the desired DFT properly ordered. The input elements can be entered in the appropriate array locations by using a simple reordering subroutine.

In general, each cycle of the algorithm involves $N/2$ butterflies, as can be seen in Fig. 13.13, and each butterfly requires one (complex) multiplication. Since there are r cycles of computation and $r = \log_2 N$, the total number of multiplications is $(N/2) \log_2 N$ as opposed to N^2 in the case of direct evaluation. This constitutes a considerable saving in computation. For example if $N \geq 512$, the number of multiplications is reduced to a fraction of 1 percent of that required by direct evaluation.

Decimation-in-Frequency Algorithm

In the preceding algorithm the given sequence is split in two by separating the even- and odd-index elements. The same procedure is then applied repeatedly on each new sequence until one-element sequences are obtained. An alternative FFT algorithm can be developed by splitting the given sequence about its midpoint and then repeating the same for each resulting sequence.

We can write

$$X(k) = \sum_{n=0}^{N/2-1} x(n)W_N^{-kn} + \sum_{n=N/2}^{N-1} x(n)W_N^{-kn}$$

$$= \sum_{n=0}^{N/2-1} \left[x(n) + W_N^{-kN/2}x\left(n + \frac{N}{2}\right) \right] W_N^{-kn}$$

and with $k \to 2k$ or $2k + 1$

$$X(2k) = \sum_{n=0}^{N/2-1} x_{10}(n)W_{N/2}^{-kn}$$

$$X(2k + 1) = \sum_{n=0}^{N/2-1} x_{11}(n)W_{N/2}^{-kn}$$

where

$$x_{10}(n) = x(n) + x\left(n + \frac{N}{2}\right) \tag{13.23}$$

$$x_{11}(n) = \left[x(n) - x\left(n + \frac{N}{2}\right)\right]W_N^{-n} \tag{13.24}$$

for $0 \le n \le N/2 - 1$. Thus the even- and odd-index values of $X(k)$ are given by the DFTs of $x_{10}(n)$ and $x_{11}(n)$, respectively. Assuming that the values of $x(n)$ are stored sequentially in an array, the values of $x_{10}(n)$ and $x_{11}(n)$ can be computed as illustrated in Fig. 13.14a, where the left-hand butterfly represents Eqs. (13.23) and (13.24).

The same cycle can now be applied to $x_{10}(n)$ and $x_{11}(n)$. For $x_{10}(n)$

$$X(2k) = \sum_{n=0}^{N/4-1} \left[x_{10}(n) + W_N^{-kN/2}x_{10}\left(n + \frac{N}{4}\right)\right]W_{N/2}^{-kn}$$

and, similarly, for $x_{11}(n)$

$$X(2k + 1) = \sum_{n=0}^{N/4-1} \left[x_{11}(n) + W_N^{-kN/2}x_{11}\left(n + \frac{N}{4}\right)\right]W_{N/2}^{-kn}$$

Hence with $k \to 2k$ or $2k + 1$

$$X(4k) = \sum_{n=0}^{N/4-1} x_{20}(n)W_{N/4}^{-kn}$$

$$X(4k + 2) = \sum_{n=0}^{N/4-1} x_{21}(n)W_{N/4}^{-kn}$$

$$X(4k + 1) = \sum_{n=0}^{N/4-1} x_{22}(n)W_{N/4}^{-kn}$$

$$X(4k + 3) = \sum_{n=0}^{N/4-1} x_{23}(n)W_{N/4}^{-kn}$$

where

$$x_{20}(n) = x_{10}(n) + x_{10}\left(n + \frac{N}{4}\right)$$

$$x_{21}(n) = \left[x_{10}(n) - x_{10}\left(n + \frac{N}{4}\right)\right]W^{-2n}$$

$$x_{22}(n) = x_{11}(n) + x_{11}\left(n + \frac{N}{4}\right)$$

$$x_{23}(n) = \left[x_{11}(n) - x_{11}\left(n + \frac{N}{4}\right)\right]W^{-2n}$$

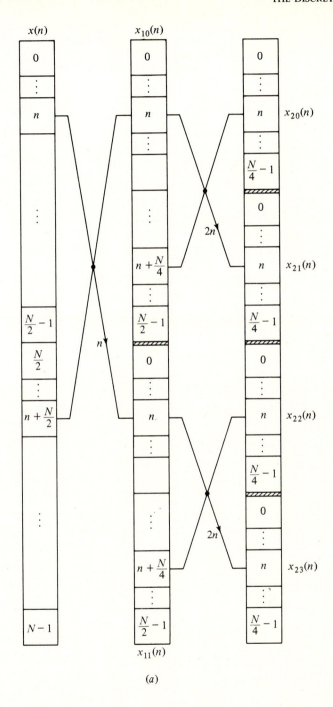

Figure 13.14 Decimation-in-frequency FFT algorithm: (a) general case;

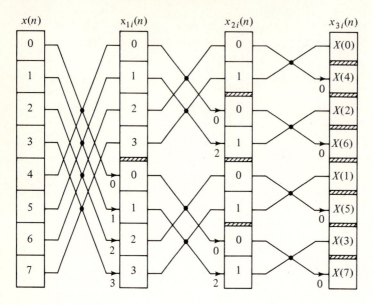

Figure 13.14 (b) $N = 8$.

for $0 \le n \le (N/4 - 1)$. The values of $x_{20}(n)$, $x_{21}(n)$, ... can be computed as in Fig. 13.14a. The DFT of each of these sequences gives one-quarter of the values of $X(k)$.

Similarly, the mth cycle yields

$$X(2^m k) = \sum_{n=0}^{N_1 - 1} x_{m0}(n) W_{N_1}^{-kn}$$

$$X(2^m k + 2^{m-1}) = \sum_{n=0}^{N_1 - 1} x_{m1}(n) W_{N_1}^{-kn}$$

. . . .

where $N_1 = N/2^m$ and

$$x_{m0}(n) = x_{(m-1)0}(n) + x_{(m-1)0}\left(n + \frac{N}{2^m}\right)$$

$$x_{m1}(n) = \left[x_{(m-1)0}(n) - x_{(m-1)0}\left(n + \frac{N}{2^m}\right) \right] W_N^{-2^{m-1}n}$$

. . . .

for $0 \le n \le N/2^m - 1$.

As in the previous case, the procedure terminates when $m = r$, at which time $x_{r0}(n)$, $x_{r1}(n)$, ... reduce the one-element sequences each giving one value of the desired DFT, i.e.,

$$X(0) = x_{r0}(0)$$
$$X(2^{r-1}) = x_{r1}(0)$$
$$\cdots\cdots\cdots\cdots$$

The complete algorithm for $N = 8$ is illustrated in Fig. 13.14*b*. As can be seen, the values of $X(k)$ appear in the output array disordered. However, reordering can be easily accomplished by reversing the r-bit binary representation of the location index at the end of computation. The advantage of this algorithm is that the values of $x(n)$ are entered in the input array sequentially.

Inverse DFT

Owing to the similarity between Eqs. (13.2) and (13.3), the preceding two algorithms can readily be employed for the computation of the IDFT. Equation (13.3) can be put in the form

$$x^*(n) = \left[\frac{1}{N} \sum_{k=0}^{N-1} X(k) W^{kn} \right]^* = \frac{1}{N} \sum_{k=0}^{N-1} X^*(k) W^{-kn}$$

or

$$x^*(n) = \mathscr{D} \frac{1}{N} X^*(k)$$

Thus if a program is available which can be used to compute the DFT of a complex signal $x(n)$, entering the complex conjugate of $X(k)/N$ as input will yield the complex conjugate of $x(n)$ as output.

13.12 DIGITAL-FILTER IMPLEMENTATION

The response of a nonrecursive filter to an excitation $x(n)$ is given by

$$y(n) = \sum_{m=-\infty}^{\infty} x(n-m)h(m)$$

and if

$$h(n) = 0 \qquad \text{for } n < 0 \text{ and } n > N - 1$$
$$x(n) = 0 \qquad \text{for } n < 0 \text{ and } n > L - 1$$

we have

$$y(n) = \sum_{m=0}^{N-1} x(n-m)h(m) \tag{13.25}$$

for $0 \leq n \leq N + L - 2$. A software implementation for the filter can readily be obtained by programming Eq. (13.25) directly. However, this approach can involve a large amount of computation since N multiplications are necessary for each sample of the response. The alternative is to use the FFT method [9].

Let us define $(L + N - 1)$-element DFTs for $h(n)$, $x(n)$, and $y(n)$, as in Sec. 13.2, which we can designate as $H(k)$, $X(k)$, and $Y(k)$, respectively. From Eqs. (13.14) and (13.25)

$$Y(k) = H(k)X(k)$$

and hence

$$y(n) = \mathscr{D}^{-1}H(k)X(k)$$

Therefore, the response of the filter can be computed by using the following procedure:

1. Compute the DFTs of $h(n)$ and $x(n)$ using an FFT algorithm.
2. Compute the product $H(k)X(k)$ for $k = 0, 1, \ldots$.
3. Compute the IDFT of $Y(k)$ using an FFT algorithm.

The evaluation of $H(k)$, $X(k)$, or $y(n)$ requires $[(L + N - 1)/2] \log_2 (L + N - 1)$ complex multiplications, and step 2 above entails $L + N - 1$ of the same. Since one complex multiplication corresponds to four real ones, the total number of real multiplications per output sample is $6 \log_2 (L + N - 1) + 4$, as opposed to N in the case of direct evaluation. Clearly, for large values of N, the FFT approach is the more efficient. For example, if $N = L = 512$, the number of multiplications is reduced to 12.5 percent of that required by direct evaluation.

In the above implementation, the entire input sequence must be available before the processing can start. Consequently, if the input sequence is long, a significant delay in computation is introduced, which is usually objectionable in real-time applications. For such applications, the input sequence can be segmented, and each segment can be processed individually.

We can write

$$x(n) = \sum_{i=0}^{q} x_i(n)$$

for $0 \leq n \leq (q + 1)L - 1$, where

$$x_i(n) = \{u(n - iL) - u[n - (i + 1)L]\}x(n)$$

or

$$x_i(n) = \begin{cases} x(n) & \text{for } iL \leq n \leq (i + 1)L - 1 \\ 0 & \text{otherwise} \end{cases}$$

as illustrated in Fig. 13.15. With this manipulation Eq. (13.25) becomes

$$y(n) = \sum_{m=0}^{N-1} \sum_{i=0}^{q} x_i(n - m)h(m)$$

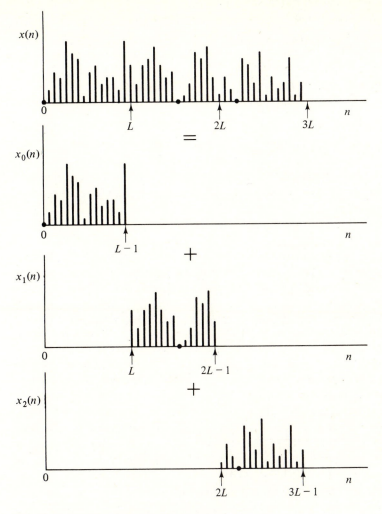

Figure 13.15 Segmentation of input sequence.

and by interchanging the order of summation

$$y(n) = \sum_{i=0}^{q} c_i(n) \tag{13.26}$$

where

$$c_i(n) = \sum_{m=0}^{N-1} x_i(n-m)h(m) \tag{13.27}$$

or

$$c_i(n) = \sum_{m=0}^{N-1} \{u(n-m-iL) - u[n-m-(i+1)L]\}x(n-m)h(m) \tag{13.28}$$

In this way, $y(n)$ can be computed by evaluating a number of partial convolutions.

For $iL - 1 \leq n \leq (i + 1)L + N - 1$, Eq. (13.28) gives

$$c_i(iL - 1) = 0$$

$$c_i(iL) = x(iL)h(0)$$

$$c_i(iL + 1) = x(iL + 1)h(0) + x(iL)h(1)$$

$$\cdots\cdots\cdots\cdots\cdots\cdots\cdots\cdots$$

$$c_i[(i + 1)L + N - 2] = x[(i + 1)L - 1]h(N - 1)$$

$$c_i[(i + 1)L + N - 1] = 0$$

Evidently, the ith partial-convolution sequence has $L + N - 1$ nonzero elements and can be stored in an array C_i, as demonstrated in Fig. 13.16. The elements of C_i can be computed as

$$c_i(n) = \mathcal{D}^{-1}H(k)X_i(k)$$

Now from Eq. (13.26) an array Y containing the values of $y(n)$ can be readily formed, as illustrated in Fig. 13.16, by entering the elements of nonoverlapping segments in C_0, C_1, ... and then adding the elements in overlapping adjacent segments. As can be seen, processing can start as soon as L input elements are received, and the first batch of L output samples is available as soon as the first input segment is processed. In this way delays in computation can be minimized.

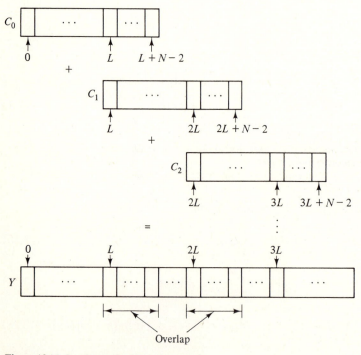

Figure 13.16 Overlap-and-add implementation.

REFERENCES

1. J. W. Cooley and J. W. Tukey, An Algorithm for the Machine Calculation of Complex Fourier Series, *Math. Comp.*, vol. 19, pp. 297–301, April 1965.
2. W. T. Cochran, J. W. Cooley, D. L. Favin, H. D. Helms, R. A. Kaenel, W. W. Lang, G. C. Maling, D. E. Nelson, C. M. Rader, and P. D. Welch, What Is the Fast Fourier Transform?, *IEEE Trans. Audio Electroacoust.*, vol. AU-15, pp. 45–55, June 1967.
3. G. D. Bergland, A Guided Tour of the Fast Fourier Transform, *IEEE Spectrum*, vol. 6, pp. 41–52, July 1969.
4. J. W. Cooley, P. A. W. Lewis, and P. D. Welch, Historical Notes on the Fast Fourier Transform, *IEEE Trans. Audio Electroacoust.*, vol. AU-15, pp. 76–79, June 1967.
5. J. W. Cooley, P. A. W. Lewis, and P. D. Welch, Application of the Fast Fourier Transform to Computation of Fourier Integrals, Fourier Series and Convolution Integrals, *IEEE Trans. Audio Electroacoust.*, vol. AU-15, pp. 79–84, June 1967.
6. H. Babic and G. C. Temes, Optimum Low-Order Windows for Discrete Fourier Transform Systems, *IEEE Trans. Acoust., Speech, Signal Process.*, vol. ASSP-24, pp. 512–517, December 1976.
7. B. Gold and K. L. Jordan, Jr., A Direct Search Procedure for Designing Finite Duration Impulse Response Filters, *IEEE Trans. Audio Electroacoust.*, vol. AU-17, pp. 33–36, March 1969.
8. L. R. Rabiner, B. Gold, and C. A. McGonegal, An Approach to the Approximation Problem for Nonrecursive Digital Filters, *IEEE Trans. Audio Electroacoust.*, vol. AU-18, pp. 83–106, June 1970.
9. H. D. Helms, Fast Fourier Transform Method of Computing Difference Equations and Simulating Filters, *IEEE Trans. Audio Electroacoust.*, vol. AU-15, pp. 85–90, June 1967.

ADDITIONAL REFERENCES

Bertram, S.: On the Derivation of the Fast Fourier Transform, *IEEE Trans. Audio Electroacoust.*, vol. AU-18, pp. 55–58, March 1970.
Bingham, C., M. D. Godfrey, and J. W. Tukey: Modern Techniques of Power Spectrum Estimation, *IEEE Trans. Audio Electroacoust.*, vol. AU-15, pp. 56–66, June 1967.
Cooley, J. W., P. A. W. Lewis, and P. D. Welch: The Fast Fourier Transform Algorithm: Programming Considerations in the Calculation of Sine, Cosine and Laplace Transforms, *J. Sound Vib.*, vol. 12, pp. 315–337, July 1970.
———, ———, and ———: The Finite Fourier Transform, *IEEE Trans. Audio Electroacoust.*, vol. AU-17, pp. 77–85, June 1969.
Glassman, J. A.: A Generalization of the Fast Fourier Transform, *IEEE Trans. Comput.*, vol. C-19, pp. 105–116, February 1970.
Oppenheim, A. V., and C. J. Weinstein: Effects of Finite Register Length in Digital Filtering and the Fast Fourier Transform, *Proc. IEEE*, vol. 60, pp. 957–976, August 1972.
Pease, M. C.: An Adaptation of the Fast Fourier Transform for Parallel Processing, *J. Ass. Comput. Mach.*, vol. 15, pp. 252–264, April 1968.
Rader, C. M.: Discrete Fourier Transforms When the Number of Data Samples Is Prime, *Proc. IEEE*, vol. 56, pp. 1107–1108, June 1968.
Stockham, T. G. Jr.: High-Speed Convolution and Correlation, *1966 Spring Joint Comput. Conf. Proc.*, vol. 28, pp. 229–233, 1966.
Weinstein, C. J.: Roundoff Noise in Floating Point Fast Fourier Transform Computation, *IEEE Trans. Audio Electroacoust.*, vol. AU-17, pp. 209–215, September 1969.
Welch, P. D.: The Use of Fast Fourier Transform for the Estimation of Power Spectra: A Method Based on Time Averaging over Short, Modified Periodograms, *IEEE Trans. Audio Electroacoust.*, vol. AU-15, pp. 70–73, June 1967.
———: A Fixed-Point Fast Fourier Transform Error Analysis, *IEEE Trans. Audio Electroacoust.*, vol. AU-17, pp. 151–157, June 1969.

PROBLEMS

13.1 Show that

$$\sum_{k=0}^{N-1} W^{k(n-m)} = \begin{cases} N & \text{for } m = n \\ 0 & \text{otherwise} \end{cases}$$

13.2 Show that
(a) $\mathscr{D}x_p(nT + mT) = W^{km}X_p(jk\Omega)$
(b) $\mathscr{D}^{-1}X_p(jk\Omega + jl\Omega) = W^{-nl}x_p(nT)$

13.3 The definition of the DFT can be extended to include complex discrete-time signals. Show that
(a) $\mathscr{D}x_p^*(nT) = X_p^*(-jk\Omega)$
(b) $\mathscr{D}^{-1}X_p^*(jk\Omega) = x_p^*(-nT)$

13.4 (a) A complex discrete-time signal is given by

$$x_p(nT) = x_{p1}(nT) + jx_{p2}(nT)$$

where $x_{p1}(nT)$ and $x_{p2}(nT)$ are real. Show that

$$\text{Re } X_{p1}(jk\Omega) = \tfrac{1}{2}\{\text{Re } X_p(jk\Omega) + \text{Re } X_p[j(N-k)\Omega]\}$$

$$\text{Im } X_{p1}(jk\Omega) = \tfrac{1}{2}\{\text{Im } X_p(jk\Omega) - \text{Im } X_p[j(N-k)\Omega]\}$$

$$\text{Re } X_{p2}(jk\Omega) = \tfrac{1}{2}\{\text{Im } X_p(jk\Omega) + \text{Im } X_p[j(N-k)\Omega]\}$$

$$\text{Im } X_{p2}(jk\Omega) = -\tfrac{1}{2}\{\text{Re } X_p(jk\Omega) - \text{Re } X_p[j(N-k)\Omega]\}$$

(b) A DFT is given by

$$X_p(jk\Omega) = X_{p1}(jk\Omega) + jX_{p2}(jk\Omega)$$

where $X_{p1}(jk\Omega)$ and $X_{p2}(jk\Omega)$ are real DFTs. Show that

$$\text{Re } x_{p1}(nT) = \tfrac{1}{2}\{\text{Re } x_p(nT) + \text{Re } x_p[(N-n)T]\}$$

$$\text{Im } x_{p1}(nT) = \tfrac{1}{2}\{[\text{Im } x_p(nT) - \text{Im } x_p[(N-n)T]\}$$

$$\text{Re } x_{p2}(nT) = \tfrac{1}{2}\{\text{Im } x_p(nT) + \text{Im } x_p[(N-n)T]\}$$

$$\text{Im } x_{p2}(nT) = -\tfrac{1}{2}\{\text{Re } x_p(nT) - \text{Re } x_p[(N-n)T]\}$$

13.5 Figure P13.5 shows four real discrete-time signals. Classify their DFTs as real, imaginary, or complex. Assume that $N = 10$ in each case.

13.6 Find the DFTs of the following periodic signals:

(a) $x_p(nT) = \begin{cases} 1 & \text{for } n = 3, 7 \\ 0 & \text{for } n = 0, 1, 2, 4, 5, 6, 8, 9 \end{cases}$

(b) $x_p(nT) = \begin{cases} 1 & \text{for } 0 \le n \le 5 \\ 2 & \text{for } 6 \le n \le 9 \end{cases}$

(c) $x_p(nT) = \begin{cases} n & \text{for } 0 \le n \le 2 \\ 0 & \text{for } 3 \le n \le 7 \\ -(10-n) & \text{for } n = 8, 9 \end{cases}$

The period is $10T$ in each case.

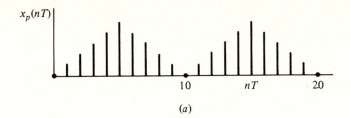

(a)

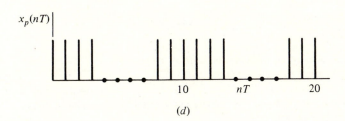

(b)

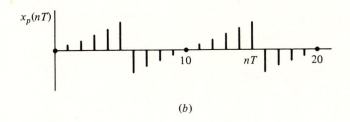

(c)

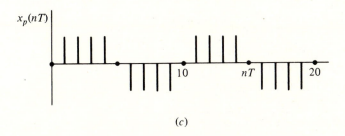

(d)

Figure P13.5

13.7 A periodic signal is given by

$$x_p(nT) = \sum_{r=-\infty}^{\infty} w_H(nT + rNT)$$

where

$$w_H(nT) = \begin{cases} \alpha + (1-\alpha) \cos \dfrac{2\pi n}{N-1} & \text{for } |n| \leq \dfrac{N-1}{2} \\ 0 & \text{otherwise} \end{cases}$$

Find $X_p(jk\Omega)$.

13.8 Obtain the IDFTs of the following:

(a) $X_p(jk\Omega) = (-1)^k \left(1 + 2\cos\dfrac{2\pi k}{10}\right)$

(b) $X_p(jk\Omega) = 1 + 2j(-1)^k \left(\sin\dfrac{3k\pi}{5} + \sin\dfrac{4k\pi}{5}\right)$

The value of N is 10.

13.9 Find the z transform of $x(nT)$ for the DFTs of Prob. 13.8. Assume that $x(nT) = 0$ outside the range $0 \le n \le 9$ in each case.

13.10 (a) Show that the first two formulas in Table 13.1 satisfy the relation

$$h(nT) = h[(N - 1 - n)T]$$

(b) Show that the last two formulas in Table 13.1 satisfy the relation

$$h(nT) = -h[(N - 1 - n)T]$$

13.11 (a) Derive the second formula in Table 13.1
(b) Derive the fourth formula in Table 13.1.

13.12 Design a nonrecursive lowpass filter with a frequency response

$$H_D(e^{j\omega T}) \simeq \begin{cases} 1 & \text{for } 0 \le |\omega| \le 12 \text{ rad/s} \\ 0 & \text{for } 13 \le |\omega| \le 32 \text{ rad/s} \end{cases}$$

Use the frequency-sampling method, and assume that $\omega_s = 64$ rad/s, $N = 64$.

13.13 (a) An optimized version of the filter in Prob. 13.12 can be obtained by assigning

$$A(13\Omega) = 0.74040381 \qquad A(14\Omega) = 0.27101526 \qquad A(15\Omega) = 0.02996826$$

(see Ref. 8). Plot the impulse and amplitude responses of the filter.
(b) Compare this design with the direct design of Prob. 13.12.

13.14 (a) Design a highpass nonrecursive filter with a frequency response

$$H(e^{j\omega T}) \simeq \begin{cases} 0 & \text{for } 0 \le |\omega| \le 20 \text{ rad/s} \\ 1 & \text{for } 21 \le |\omega| \le 32 \text{ rad/s} \end{cases}$$

Use the frequency-sampling method and assume that $\omega_s = 64$ rad/s, $N = 64$.
(b) Repeat part (a) if $\omega_s = 65$ rad/s and $N = 65$.
(c) Find the minimum stopband attenuation in each case.

13.15 (a) Show that a nonrecursive transfer function obtained through the frequency-sampling method can be realized by using a set of parallel, second-order recursive sections in cascade with an elementary Nth-order nonrecursive section.
(b) Using the approach in part (a), realize the transfer function of Prob. 13.12.

13.16 Periodic signals $x(n)$ and $h(n)$ are given by

$$x(n) = \begin{cases} 1 & \text{for } 0 \le n \le 4 \\ 2 & \text{for } 5 \le n \le 9 \end{cases}$$

$$h(n) = n \qquad \text{for } 0 \le n \le 9$$

Find

$$y(n) = \sum_{m=0}^{9} x(m)h(n - m)$$

for $0 \le n \le 9$.

13.17 Show that

$$\mathscr{D}x(n)h(n) = \frac{1}{N} \sum_{m=0}^{N-1} X(m)H(k-m)$$

where $X(k) = \mathscr{D}x(n)$ and $H(k) = \mathscr{D}h(n)$.

13.18 Construct the flow graph for a 16-element decimation-in-time FFT algorithm.

13.19 Construct the flow graph for a 16-element decimation-in-frequency FFT algorithm.

13.20 (a) Compute the Fourier-series coefficients for the periodic signal depicted in Fig. P13.20 by using a 32-element FFT algorithm.

(b) Repeat part (a) using a 64-element FFT algorithm.

(c) Repeat part (a) using an analytical method.

(d) Compare the results obtained.

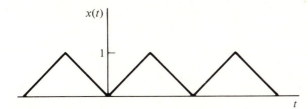

Figure P13.20

13.21 Repeat Prob. 13.20 for the signal of Fig. P13.21.

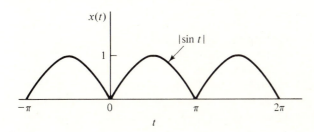

Figure P13.21

13.22 (a) Compute the Fourier transform of

$$x(t) = \begin{cases} \frac{1}{2}(1 + \cos t) & \text{for } 0 \le |t| \le \pi \\ 0 & \text{otherwise} \end{cases}$$

by using a 64-element FFT algorithm. The desired resolution in the frequency domain is 0.5 rad/s.

(b) Repeat part (a) for a frequency domain resolution of 0.25 rad/s.

(c) Repeat part (a) by using an analytical method.

(d) Compare the results in parts (a) to (c).

13.23 Repeat Prob. 13.22 for the signal

$$x(t) = \begin{cases} 1 - |t| & \text{for } |t| < 1 \\ 0 & \text{otherwise} \end{cases}$$

The desired frequency-domain resolutions for parts (a) and (b) are $\pi/4$ and $\pi/8$ rad/s, respectively.

13.24 An FFT program is available, which allows for a maximum of 64 complex input elements. Show that this program can be used to process a real 128-element sequence.

FOURTEEN

HARDWARE IMPLEMENTATION

14.1 INTRODUCTION

The hardware implementation of digital filters can assume various forms, depending on the desired degree of dedication or specialization. It may range from a dedicated computer designed to handle efficiently the relatively simple algorithms encountered in digital filters [1] to a specialized piece of hardware designed to perform a specific filtering task like the many found in communication systems.

In this chapter we consider the latter type of implementation. We start with a brief review of the necessary mathematical tool, namely boolean algebra, and then proceed to examine the various types of devices and subsystems that can be used as components in the implementation. Subsequently, in Sec. 14.7, we describe three specific approaches to implementation and discuss their merits and demerits.

14.2 BOOLEAN ALGEBRA

The universal mathematical tool for the implementation of digital hardware is the binary boolean algebra.

Definition

This is a mathematical system consisting of a set S which comprises two distinct elements designated by 0 and 1, and two operators designated by $+$ and $\cdot$. An indeterminate element of S can be represented by a variable, for example, A, and a specific element of S is said to be a constant. Boolean variables and constants can be interconnected by the operators $+$ and/or $\cdot$ to form boolean expressions, for example, $A + B$, $A \cdot B + C$. A boolean expression, like a single boolean variable, represents an indeterminate element of S. Two boolean expressions are said to be equal if they represent the same element of S for each possible assignment of constants to their variables. Equality is designated by the conventional equality sign $=$, for example, $A = A + A \cdot B$.

Postulates

The binary boolean algebra is based on the following consistent and mutually independent postulates:

1. Two combination rules are defined as

$$A + B \in S$$

$$A \cdot B \in S \qquad \text{(abbreviated } AB \in S)$$

referred to as OR and AND rules, respectively.

2. The OR and AND rules satisfy the following laws:
   a. Commutative law

   $$A + B = B + A$$

   $$A \cdot B = B \cdot A$$

   b. Associative law

   $$(A + B) + C = A + (B + C)$$

   $$(A \cdot B) \cdot C = A \cdot (B \cdot C)$$

   c. Distributive law

   $$A + (B \cdot C) = (A + B) \cdot (A + C)$$

   $$A \cdot (B + C) = (A \cdot B) + (A \cdot C)$$

3. For any element $A \in S$

   $$A + 0 = A$$

   $$A \cdot 1 = A$$

4. For any element $A \in S$ there exists a related element in S denoted as $\bar{A}$ and referred to as the *complement* of A such that

   $$A + \bar{A} = 1 \qquad \text{and} \qquad A \cdot \bar{A} = 0$$

Theorems

The preceding postulates lead directly to the following theorems.

Theorem 14.1

$$A + 1 = 1$$

$$A \cdot 0 = 0$$

Theorem 14.2

$$A + A = A$$

$$A \cdot A = A$$

Theorem 14.3

$$A + A \cdot B = A$$

$$A \cdot (A + B) = A$$

Theorem 14.4

$$A + \bar{A} \cdot B = A + B$$

$$A \cdot (\bar{A} + B) = A \cdot B$$

Theorem 14.5

$$\overline{A + B} = \bar{A} \cdot \bar{B}$$

$$\overline{A \cdot B} = \bar{A} + \bar{B}$$

Boolean Functions

A variable F may depend on other variables, say $A, B, C, \ldots$. We can indicate this fact by the notation

$$F = f(A, B, C, \ldots)$$

A relation like this is referred to as a *boolean* (or *switching*) *function*. It can be represented by a truth table, which is an exhaustive tabular evaluation of F, or by a Karnaugh map, which is a corresponding pictorial representation [2–3].

A list of the most important two-variable boolean functions is found in Table 14.1. The truth tables of these functions can be deduced as in Table 14.2 by using the boolean postulates and theorems.

Table 14.1 Two-variable boolean functions

$f(A, B)$	Name	Symbolic representation
$A + B$	OR	$A + B$
$A \cdot B$	AND	$A \cdot B$
$\overline{A + B}$	NOR	$A \downarrow B$
$\overline{A \cdot B}$	NAND	$A \uparrow B$
$\bar{A} \cdot B + A \cdot \bar{B}$	EXCLUSIVE-OR	$A \oplus B$
$\bar{A} \cdot \bar{B} + A \cdot B$	EXCLUSIVE-NOR	$A \odot B$

Table 14.2 Truth tables of two-variable boolean functions

A	B	A + B	A · B	A ↓ B	A ↑ B	A ⊕ B	A ⊙ B
0	0	0	0	1	1	0	1
0	1	1	0	0	1	1	0
1	0	1	0	0	1	1	0
1	1	1	1	0	0	0	1

Gates

Switching functions can be implemented by using electronic devices like diodes, bipolar transistors, or metal oxide–silicon (MOS) transistors. The general approach is to obtain implementations for the basic functions given in Table 14.1 and then use them as basic elements for the implementation of any desired switching function. Implementations of the basic functions are referred to collectively as *gates*.

An elementary gate using resistors and bipolar transistors is illustrated in Fig. 14.1a. Its operation can be deduced by examining its response to voltage levels. With $V_A = V_B = V_L$, where $V_L < 0.7$ V, the two base-emitter junctions are

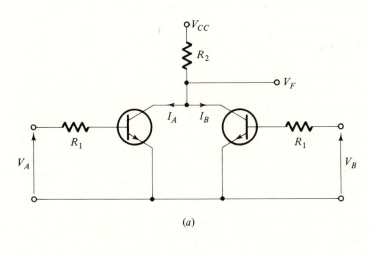

(a)

V_A	V_B	V_F
V_L	V_L	V_{CC}
V_L	V_H	V_{SAT}
V_H	V_L	V_{SAT}
V_H	V_H	V_{SAT}

(b)

A	B	F
0	0	1
0	1	0
1	0	0
1	1	0

(c)

Figure 14.1 (a) Resistor-transistor gate; (b) voltage truth table; (c) binary truth table.

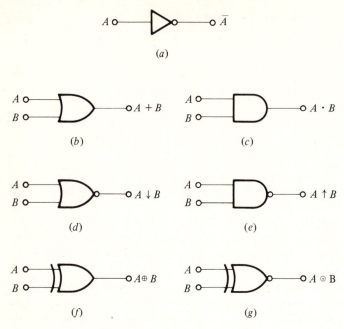

Figure 14.2 Standard gate symbols: (*a*) Inverter, (*b*) OR, (*c*) AND, (*d*) NOR, (*e*) NAND, (*f*) EXCLUSIVE-OR, (*g*) EXCLUSIVE-NOR.

reverse-biased, and hence the two collector currents are approximately zero. Thus the voltage drop across R_2 is negligible, and therefore $V_F = V_{CC}$. On the other hand, if V_A or $V_B = V_H$ or V_A and $V_B = V_H$, where $V_H > 0.7$ V, one or both transistors will be driven into saturation and so $V_F = V_{SAT} \approx 0.2$ V. The operation of the circuit is summarized in Fig. 14.1*b*. Now by identifying low and high voltages with the binary elements 0 and 1, respectively, the truth table of Fig. 14.1*c* can be constructed. This is recognized as the truth table of the NOR function (see Tables 14.1 and 14.2), and hence the device operates as a NOR gate.

There are many other circuits that can be used as gates. The standard symbols used to represent the various types of gates are shown in Fig. 14.2.

A gate, like any other electronic device, will take a finite time to react to changes in input levels because of capacitive effects and because of charge accumulation in the various regions of the transistors. The delay introduced, referred to as the *propagation delay*, is the single most important parameter of a gate. Its magnitude depends on the gate configuration.

14.3 COMBINATIONAL CIRCUITS

Digital hardware can be designed by using combinational and/or sequential circuits. We consider here the former.

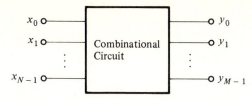

Figure 14.3 Multi-input multioutput combinational circuit.

Characterization

A combinational circuit is a collection of interconnected gates. It has a number of input terminals and one or more output terminals. It can be represented by a block diagram, as depicted in Fig. 14.3. Inputs x_0, x_1, ..., x_{N-1} constitute the excitation, and outputs y_0, y_1, ..., y_{M-1} constitute the response of the device. Each input x_i can assume two binary values, namely 0 or 1, and hence $x_{N-1} x_{N-2} \cdots x_0$ can assume 2^N binary values, referred to as *input combinations*. Similarly, with M outputs 2^M output combinations are possible, which are related to the input combinations by some rule of correspondence.

The distinctive property of a combinational circuit is that the output combination at a given instant depends only on the input combination at that instant (assuming steady-state conditions). That is, a combinational circuit has no memory of previous input or output combinations.

An N-input, M-output combinational circuit can be characterized by a set of M switching functions such as

$$y_0 = f_0(x_0, x_1, \ldots, x_{N-1})$$
$$y_1 = f_1(x_0, x_1, \ldots, x_{N-1})$$
$$\cdots\cdots\cdots\cdots\cdots\cdots\cdots\cdots\cdots\cdots\cdots$$
$$y_{M-1} = f_{M-1}(x_0, x_1, \ldots, x_{N-1})$$

Analysis

The analysis of a combinational circuit can be accomplished by deducing its switching functions or its truth table.

Example 14.1 Analyze the circuit of Fig. 14.4*a*.

SOLUTION From Fig. 14.4*a*

$$y_0 = \bar{x}_1 \bar{x}_0 \qquad y_1 = \bar{x}_1 x_0 \qquad y_2 = x_1 \bar{x}_0 \qquad y_3 = x_1 x_0$$

The truth table of the circuit is shown in Fig. 14.4*b*. The circuit is referred to as a two-input, four-output decoder.

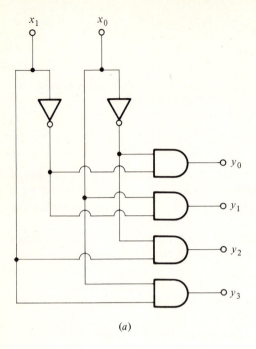

(a)

x_1	x_0	y_3	y_2	y_1	y_0
0	0	0	0	0	1
0	1	0	0	1	0
1	0	0	1	0	0
1	1	1	0	0	0

(b)

Figure 14.4 (a) Two-input decoder (Example 14.1); (b) truth table.

Multiplexers

N-input, M-output combinational circuits $(M = 2^N)$ in which

$$y_i = \begin{cases} 1 & \text{for } i = \sum_{j=0}^{N-1} x_j 2^j \\ 0 & \text{otherwise} \end{cases}$$

for $i = 0, 1, \ldots, M - 1$ are known as *decoders*, e.g., the circuit of Fig. 14.4a. Circuits of this type can be used to design multiplexers, read-only memories, and other devices.

A four-input multiplexer using the decoder of Fig. 14.4a is illustrated in Fig. 14.5a. The switching function of this circuit is

$$F = a_0 \bar{x}_1 \bar{x}_0 + a_1 \bar{x}_1 x_0 + a_2 x_1 \bar{x}_0 + a_3 x_1 x_0$$

and hence
$$F = \begin{cases} a_0 \\ a_1 \\ a_2 \\ a_3 \end{cases} \quad \text{for } x_1 x_0 = \begin{cases} 00 \\ 01 \\ 10 \\ 11 \end{cases}$$

The circuit can thus be used as a four-input selection switch like that depicted in Fig. 14.5b.

A four-output distribution switch of the type shown in Fig. 14.6a can be constructed by using a decoder as depicted in Fig. 14.6b. For this device

$$F_0 = F\bar{x}_1\bar{x}_0 \qquad F_1 = F\bar{x}_1 x_0 \qquad F_2 = Fx_1\bar{x}_0 \qquad F_3 = Fx_1 x_0$$

and so
$$\begin{aligned} F_0 &= F \\ F_1 &= F \\ F_2 &= F \\ F_3 &= F \end{aligned} \quad \text{if } x_1 x_0 = \begin{cases} 00 \\ 01 \\ 10 \\ 11 \end{cases}$$

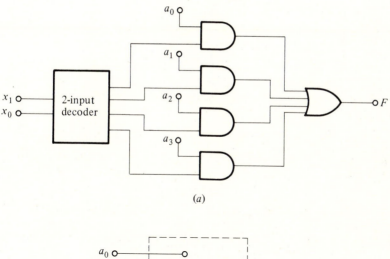

(a)

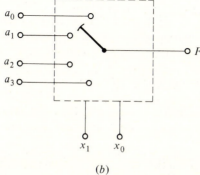

(b)

Figure 14.5 (a) Four-input multiplexer; (b) four-input selection switch.

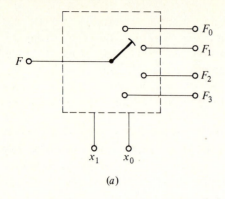

(a)

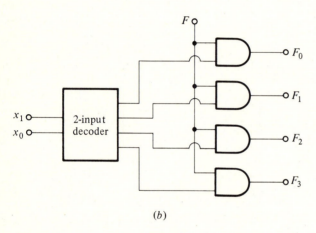

(b)

Figure 14.6 (a) Four-output distribution switch; (b) implementation using a two-input decoder.

An M-input multiplexer is characterized by the switching function

$$F = a_0(\bar{x}_{N-1} \cdots \bar{x}_1 \bar{x}_0) + a_1(\bar{x}_{N-1} \cdots \bar{x}_1 x_0) + \cdots + a_{M-1}(x_{N-1} \cdots x_1 x_0)$$

where $N = \log_2 M$. This is the most general N-variable switching function, and since the ith term can be deleted or retained by setting a_i to 0 or 1, the multiplexer can be used to realize any one of the possible N-variable switching functions. One need only assign appropriate values to a_0, a_1, ..., a_{N-1}. For example, the EXCLUSIVE-OR can be realized by assigning $a_0 = a_3 = 0$, $a_1 = a_2 = 1$ in the two-input multiplexer of Fig. 14.5a.

Parallel Addition

The basic units for the construction of arithmetic devices are the half and full adders. The half adder is a combinational circuit which will add an augend bit and an addend bit to produce a sum bit and a carry bit. The full adder is similar to the half adder except that it will add an augend bit, an addend bit, and a previous carry bit to produce a sum bit and a new carry bit. The half adder can readily be implemented by using an EXCLUSIVE-OR and an AND gate, as in Fig. 14.7a. The full adder can be implemented by interconnecting two half adders, as in Fig. 14.7b.

Parallel addition of two N-bit binary numbers $a_{N-1} \cdots a_1 a_0$ and $b_{N-1} \cdots b_1 b_0$ can be carried out by using a half adder and $N-1$ full adders, as depicted in Fig. 14.8.

The most important parameter of a parallel adder is its addition time. This is the time taken by the circuit under worst-case conditions to reach a steady state after a change in its input levels. In effect, this is the cumulative delay of the longest path between input and output. The longest path in Fig. 14.8 is between input b_0 and output c_{N-1}. Assuming a gate delay of T_gs for AND and OR

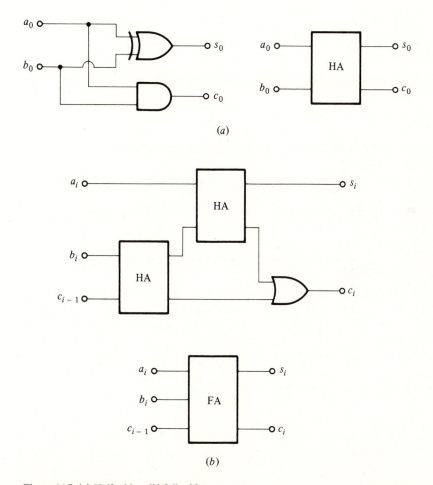

Figure 14.7 (a) Half adder; (b) full adder.

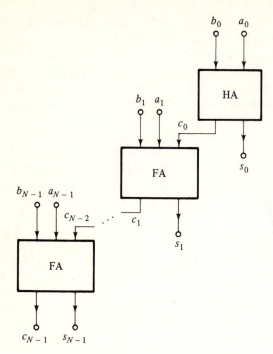

Figure 14.8 N-bit parallel adder.

gates, and $3T_g$s for EXCLUSIVE OR gates, the addition time for the adder of Fig. 14.8 for $N \geq 4$ is

$$T_a = [2(N - 1) + 1]T_g$$

The addition time can be reduced by using carry-look-ahead circuits [4]. The use of these circuits, however, tends to increase the cost of the implementation and also the power dissipation.

Parallel Multiplication

Like parallel addition, parallel multiplication can be performed by employing half and full adders. Consider a (4×4)-bit multiplication like the following:

$$
\begin{array}{cccc}
x_3 & x_2 & x_1 & x_0 \\
m_3 & m_2 & m_1 & m_0 \\
\hline
p_3 & p_2 & p_1 & p_0 \\
q_3 & q_2 & q_1 & q_0 \\
r_3 & r_2 & r_1 & r_0 \\
t_3 & t_2 & t_1 & t_0 \\
\hline
y_7 & y_6 & y_5 & y_4 & y_3 & y_2 & y_1 & y_0
\end{array}
$$

We can write

$$p_i = m_0 x_i \qquad q_i = m_1 x_i \qquad r_i = m_2 x_i \qquad t_i = m_3 x_i$$

for $i = 0$ to 3. Hence bits p_0 through t_3 can readily be generated by using AND gates. Now by adding the above partial products according to the scheme of Table 14.3 the desired product can be formed. In this table c_1 to c_{12} represent carry bits whereas s_1 to s_6 represent partial-sum bits. The implementation of this algorithm is shown in Fig. 14.9.

The amount of necessary hardware for the implementation of parallel multipliers tends to increase very rapidly as the number of bits in the operands is increased. For example, a (10×10)-bit multiplier of the type shown in Fig. 14.9 would require 100 AND gates and 90 full adders besides a complicated interconnection network. However, an advantage is gained at this price, namely high speed of operation.

Read-only Memories

A read-only memory (ROM) is essentially a multi-input, multioutput combinational circuit. It is referred to as a *memory* in the sense that each input combination or address results in the recovery of an output combination or stored word. It is referred to as *read-only* memory because its data content is fixed.

A ROM differs from a conventional combinational circuit only in its construction. It generally comprises a decoder and an encoder, which are con-

Table 14.3 (4 × 4)-bit multiplication algorithm

c_{12}	y_7	c_8	c_4		p_3	p_2	p_1	p_0
				q_3	q_2	q_1	q_0	
			$c_4 \leftarrow$	$c_3 \leftarrow$	$c_2 \leftarrow$	$c_1 \leftarrow$		
			c_4	s_3	s_2	s_1		
			r_3	r_2	r_1	r_0		
		$c_8 \leftarrow$	$c_7 \leftarrow$	$c_6 \leftarrow$	$c_5 \leftarrow$			
		c_8	s_6	s_5	s_4			
		t_3	t_2	t_1	t_0			
$c_{12} \leftarrow$	$c_{11} \leftarrow$	$c_{10} \leftarrow$	$c_9 \leftarrow$					
y_7	y_6	y_5	y_4	y_3	y_2	y_1	y_0	

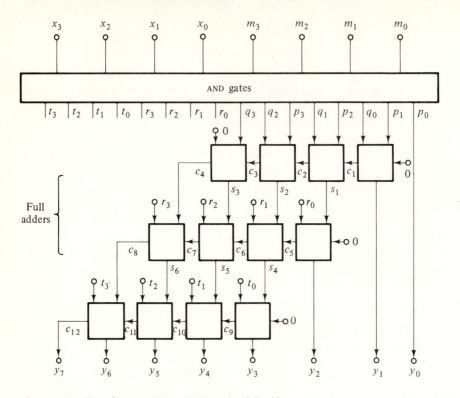

Figure 14.9 A (4×4)-bit parallel multiplier using full adders.

nected in cascade, as in Fig. 14.10*a*. The function of the decoder is to convert each input combination into a distinct intermediate word in which all bits except one are zeros. The function of the encoder, on the other hand, is to convert the intermediate words into some prescribed set of words. A typical truth table for a two-input, four-output ROM is given in Fig. 14.10*b*.

The encoder can be implemented by using diodes, bipolar transistors, or MOS transistors. Two possibilities are illustrated in Fig. 14.11.

The main parameters of a ROM are its *memory size* (or *capacity*) and its *access time*. The memory size is the product of the number of output words times the number of bits per word. The access time is the propagation delay of the device.

Three types of ROMs are available: standard ROMs, which are designed to perform standard functions; programmable ROMs (or PROMs), which can be programmed after fabrication according to the customer's specifications; and erasable ROMs, which can be reprogrammed by using special instruments.

ROMs can be used as code converters, look-up tables, or components in arithmetic devices. The design of a parallel multiplier through the use of ROMs is illustrated by the following example.

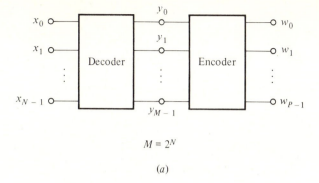

$$M = 2^N$$

(a)

x_1	x_0	y_3	y_2	y_1	y_0	w_3	w_2	w_1	w_0
0	0	0	0	0	1	0	1	0	1
0	1	0	0	1	0	0	1	0	0
1	0	0	1	0	0	1	0	1	0
1	1	1	0	0	0	0	1	1	0

(b)

Figure 14.10 (a) Read-only memory; (b) typical truth table.

Example 14.2 Design a (4×4)-bit multiplier using ROMs and 2-bit adders.

SOLUTION Let the multiplicand and multiplier be $x_3 x_2 x_1 x_0$ and $m_3 m_2 m_1 m_0$, respectively. By writing

$$x_3 x_2 x_1 x_0 = x_3 x_2 00 + 00 x_1 x_0$$

$$m_3 m_2 m_1 m_0 = m_3 m_2 00 + 00 m_1 m_0$$

the desired product can be expressed as

$$y_7 y_6 y_5 y_4 y_3 y_2 y_1 y_0 = p_3 p_2 p_1 p_0 + q_3 q_2 q_1 q_0 00 + r_3 r_2 r_1 r_0 00 + t_3 t_2 t_1 t_0 0000$$

where $p_3 p_2 p_1 p_0 = (x_1 x_0)(m_1 m_0)$ $\quad q_3 q_2 q_1 q_0 = (x_3 x_2)(m_1 m_0)$

$\quad\quad\quad r_3 r_2 r_1 r_0 = (x_1 x_0)(m_3 m_2)$ $\quad t_3 t_2 t_1 t_0 = (x_3 x_2)(m_3 m_2)$

These partial products can be generated by using four (16×4)-bit ROMs. The desired product can be formed according to the algorithm of Table 14.4. In this table bits are added two by two, as indicated by the solid lines and, as in the algorithm of Table 14.3, c_1 to c_4 and s_1 to s_4 represent carry and partial-sum bits, respectively. The corresponding multiplier implementation is depicted in Fig. 14.12.

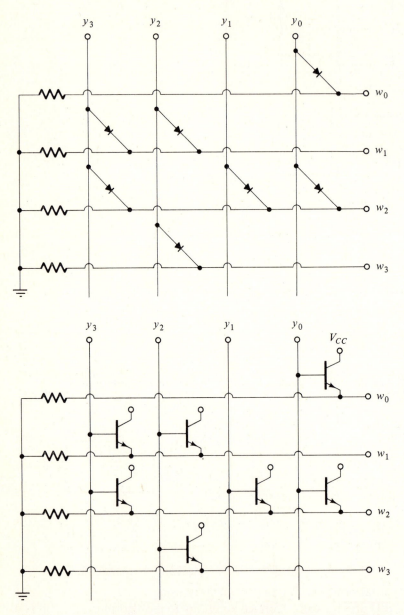

Figure 14.11 Implementation of encoders by means of diodes or bipolar transistors.

Table 14.4 (4 × 4)-bit multiplication algorithm

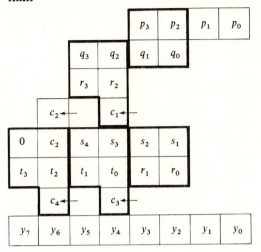

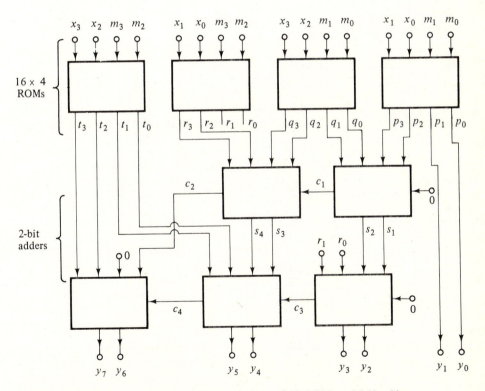

Figure 14.12 A (4 × 4)-bit parallel multiplier using (16 × 4)-bit ROMs and 2-bit adders.

14.4 FLIP-FLOPS, REGISTERS, AND COUNTERS

Sequential circuits, unlike combinational circuits, must necessarily incorporate memory elements such as flip-flops.

A flip-flop is a device with two outputs y_1 and y_2 and one or more inputs. Of the four possible values of $y_1 y_2$ only two are allowed, namely 01 and 10. In effect, a flip-flop is a two-state device. Various types of flip-flops are in current use. The most frequently used ones are the following:

1. Set-reset (SR) flip-flop (also known as SR latch)
2. Clocked SR flip-flop
3. JK flip-flop
4. Delay (D) flip-flop

The most basic of these is the SR flip-flop.

SR Flip-Flop

An SR flip-flop can be constructed by using two NOR gates, as depicted in Fig. 14.13a. The analysis of this device consists of finding the state transitions brought about by the four possible values of SR, namely 00, 01, 10, and 11.

From Fig. 14.13a

$$y_1 = \bar{R}\bar{y}_2 \qquad y_2 = \bar{S}\bar{y}_1 \qquad (14.1)$$

Hence with $SR = 00$

$$y_1 = \bar{y}_2 \qquad y_2 = \bar{y}_1$$

and if $y_1 y_2 = 01$ or 10 initially, the steady-state response will be $y_1 y_2 = 01$ or 10. That is, in either case no state transition is brought about by the excitation. With $SR = 01$, Eq. (14.1) gives

$$y_1 = 0 \qquad y_2 = 1$$

and if $y_1 y_2 = 10$ initially, the flip-flop will be reset (or cleared) such that $y_1 y_2 = 01$ immediately after excitation. On the other hand, if $y_1 y_2 = 01$ initially, no action will take place. With $SR = 10$ we have

$$y_2 = 0 \qquad y_1 = 1$$

and if $y_1 y_2 = 01$ initially, the flip-flop will be set such that $y_1 y_2 = 10$ at steady state. If $y_1 y_2 = 10$ initially, no action will take place. Finally, with $SR = 11$ Eq. (14.1) yields

$$y_1 = y_2 = 0$$

However, this equation holds only if S and R are changed simultaneously from 0 to 1. In practice, simultaneous changes in S and R are highly improbable, and as a result the steady-state values of y_1 and y_2 for this case are difficult to predict. Because of this reason, input $SR = 11$ is normally avoided.

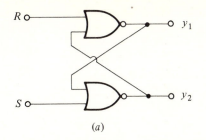

(a)

S	R	$y(n)$	$y(n+1)$
0	0	0	0
0	0	1	1
0	1	0	0
0	1	1	0
1	0	0	1
1	0	1	1
1	1	0	X
1	1	1	X

Not allowed $\begin{bmatrix} & & \\ & & \end{bmatrix}$ for last two rows

(b)

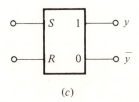

(c)

Figure 14.13 (a) SR flip-flop, (b) transition table, (c) symbol.

The operation of the flip-flop is summarized by the table of Fig. 14.13b, where $y(n)$ and $y(n+1)$ represent y_1 (or $\bar{y}_2$) before and after excitation, respectively. Tables like this are referred to as *transition tables*. The usual symbol for the SR flip-flop is shown in Fig. 14.13c.

Clocked SR Flip-Flop

An SR flip-flop might be driven by a combinational circuit like that in Fig. 14.14a, where signals X and Y may have the waveforms shown in Fig. 14.14b. Let us assume that each gate has a propagation delay T_g and that the flip-flop is initially reset ($y = 0$). At $t = 0$, $Z = \bar{X} = 1$ and $S = YZ = 0$. At t_1, X and Y change simultaneously from 0 to 1. Hence T_g s later at t_2, Z will change to 0, as depicted in Fig. 14.14b. At $t_1(+)$ $Z = Y = 1$, and so T_g seconds later at t_2, S will change from 0 to 1. Since $SR = 10$, the flip-flop will be set after a brief delay. At $t_2(+)$ $Y = 1$, $Z = 0$, and as a result S will change again T_g seconds later at t_3 from 1 to 0.

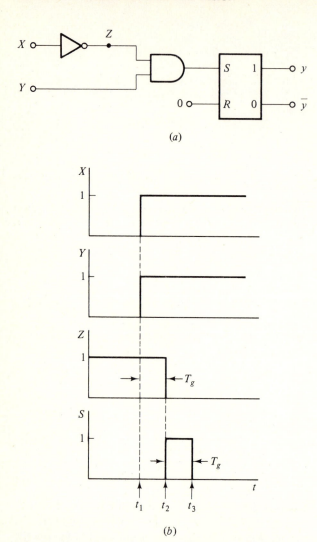

Figure 14.14 (a) SR flip-flop driven by a combinational circuit; (b) timing diagram.

According to the table of Fig. 14.13b, the flip-flop will remain set. Had the two gates been ideal with zero delay, S would have remained 0 before, during, and after the signal transition and the flip-flop would remain reset. Clearly, a combinational circuit driving an SR flip-flop may generate spurious pulses which may cause undesirable flip-flop transitions.

The above problem can be overcome by isolating the SR flip-flop from its driving circuit during signal transitions by using two AND gates and a control (or clock) signal C, as depicted in Fig. 14.15. Since $S = S'C$ and $R = R'C$, $SR = 00$ if $C = 0$ or $SR = S'R'$ if $C = 1$. Thus by keeping $C = 0$ during signal transitions,

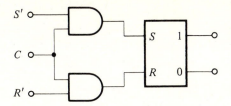

Figure 14.15 Clocked SR flip-flop.

spurious pulses can be prevented from initiating undesirable state transitions in the flip-flop.

The modified flip-flop is said to be a clocked SR flip-flop. Its transition table is the same as that in Fig. 14.13b except that state transitions will take place only if a clock pulse is applied.

JK Flip-Flop

The JK flip-flop is a variation of the SR flip-flop in which the input combination 11 is permissible. The structure of a clocked JK flip-flop is shown in Fig. 14.16a. The transition table of the JK flip-flop can be deduced from that of the SR flip-flop, as depicted in Fig. 14.16b, by noting that $S = J\bar{y}$, $R = Ky$.

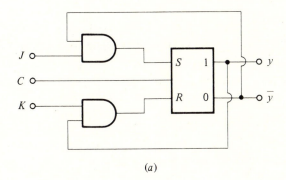

(a)

J	K	$y(n)$	S	R	$y(n+1)$
0	0	0	0	0	0
0	0	1	0	0	1
0	1	0	0	0	0
0	1	1	0	1	0
1	0	0	1	0	1
1	0	1	0	0	1
1	1	0	1	0	1
1	1	1	0	1	0

(b)

Figure 14.16 (a) Clocked JK flip-flop; (b) transition table.

Delay Flip-Flop

The delay flip-flop can be constructed by using a JK flip-flop and an inverter, as depicted in Fig. 14.17a. Its transition table is shown in Fig. 14.17b. As can be seen,

$$y(n + 1) = D$$

and hence this flip-flop will record the value of its input. The delay flip-flop is the basic component for the construction of registers.

Registers

Registers are devices which can be used to store binary words. The most common type is the shift register, which is designed with a shift capability for loading and removing the data. Depending on the way bits are loaded and removed, four types of registers can be identified, as follows:

1. Serial-input, serial-output
2. Serial-input, parallel-output
3. Parallel-input, serial-output
4. Parallel-input, parallel-output

The four possibilities are illustrated in Fig. 14.18a.
   A 5-bit serial-input, serial-output shift register can readily be constructed by cascading five delay flip-flops as in Fig. 14.18b. The operation of this device is easy to explain.
   Let us assume that signal x in Fig. 14.18b is of the form depicted in Fig. 14.19 and that $y_0 y_1 y_2 y_3 y_4 = 00000$ initially. Also let C be a clock signal of the type

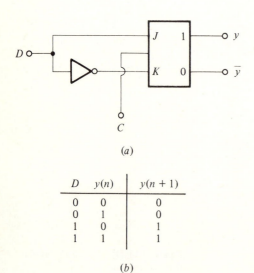

(a)

D	$y(n)$	$y(n + 1)$
0	0	0
0	1	0
1	0	1
1	1	1

(b)

Figure 14.17 (a) Delay flip-flop; (b) transition table.

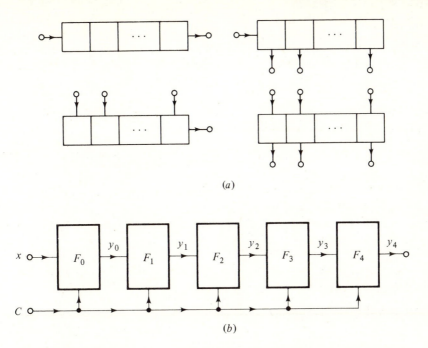

(a)

(b)

Figure 14.18 (a) The four types of registers; (b) serial-input, serial-output shift register.

shown in Fig. 14.19 and assume that the pulse duration T_p is less than T_f, the propagation delay of a flip-flop. Each flip-flop will record the value of its input during each clock pulse. At instant $t_1 +$, $x = 1$, $y_0 = \cdots = y_4 = 0$, $C = 1$, and so T_f s later y_0 will change to 1 whereas $y_1, \ldots, y_4$ will remain 0, as shown in Fig. 14.19. At instant $t_2 +$, $x = 1$, $y_0 = 1$, $y_1 = \cdots = y_4 = 0$, $C = 1$, and so y_0 will remain 1, and T_f s later y_1 will change to 1 whereas $y_2, \ldots, y_4$ will remain 0. Proceeding in the same way, we can complete the timing diagram of Fig. 14.19. The activity of the register is summarized in Table 14.5. As can be seen, bits enter at the left, proceed along the register in bucket-brigade fashion, and eventually exit at the right.

Table 14.5 Operation of a shift register

x	y_0	y_1	y_2	y_3	y_4
0	0	0	0	0	0
1	1	0	0	0	0
1	1	1	0	0	0
0	0	1	1	0	0
0	0	0	1	1	0
1	1	0	0	1	1

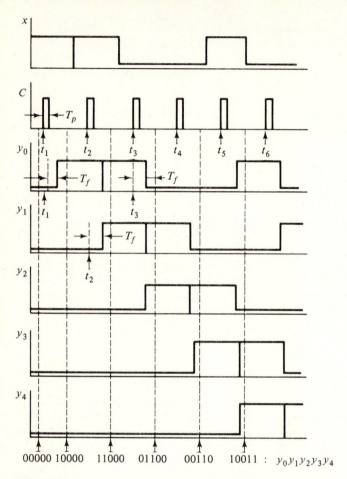

Figure 14.19 Typical timing diagram for a 5-bit serial-input, serial-output shift register.

Counters

Counters are devices which can be used to count pulses. They may be up or down counters, generating binary numbers in ascending or descending order; ripple counters, generating sequences such as 0001, 0010, 0100, ... ; modulo-N counters, and so on [2–4]. Counters are used extensively for the generation of control signals in sequential circuits. Their basic building block is the flip-flop.

14.5 SEQUENTIAL CIRCUITS

Two types of signals can be employed in sequential circuits, namely level and pulse signals. A level signal is one that remains at 0 or 1 for arbitrary periods of time and changes values at intervals which are long relative to the delay of a gate.

A pulse signal, on the other hand, is one that remains at 0 and changes to 1 for very short periods by comparison with the delay of a gate. A periodic pulse signal is said to be a *clock signal*.

Depending on the types of signals allowed, three general classes of sequential circuits can be identified [2]:

1. Clock mode
2. Pulse mode
3. Level mode

The most suitable of these for the design of digital filters is the class of clock-mode circuits. The general model for these circuits is illustrated in Fig. 14.20. The memory comprises a set of clocked flip-flops, and all signals except for the clock signal are level signals.

A basic requirement of clock-mode circuits is that flip-flops may change state only once for each clock pulse. This means that the duration of the clock pulse must be shorter than the delay of the fastest path between memory input and combinational-circuit output. At the same time, the interval between clock pulses must be longer than the delay of the slowest path between memory input and combinational-circuit output.

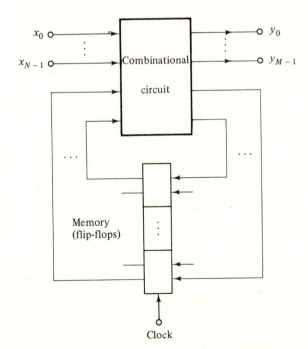

Figure 14.20 Model for clock-mode sequential circuits.

Sequential circuits are generally slower than combinational circuits since operations are performed in sequence. However, they are also more economical since subsystems are time-shared.

In the following sections we discuss a number of clock-mode, sequential circuits that can be used as components in digital filters.

Two's-Complement Serial Addition

Two's-complement serial addition can be performed by using a full adder and a delay flip-flop, as depicted in Fig. 14.21. Signal R is 0 during the first clock pulse and is set to 1 subsequently. Its purpose is to prevent a carry bit stored in D in a previous addition from activating the full adder. The two numbers to be added are funneled into the adder bit by bit with their least significant bits first, and each time the augend, addend, and previous carry bit are added to form a sum bit and a new carry bit. The sum bit is passed on to the sum register, whereas the new carry bit is passed on to the delay flip-flop.

Two's-Complement Serial Subtraction

Two's-complement subtraction can be performed by adding the complemented subtrahend to the minuend. The two's complement of any number can be formed by complementing all bits and then adding a 1 at the least significant position. Hence a two's-complement serial subtractor can be constructed by modifying the adder of Fig. 14.21, as illustrated in Fig. 14.22a. As before, R is 0 during the first clock pulse, in which case $F = 1$. In this way a 1 is added at the least significant

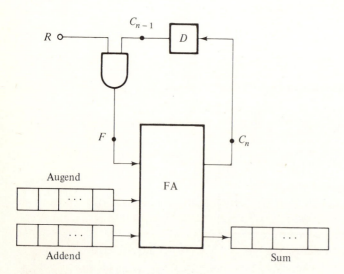

Figure 14.21 Two's-complement serial adder.

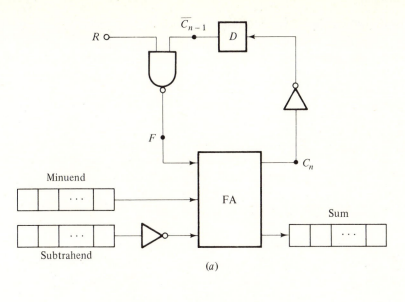

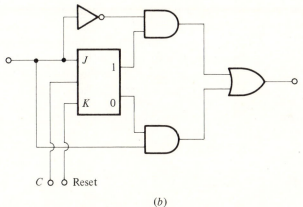

Figure 14.22 (*a*) Two's-complement serial subtractor; (*b*) two's complementer.

position of the sum. On the second and subsequent clock pulses R is set to 1, and so

$$F = \overline{R\overline{C}}_{n-1} = C_{n-1}$$

as in the two's-complement adder.

An alternative approach to two's-complement subtraction is to use a two's complementer in series with the addend input in Fig. 14.21. The two's complement of a number can also be formed by complementing all bits except the last 1 to the right and all subsequent zeros, e.g., for 1.1001100 we have 0.0110100. Hence a serial two's complementer can readily be designed by using a JK flip-flop as in Fig. 14.22*b*.

Serial Signed-Magnitude Multiplication

A multiplication like

$$x_0. \; x_1 \; x_2 \quad x_3$$
$$m_0. m_1$$

$$\overline{}$$

$$p_0 \; p_1 \; p_2 \quad p_3$$
$$q_0 \, q_1 \quad q_2 \; q_3$$

$$\overline{}$$

$$y_0. \, y_1 \; y_2 \; y_3 \quad y_4$$

where $\qquad x = x_0.\,x_1 x_2\,x_3 \qquad$ and $\qquad m = m_0.\,m_1$

are positive signed-magnitude numbers, can be performed by employing the circuit of Fig. 14.23a. The two AND gates are used to generate the bits of the partial products one by one, whereas flip-flop D_1 is used to delay the generation of q_i by one clock pulse so that p_3 can be added to 0, p_2 to q_3, and so on. The carry flip-flop is assumed to be reset at the start of the multiplication cycle.

The above multiplier will receive a 4-bit multiplicand and produce a 5-bit product. A modified version of this circuit incorporating truncation is illustrated in Fig. 14.23b. With D_1 and D_2 initially reset and $T = 0$ during the first clock pulse, $F = Q = P' = 0$, and so the least significant bit of the product (namely y_4) will be reset to 0. On the second and subsequent clock pulses $T = 1$, $P' = P$, and the circuit will operate as before.

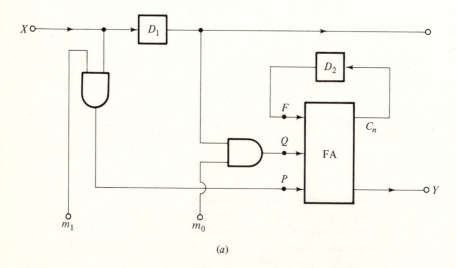

(a)

Figure 14.23 Rudimentary serial magnitude multiplier: (a) basic circuit;

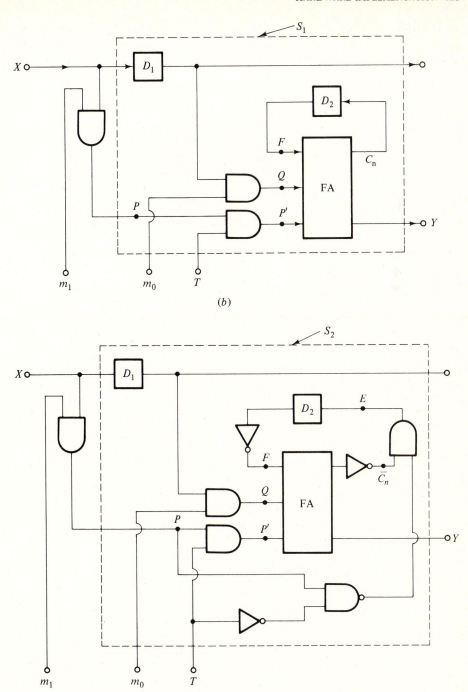

Figure 14.23 (*b*) circuit incorporating truncation, (*c*) circuit incorporating rounding.

Rounding can be incorporated by modifying the circuit of Fig. 14.23b, as in Fig. 14.23c. Signal T is again 0 during the first clock pulse and is set to 1 subsequently. With D_1 initially reset and D_2 set, $F = Q = P' = 0$ during the first clock pulse and so $\bar{C}_n = 1$, $Y = 0$. Also

$$E = \overline{P}\overline{T} \cdot \bar{C}_n = \bar{p}_3$$

and as a consequence $\bar{p}_3$ will be stored in D_2 so that p_3 can be added to the product during the following clock pulse. On the second and subsequent clock pulses $E = \bar{C}_n$, and the circuit will operate as before.

A serial multiplier which will multiply any two positive numbers

$$x = x_0.x_1 x_2 \cdots x_{N-1} \qquad m = m_0.m_1 m_2 \cdots m_K$$

can now be implemented by repeating section S_1 in Fig. 14.23b and using section S_2 of Fig. 14.23c at the output if quantization by rounding is desired. Such a device is illustrated in Fig. 14.24. Two features of this device are of interest:

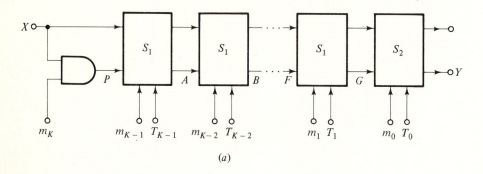

(a)

$$
\begin{aligned}
&x_0.x_1 x_2 \cdots x_{N-1}\\
&m_0.\, m_1 m_2 \cdots m_K\\
\hline
&p_0 p_1 p_2 \cdots p_{N-1}\\
&q_0 q_1 q_2 \cdots q_{N-1}\\
\hline
&a_0 a_1 a_2 \cdots a_{N-1}\,0\\
&r_0 r_1 r_2 \quad\cdots r_{N-1}\\
\hline
&b_0 b_1 b_2 \cdots b_{N-1}\,00\\
&\cdots \cdots \cdots \cdots\\
\hline
&f_0 f_1 f_2 \cdots f_{N-1}\,0\cdots 00\\
&v_0 v_1 v_2 \cdots v_{N-1}\\
\hline
&g_0 g_1 g_2 \cdots g_{N-1}\,00\cdots 00\\
&w_0 w_1 w_2 \cdots w_{N-1}\\
&\qquad\qquad g_{N-1}\\
\hline
&y_0 \cdot y_1 y_2 \cdots y_{N-1}\underbrace{000\cdots 00}_{K\ \text{zeros}}
\end{aligned}
$$

(b)

Figure 14.24 (a) General magnitude multiplier; (b) multiplication algorithm.

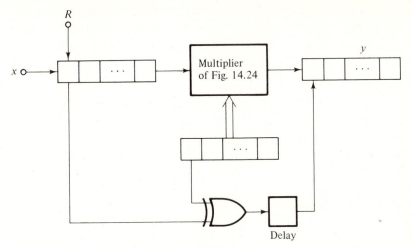

R

x

y

Multiplier
of Fig. 14.24

Delay

Figure 14.25 Serial signed-magnitude multiplier.

(1) with $x_0 = m_0 = 0$ all carry flip-flops will be put in their initial state at the end of the multiplication cycle; i.e., the multiplier is self-resetting. (2) The multiplier will introduce a delay of K bits, as can be seen in Fig. 14.24b.

A signed-magnitude multiplier which will multiply positive as well as negative numbers can be obtained by modifying the circuit of Fig. 14.24a as depicted in Fig. 14.25. The EXCLUSIVE-OR is used to generate the sign of the product while signal R is used to reset the sign bit in the case of a negative multiplicand.

Two's-Complement Multiplication

A two's-complement multiplier is a device that will receive two numbers in two's-complement representation and produce the correct product, again in two's-complement representation. The first step in the design of such a device is to develop a suitable multiplication algorithm.

The two's complement of a number x, designated as $\tilde{x}$, is given by

$$\tilde{x} = \begin{cases} x & \text{if } x \geq 0 \\ 2 - |x| & \text{if } x < 0 \end{cases}$$

Conversely, if

$$\tilde{x} = x_0 . x_1 x_2 \cdots x_L$$

we have

$$x = -x_0 + \sum_{i=1}^{L} x_i 2^{-i} \tag{14.2}$$

For example, if $\tilde{x} = 0.010101$,

$$x = \tfrac{1}{4} + \tfrac{1}{16} + \tfrac{1}{64} = 0.328125$$

and if $\tilde{x} = 1.101011$,

$$x = -1 + \tfrac{1}{2} + \tfrac{1}{8} + \tfrac{1}{32} + \tfrac{1}{64} = -0.328125$$

Consider the product

$$y = xm \tag{14.3}$$

and let $\tilde{y}$, $\tilde{x}$, and $\tilde{m}$ be the two's complements of y, x, and m, respectively. From Eqs. (14.2) and (14.3)

$$y = -x_0 m + \sum_{i=1}^{L} x_i m 2^{-i}$$

and so

$$\tilde{y} = \text{two's complement } [-x_0 m + 2^{-1}(x_1 m + \cdots + 2^{-1}(x_{L-1} m + 2^{-1}(x_L m)))] \tag{14.4}$$

Now consider the two's complement of $2^{-1}u$, where

$$\tilde{u} = u_0.u_1 u_2 \cdots u_M$$

For $u \geq 0$ (or $u_0 = 0$)

$$\text{Two's complement } (2^{-1}u) = 2^{-1}\tilde{u}$$

and for $u < 0$ (or $u_0 = 1$)

$$\text{Two's complement } (2^{-1}u) = 2 - |2^{-1}u| = 1 + 2^{-1}(2 - |u|) = 1 + 2^{-1}\tilde{u}$$

that is, $\text{Two's complement } (2^{-1}u) = \begin{cases} 2^{-1}\tilde{u} & \text{if } u_0 = 0 \\ 1 + 2^{-1}\tilde{u} & \text{if } u_0 = 1 \end{cases}$

This operation, referred to as *two's-complement right shift*, may be designated as

$$\text{Two's complement } (2^{-1}u) = 2_2^{-1}\tilde{u}$$

By using this notation in Eq. (14.4), we obtain

$$\tilde{y} = -x_0 \tilde{m} + 2_2^{-1}(x_1 \tilde{m} + \cdots + 2_2^{-1}(x_{L-1} \tilde{m} + 2_2^{-1}(x_L \tilde{m}))) \tag{14.5}$$

where the $+$ and $-$ signs designate two's-complement addition and subtraction, respectively. Therefore, the two's complement of the desired product can be formed by using the following algorithm:

1. Clear accumulator register.
2. Add $x_L \tilde{m}$ (two's-complement addition) to accumulator content.
3. Shift accumulator content to right by 1 bit (two's-complement shift).
4. Repeat steps 2 and 3 for $x_{L-1}, \ldots, x_1$.
5. Subtract $x_0 \tilde{m}$ from accumulator content (two's-complement subtraction).

The algorithm is illustrated by the following example.

Example 14.3 Form the product

$$y = 0.8125(-0.390625)$$

by using two's-complement multiplication. Assume a 12-bit accumulator.

Table 14.6 Two's-complement multiplication (Example 14.3)

Operation	Accumulator
Clear	0.0 0 0 0 0 0 0 0 0 0 0
Add $x_4 \tilde{m} = 1.100111$	1.1 0 0 1 1 1 0 0 0 0 0
Shift	1.1 1 0 0 1 1 1 0 0 0 0
Add $x_3 \tilde{m} = 0.000000$	1.1 1 0 0 1 1 1 0 0 0 0
Shift	1.1 1 1 0 0 1 1 1 0 0 0
Add $x_2 \tilde{m} = 1.100111$	1.1 0 0 0 0 0 1 1 0 0 0
Shift	1.1 1 0 0 0 0 0 1 1 0 0
Add $x_1 \tilde{m} = 1.100111$	1.0 1 0 1 1 1 0 1 1 0 0
Shift	1.1 0 1 0 1 1 1 0 1 1 0
Subtract $x_0 \tilde{m} = 0.000000$	1.1 0 1 0 1 1 1 0 1 1 0

SOLUTION The two's complements of the operands are

$$\tilde{x} = 0.1101 \qquad \tilde{m} = 1.100111$$

According to Table 14.6, the above algorithm yields

$$\tilde{y} = 1.1010111011 \qquad \text{or} \qquad y = -0.0101000101 = -0.3173828125$$

An implementation of the above algorithm can readily be obtained, as illustrated in Fig. 14.26.

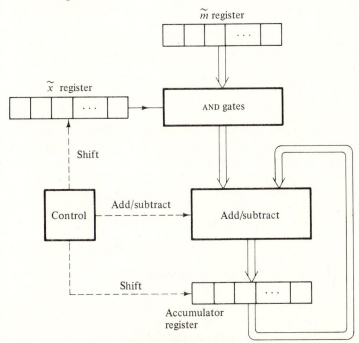

Figure 14.26 Serial two's-complement multiplier.

14.6 IC FAMILIES

The basic component in the design of modern digital systems, the *integrated circuit* (IC), is a complete circuit fabricated on a single chip of silicon; its complexity may range from a few gates to a few thousand gates. ICs containing between 12 and 100 gates are generally referred to as *medium-scale integrated* (MSI) *circuits*, whereas ICs containing in excess of 100 gates are referred to as *large-scale integrated* (LSI) *circuits*. Several families of ICs have evolved in recent years the most important of which are the following:

1. Transistor-transistor logic (T^{2}L)
2. Emitter-coupled logic (ECL)
3. Metal oxide–semiconductor logic (MOS logic)
4. Integrated injection logic (I^{2}L)

Since each of these families is based on a specific gate configuration which has certain fundamental characteristics, each family has its own distinctive features.

Desirable Features

The desirable features in an IC family are as follows [5–7]:

1. High speed
2. Low power dissipation
3. High noise immunity
4. Low generated noise
5. Logic flexibility
6. Low cost

High speed is particularly important in the implementation of digital filters, primarily for economical reasons. With the availability of high-speed ICs, a parallel design can often be replaced by a serial one where fewer ICs are time-shared to perform the same filtering task. Alternatively, a high-speed parallel design can be time-shared to perform many filtering tasks at the same time.

Low power dissipation is a desirable feature because it leads to low cooling costs as well as low power-supply costs. However, low power dissipation is usually associated with other inferior characteristics, e.g., low speed, and a compromise must be made by the designer.

Digital systems are often subjected to spurious signals produced by switching transiences and excessive coupling between signal leads and power lines. Signals of this type, referred to collectively as noise, are often indistinguishable from valid logic signals and hence can introduce malfunctions or unpredictable operation. Noise can be eliminated by careful layout and adequate shielding, but it is usually more economical to design the gates themselves with a certain degree of noise immunity. The higher the level of noise immunity, the fewer the necessary precautions to prevent false logic signals.

When a gate changes state, the current drain from the power supply can change rapidly to a new level, and if a power line is close to signal lines, undesirable noise will be induced. This problem can be overcome by using short signal leads that are well separated from power lines and by adequately bypassing the power supplies. However, it is often more economical to prevent this problem by using ICs in which the noise generated is low. In such ICs the current drain in the basic gate tends to remain constant or changes at a slow rate during state transitions.

Logic flexibility (or versatility) reflects the capability of the IC family to meet as many system demands as possible. It is based on the following factors:

1. Good selection of standard functional blocks
2. Compatibility with other IC families
3. Availability of wired functions and complementary outputs
4. Large fanout

Each IC manufacturer provides a selection of standard functional blocks such as multiplexers, registers, ROMs, and multipliers. With a good selection of such devices available, the design can often be simplified and the costs reduced.

Two IC families are said to be *compatible* if the basic gates of the two families have similar characteristics under the same operating conditions. A typical system of moderate complexity is almost invariably designed with more than one IC family. If compatible families are employed throughout the system, the interfacing of the various subsystems is considerably simplified.

Wired functions, e.g., WIRED-AND, can sometimes be obtained by simply connecting together the outputs of two ICs. Since these functions are obtained without the need for additional hardware, they can often reduce the power dissipation as well as the costs. The availability of complementary outputs simplifies the interconnection of subsystems.

If a load is connected at the output of a gate, current will flow to the load, and since the gate has a finite output resistance, the output voltage level will be reduced. If the loading becomes excessive, the specified output voltage levels cannot be maintained and the gate will either malfunction or become too sensitive to electrical noise. Normally, a gate is required to drive a number of other gates. Hence it is convenient to measure the load on a gate in terms of *unit loads*, where one unit load represents the loading produced by a single gate input. The number of unit loads that a gate can drive and still maintain the specified output voltage levels is said to be the *fanout* of the gate. A large fanout is a desirable feature since it eliminates the need for power or buffer amplifiers.

Finally, the cost of the IC family should be as low as possible, for obvious reasons. It should be mentioned, however, that layout, shielding, cooling, and interfacing costs must be included in the cost estimate on a pro rata basis because they tend to depend to some extent on the IC family.

T²L Family

The most frequently used IC family is T²L because of its good allround character-istics. Its features are good load-drive capability, high noise immunity, good speed–power-dissipation product, and availability of a good selection of standard functional blocks. On the negative side, T²L devices tend to be somewhat noisy.

Four subfamilies of T²L are available as follows:

1. Standard T²L
2. Standard low-power T²L
3. Schottky-clamped T²L
4. Schottky-clamped low-power T²L.

The Schottky-clamped subfamilies are designed specifically for high-speed appli-cations and are more expensive than standard T²L.

ECL Family

ECL is the fastest and as is to be expected the most expensive form of logic. It features good noise immunity, low noise generation, excellent load-drive capabil-ity, and availability of WIRED-OR and complementary outputs. Its main disadvan-tages are noncompatibility with other forms of logic and high power dissipation.

MOS Family

MOS logic is the cheapest form of logic but also the slowest. It consists of three subfamilies:

1. p-channel MOS (PMOS)
2. n-channel MOS (NMOS)
3. Complementary MOS (CMOS)

The first two are similar except that NMOS is somewhat faster. CMOS features extremely low power dissipation and is faster than NMOS but also more expensive.

I²L Family

I²L is a relatively recent form of logic and its impact is yet to be assessed. It can be as fast as T²L and as inexpensive as NMOS, and its power dissipation can be as low as that of CMOS. Furthermore, it can be mixed with other forms of bipolar logic such as T²L for the fabrication of LSI devices. The expectations are therefore high.

The features of the various IC families are summarized in Table 14.7.

Table 14.7 Comparison of IC families [5–9]

Type	Gate delay ns	Gate power dissipation, mW	Noise immunity†	Generated noise	Cost
NMOS	100–300	1.0	L	M	L
CMOS	30–100	1.0	H	LM	L
I²L	10–35	1.0	M	M	L
Standard T²L	10–12	12	H	MH	M
Schottky T²L	3–6	22	H	H	MH
ECL	1–3	30–55	M	LM	H

† L = low, M = medium, H = high.

14.7 DIGITAL-FILTER IMPLEMENTATIONS

The hardware implementation of digital filters can be accomplished by using the following design procedure:

1. Choose filter structure.
2. Choose between fixed-point and floating-point arithmetic.
3. Choose number representation, e.g. signed magnitude, two's complement.
4. Choose between serial and parallel processing.
5. Choose arithmetic devices.

Numerous options are obviously open to the designer, and as a consequence the implementation can assume a variety of forms.

We discuss here three specific fixed-point implementations due to Jackson, Kaiser, and McDonald [10], Peled and Liu [11], and Monkewich and Steenaart [12–13].

Jackson-Kaiser-McDonald Implementation

In the Jackson-Kaiser-McDonald (JKM) approach the filter is implemented as a connection of cascade (or parallel) second-order canonic sections. Signals are assumed to be in serial two's-complement representation with the least significant bit leading in the time slot. The multiplier coefficients, however, are stored in signed-magnitude form. Addition is carried out by using the two's-complement adder of Fig. 14.21. Multiplication, on the other hand, is carried out by using the magnitude multiplier of Fig. 14.24. The necessary interfacing between adders and multipliers can be implemented by employing two two's complementers, as illustrated in Fig. 14.27. The unit delays can be implemented by using serial-input, serial-output shift registers. As can be seen in Fig. 14.27, the N bits of the multiplicand must be available before multiplication can begin, and since the multiplier will introduce a delay of K bits (if $m = m_0 . m_1 \cdots m_K$), the unit-delay registers must

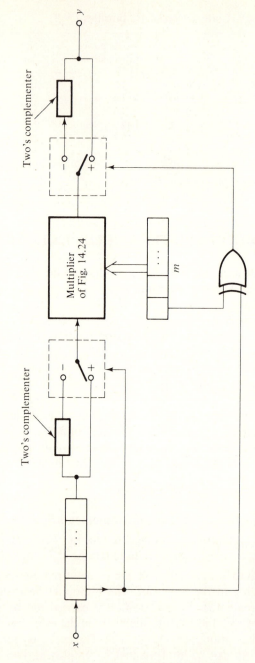

Figure 14.27 Multiplication of two's-complement numbers by using a magnitude multiplier.

424

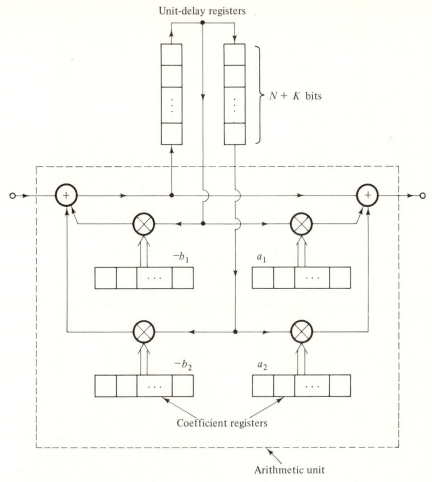

Figure 14.28 The basic JKM section.

have a length of at least $N + K$ bits so that data can be routed around the filter section in synchronism. The implementation of the basic section is depicted in Fig. 14.28.

If the input bit rate (sampling rate times the number of bits per sample) is below the capability of the digital circuits, multiplexing can be used to increase the cost effectiveness of the implementation. Three types of multiplexing are possible:

1. The same multisection implementation can be used to process a number of distinct signals simultaneously.
2. A single section can be used to effect a high-order cascade (or parallel) filter.
3. A single section can be used to effect a high-order filter which can be used to process a number of distinct signals.

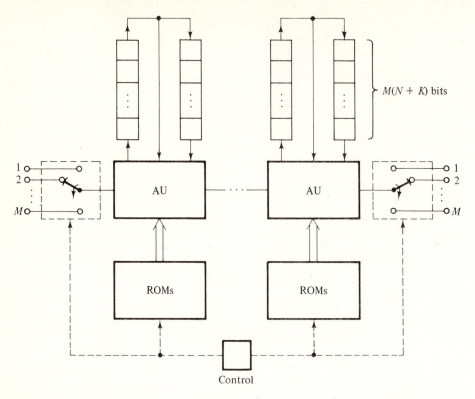

Figure 14.29 Type 1 multiplexing.

Type 1 multiplexing is illustrated in Fig. 14.29. Each section is essentially of the type shown in Fig. 14.28 except that the length of each register is increased to $M(N + K)$ bits and each coefficient register is replaced by a ROM. The input samples from the M sources are interleaved sample by sample and fed serially into the filter. The output samples emerge in the same interleaved order and can thus be easily separated.

Type 2 multiplexing can be achieved by using the scheme of Fig. 14.30. In this case the section output is routed back to the input repeatedly, and each time a new set of coefficients is loaded into the arithmetic unit. This routine ends when the Mth section is effected, at which time the processed signal is routed to the output.

Type 3 multiplexing can be achieved by replacing the multisection filter of Fig. 14.29 by the scheme of Fig. 14.30.

Peled-Liu Implementation

The Peled-Liu (PL) implementation, like the preceding one, is based on two's-complement serial representation. Its essential feature is that all multiplications and additions of each section are performed simultaneously through the use of a

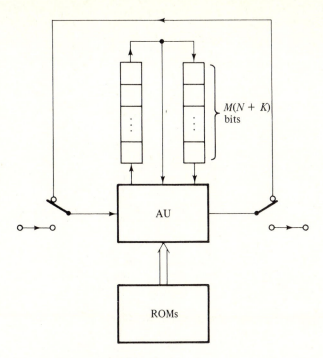

Figure 14.30 Type 2 multiplexing.

ROM together with an adder-subtractor circuit. The basis of this implementation is as follows.

Consider the second-order difference equation

$$y(n) = a_0 x(n) + a_1 x(n-1) + a_2 x(n-2) - b_1 y(n-1) - b_2 y(n-2) \quad (14.6)$$

and assume that

$$\tilde{x}(n) = x_0(n).x_1(n)x_2(n) \cdots x_L(n)$$

$$\tilde{y}(n) = y_0(n).y_1(n)y_2(n) \cdots y_L(n)$$

are the two's complements of $x(n)$ and $y(n)$, respectively. Also let $\tilde{a}_i$ and $\tilde{b}_i$ be the two's complements of a_i and b_i. Numbers $x(n)$ and $y(n)$ can be expressed as

$$x(n) = -x_0(n) + \sum_{i=1}^{L} x_i(n)2^{-i}$$

$$y(n) = -y_0(n) + \sum_{i=1}^{L} y_i(n)2^{-i}$$

and hence Eq. (14.6) can be put in the form

$$y(n) = \sum_{i=1}^{L} 2^{-i}[a_0 x_i(n) + a_1 x_i(n-1) + a_2 x_i(n-2)$$

$$- b_1 y_i(n-1) - b_2 y_i(n-2)] - [a_0 x_0(n) + a_1 x_0(n-1)$$

$$+ a_2 x_0(n-2) - b_1 y_0(n-1) - b_2 y_0(n-2)]$$

Therefore, by analogy with Eqs. (14.4) and (14.5)

$$\tilde{y}(n) = \sum_{i=1}^{L} 2_2^{-i} F_i - F_0$$

where

$$F_i = \tilde{a}_0 x_i(n) + \tilde{a}_1 x_i(n-1) + \tilde{a}_2 x_i(n-2) + (-\tilde{b}_1) y_i(n-1) + (-\tilde{b}_2) y_i(n-2)$$

$$(14.7)$$

The two's complement of $y(n)$ can thus be formed by using the following algorithm:

1. Clear accumulator register.
2. Evaluate F_i for $i = L$.
3. Add F_i(two's-complement addition) to accumulator content.
4. Shift accumulator content to the right by one bit (two's-complement shift).
5. Repeat steps 2 to 4 for $i = L-1, L-2, \ldots, 1$.
6. Evaluate F_0.
7. Subtract F_0 from accumulator content (two's-complement subtraction).

The 32 possible values of F_i can readily be generated by using a 32xL ROM, and hence the above algorithm can be implemented as depicted in Fig. 14.31. This implementation was first reported in a patent by Croisier, Esteban, Levilion, and Rizo [14].

Example 14.4 Design the ROM of Fig. 14.31 if

$$H(z) = \frac{a_0 z^2 + a_1 z + a_2}{z^2 + b_1 z + b_2}$$

where $\quad a_0 = a_2 = 0.2222545 \qquad a_1 = -0.4445091$

$$b_1 = 0.1561833 \qquad b_2 = 0.04520149$$

Assume that $L = 7$ and quantize the coefficients by rounding.

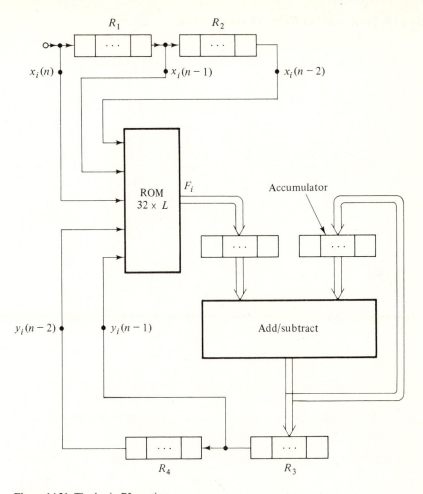

Figure 14.31 The basic PL section.

SOLUTION

$$a_0 = a_2 = 0.0011100 \qquad \tilde{a}_0 = \tilde{a}_2 = 0.0011100$$

$$a_1 = -0.0111001 \qquad \tilde{a}_1 = 1.1000111$$

$$b_1 = 0.0010100 \qquad (-\tilde{b}_1) = 1.1101100$$

$$b_2 = 0.0000110 \qquad (-\tilde{b}_2) = 1.1111010$$

The truth table of the ROM can be constructed as in Table 14.8 by using Eq. (14.7).

Table 14.8 Truth table of ROM (Example 14.4)

$x_i(n) \cdots y_i(n-2)$	F_i	$x_i(n) \cdots y_i(n-2)$	F_i
0 0 0 0 0	0 0 0 0 0 0 0	1 0 0 0 0	0 0 0 1 1 1 0 0
0 0 0 0 1	1 1 1 1 1 0 1 0	1 0 0 0 1	0 0 0 1 0 1 1 0
0 0 0 1 0	1 1 1 0 1 1 0 0	1 0 0 1 0	0 0 0 0 1 0 0 0
0 0 0 1 1	1 1 1 0 0 1 1 0	1 0 0 1 1	0 0 0 0 0 0 1 0
0 0 1 0 0	0 0 0 1 1 1 0 0	1 0 1 0 0	0 0 1 1 1 0 0 0
0 0 1 0 1	0 0 0 1 0 1 1 0	1 0 1 0 1	0 0 1 1 0 0 1 0
0 0 1 1 0	0 0 0 0 1 0 0 0	1 0 1 1 0	0 0 1 0 0 1 0 0
0 0 1 1 1	0 0 0 0 0 0 1 0	1 0 1 1 1	0 0 0 1 1 1 1 0
0 1 0 0 0	1 1 0 0 0 1 1 1	1 1 0 0 0	1 1 1 0 0 0 1 1
0 1 0 0 1	1 1 0 0 0 0 0 1	1 1 0 0 1	1 1 0 1 1 1 0 1
0 1 0 1 0	1 0 1 1 0 0 1 1	1 1 0 1 0	1 1 0 0 1 1 1 1
0 1 0 1 1	1 0 1 0 1 1 0 1	1 1 0 1 1	1 1 0 0 1 0 0 1
0 1 1 0 0	1 1 1 0 0 0 1 1	1 1 1 0 0	1 1 1 1 1 1 1 1
0 1 1 0 1	1 1 0 1 1 1 0 1	1 1 1 0 1	1 1 1 1 1 0 0 1
0 1 1 1 0	1 1 0 0 1 1 1 1	1 1 1 1 0	1 1 1 0 1 0 1 1
0 1 1 1 1	1 1 0 0 1 0 0 1	1 1 1 1 1	1 1 1 0 0 1 0 1

The above approach can be extended to the implementation of parallel or cascade structures. For the sixth-order filter of Fig. 14.32a

$$y(n) = \lambda_1 u(n) + \lambda_2 v(n) + \lambda_3 w(n)$$

and so

$$\tilde{y}(n) = \sum_{i=1}^{L} 2\bar{2}^i F_i - F_0$$

where

$$F_i = \tilde{\lambda}_1 u_i(n) + \tilde{\lambda}_2 v_i(n) + \tilde{\lambda}_3 w_i(n)$$

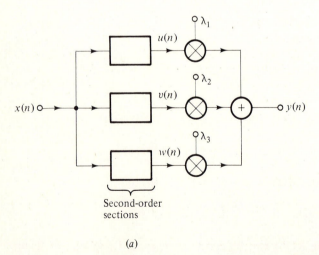

(a)

Figure 14.32 Parallel implementation of a sixth-order filter: (a) parallel structure;

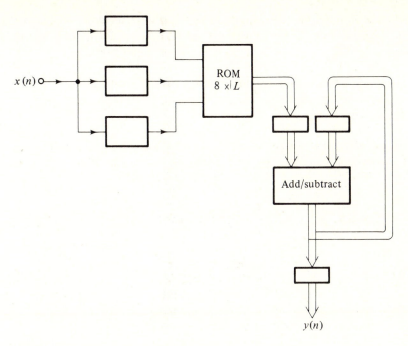

Figure 14.32 (*b*) corresponding implementation.

Hence the three scaling multipliers as well as the output adder can be implemented by using an $8 \times L$ ROM together with an adder-subtractor circuit. This possibility is depicted in Fig. 14.32*b*.

A cascade implementation of a fourth-order filter [15] is illustrated in Fig. 14.33. At the start of the cycle, the switches are in position 1. Hence ROM 1 generates the values of F_i for the first section, which are added to the accumulator in sequence. At the same time the contents of R_3 and R_4 are recirculated, while the content of R_0, namely $y(n)$, is transferred to the output in serial fashion. Once the first section is effected, the content of the accumulator, namely $v(n)$, is transferred to R_0 in parallel fashion and the switches change to position 2. The same process is repeated for the second section except that $v(n)$ is used to update the contents of R_3.

Monkewich-Steenaart Implementation

In the Monkewich-Steenaart (MS) implementation as in the Peled-Liu implementation, the need for multipliers is eliminated through the use of ROMs. The basic second-order implementation is depicted in Fig. 14.34, where the adder may operate in serial or parallel mode. The ROMs are required to store all the possible signal-coefficient products, and consequently the necessary memory capacity is considerable. For example, if signals and coefficients were in terms of a 13-bit representation, the required size of each ROM would be $13 \times 2^{13} = 106{,}496$ bits. However, by subjecting the input signal to logarithmic quantization, the number of possible signal values can be decreased to 2^7 or 2^8 without degrading the signal-to-noise ratio beyond that encountered in standard PCM systems. In this way the required memory capacity can be reduced to a feasible level.

Logarithmic quantization can be incorporated by including quantizers Q_A and Q_B in Fig. 14.34. The ROMs can be programmed to generate the true signal-coefficient products so that a standard adder can be used. The quantizer at the output of the adder converts the feedback signals back into quantized representation.

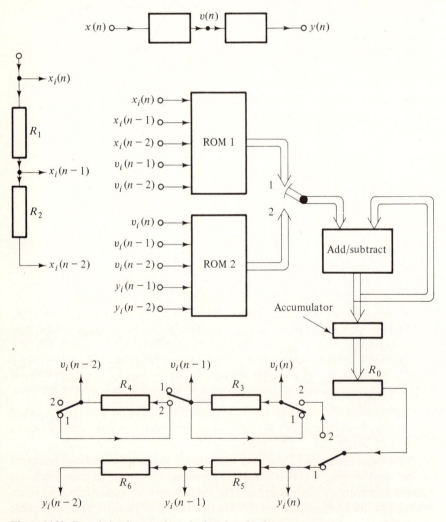

Figure 14.33 Cascade implementation of a fourth-order filter.

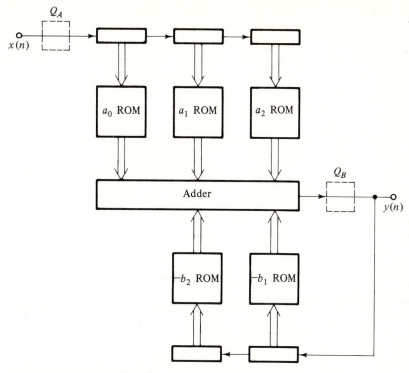

Figure 14.34 The basic MS section.

Comparison

Some comparisons between the JKM and PL implementations reported in [11] show the latter to be superior with respect to power dissipation, speed, and costs, at least on the basis of present-day technology. However, the various improvements are brought about by sacrificing the flexibility inherent in the JKM implementation. For applications where fixed filtering tasks are needed, flexibility is not an important issue, and the choice between the two should be a PL implementation. On the other hand, for applications where time-variable or programmable filters are needed, the obvious choice is a JKM implementation.

The MS implementation is essentially an alternative to the PL implementation. It is generally faster because the number of addition cycles (assuming a serial adder) is reduced to the number of coefficients. However, it is also more demanding in terms of hardware.

An advantage of the JKM implementation over the other two is that a universal arithmetic unit can be constructed in LSI form, which can be employed in a multiplicity of widely differing applications (like an operational amplifier). Hence in the long run the JKM implementation may turn out to be the more economical one.

14.8 APPLICATIONS OF DIGITAL FILTERS

Digital filters in the form of software have been used extensively in the past and will no doubt continue to be used in the future at a progressively increasing rate. Typical applications are:

1. Data smoothing and prediction
2. Image enhancement
3. Pattern recognition
4. Speech processing
5. Processing of telemetry signals
6. Processing of biomedical signals
7. Simulation of analog systems

Programmable hardware digital filters have already made their appearance in the form of digital-signal processors such as FFT processors, frequency synthetizers, and wave analyzers. Many other applications are anticipated for the future, especially in the domain of instrumentation.

Nonprogrammable digital filters are currently considered as a possible replacement of analog filters in many communication subsystems [10, 12, 16–18]. With present-day technology, digital filters are still more expensive than analog filters except for some low-frequency applications, where extensive multiplexing of hardware is possible. Nevertheless, the gains in accuracy and stability of operation may sometimes justify the extra expense. With the present trends in the fabrication of LSI circuits continuing, the cost of digital hardware is bound to drop drastically in the not too distant future. At that time digital filters will become more attractive than analog filters in many more applications. It is not expected that digital filters will replace analog filters altogether, e.g., microwave filters! Instead, like *LC*, crystal, mechanical, monolithic, and active filters, digital filters will become an invaluable addition to the bag of tricks available to the filter designer.

REFERENCES

1. L. R. Rabiner and B. Gold, "Theory and Application of Digital Signal Processing," chap. 11, Prentice-Hall, Englewood Cliffs, N.J., 1975.
2. F. J. Hill and G. R. Peterson, "Introduction to Switching Theory and Logical Design," Wiley, New York, 1974.
3. H. T. Nagle Jr., B. D. Carroll, and J. D. Irwin, "An Introduction to Computer Logic," Prentice-Hall, Englewood Cliffs, N.J., 1975.
4. G. K. Kostopoulos, "Digital Engineering," Wiley, New York, 1975.
5. L. S. Garrett, Integrated-Circuit Digital Logic Families, *IEEE Spectrum*, vol. 7, pp. 46–58, October 1970.
6. L. S. Garrett, Integrated-Circuit Digital Logic Families, *IEEE Spectrum*, vol. 7, pp. 63–72, November 1970.
7. L. S. Garrett, Integrated-Circuit Digital Logic Families, *IEEE Spectrum*, vol. 7, pp. 30–42, December 1970.

8. R. L. Horton, I^2L Takes Bipolar Integration a Significant Step Forward, *Electronics*, pp. 83–90, February 6, 1975.

9. L. Altman, The New LSI, *Electronics*, pp. 81–92, July 10, 1975.

10. L. B. Jackson, J. F. Kaiser, and H. S. McDonald, An Approach to the Implementation of Digital Filters, *IEEE Trans. Audio Electroacoust.*, vol. AU-16, pp. 413–421, September 1968.

11. A. Peled and B. Liu, A New Hardware Realization of Digital Filters, *IEEE Trans. Acoust., Speech, Signal Process.*, vol. ASSP-22, pp. 456–462, December 1974.

12. O. Monkewich and W. Steenaart, Companding for Digital Filters, *Proc. 1975 IEEE Int. Symp. Circuits Syst.*, pp. 68–71.

13. O. Monkewich and W. Steenaart, Stored Product Digital Filtering with Nonlinear Quantization, *Proc. 1976 IEEE Int. Symp. Circuits Syst.*, pp. 157–160.

14. A. Croisier, D. J. Esteban, M. E. Levilion, and V. Rizo, Digital Filter for PCM Encoded Signals, U.S. Patent 3777130, December 3, 1973.

15. A. Peled and B. Liu, " Digital Signal Processing," app. 5.1, Wiley, New York, 1976.

16. C. F. Kurth, SSD/FDM Utilizing TDM Digital Filters, *IEEE Trans. Commun. Technol.*, vol. COM-19, pp. 63–71, February 1971.

17. S. L. Freeny, R. B. Kieburtz, K. V. Mina, and S. K. Tewksbury, Design of Digital Filters for an All Digital Frequency Division Multiplex-Time Division Multiplex Translator, *IEEE Trans. Circuit Theory*, vol. CT-18, pp. 702–711, November 1971.

18. S. K. Tewksbury, Special Purpose Hardware Implementation of Digital Filters, *Proc. 1973 IEEE Int. Symp. Circuits Syst.*, pp. 418–421.

ADDITIONAL REFERENCES

Bergland, G. D.: Fast Fourier Transform Hardware Implementations: An Overview, *IEEE Trans. Audio Electroacoust.*, vol. AU-17, pp. 104–108, June 1969.

Cho, Y., and F. G. Popp: A High-Speed Arithmetic Unit for Digital Filter Implementation, *1967 NEREM Rec.*, pp. 114–115, November 1967.

Gold, B., and T. Bially: Parallelism in Fast Fourier Transform Hardware, *IEEE Trans. Audio Electroacoust.*, vol. AU-21, pp. 5–16, February 1973.

Groginsky, H. L., and G. A. Works: A Pipeline Fast Fourier Transform, *IEEE Trans. Comput.*, vol. C-19, pp. 1015–1019, November 1970.

Heute, U.: Hardware Considerations for Digital FIR Filters Especially with Regard to Linear Phase, *Arch. Elektron. Uebertrag*, vol. 29, pp. 116–120, March 1975.

Jackson, L. B., J. F. Kaiser, and H. S. McDonald: Implementation of Digital Filters, *1968 IEEE Int. Conv. Dig.*, p. 213.

Kaul, P.: An All Digital Telephony Signalling Module, *Proc. 1975 IEEE Int. Symp. Circuits Syst.*, pp. 392–395.

Liu, B., and A. Peled: A New Hardware Realization of High-Speed Fast Fourier Transformers, *IEEE Trans. Acoust., Speech, Signal Process.*, vol. ASSP-23, pp. 543–547, December 1975.

McDonald, H. S.: Impact of Large-Scale Integrated Circuits on Communication Equipment, *Proc. Natl. Electron. Conf.*, pp. 569–572, December 1968.

Nouta, R.: On the Efficient Wave Digital Filter Realization Using Programmable Hardware, *1977 IEEE Int. Symp. Circuits Syst.*, pp. 49–51.

Peled, A., and B. Liu: A New Approach to the Realization of Nonrecursive Digital Filters, *IEEE Trans. Audio Electroacoust.*, vol. AU-21, pp. 477–484, December 1973.

——, ——, and K. Steiglitz: A Note on Implementation of Digital Filters, *IEEE Trans. Acoust., Speech, Signal Process.*, vol. ASSP-23, pp. 387–389, August 1975.

Tierney, J., C. M. Rader, and B. Gold: A Digital Frequency Synthesizer, *IEEE Trans. Audio Electroacoust.*, vol. AU-19, pp. 48–56, March 1971.

PROBLEMS

14.1 Prove Theorems 14.4 and 14.5.

14.2 Construct the truth tables for the following switching functions:

(a) $F_1 = (A + \bar{B})(A + C + D)\bar{D}$

(b) $F_2 = \bar{A}\bar{B}(AC + \bar{B}) + (A + B)[(\overline{\bar{A}\bar{B}C})\downarrow(\overline{\bar{A}BC})]$

14.3 Figure P14.3 depicts a T^{2}L gate. Explain its operation.

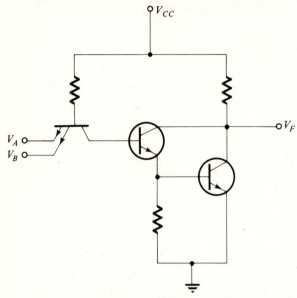

Figure P14.3

14.4 The circuit of Fig. P14.4 is called a *voter circuit*. Deduce the reason for the name.

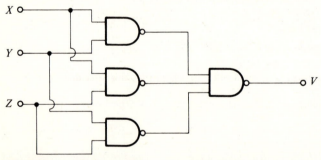

Figure P14.4

14.5 Binary numbers $x_1 x_0$ and $y_1 y_0$ are applied at the inputs of the circuit in Fig. P14.5. Show that $W = 1$ if $x_1 x_0 = y_1 y_0$.

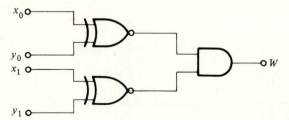

Figure P14.5

14.6 (*a*) Design a three-input decoder.

(*b*) By using the decoder in part (*a*) construct an eight-input selection switch.

(*c*) Now construct an eight-output distribution switch.

14.7 Realize the following switching functions by using an eight-input multiplexer in each case.

(*a*) $F_1 = (x_2 + x_1)\bar{x}_0 + x_2 x_1 + \bar{x}_2 \bar{x}_1 x_0$

(*b*) $F_2 = (x_2 + \bar{x}_0)(\bar{x}_1 + x_0)(\bar{x}_2 + x_1 + \bar{x}_0)$

14.8 Design a full adder by using a minimum number of NAND gates.

14.9 (*a*) Design a (3×3)-bit parallel multiplier using the full adder of Fig. 14.7*b*.

(*b*) Estimate the multiplication time if $T_g = 10$ ns.

14.10 A ROM is required to perform the following code conversion:

Input	Output
0 0 0	0 0 1 0
0 0 1	0 1 1 0
0 1 0	0 1 1 1
0 1 1	0 1 0 1
1 0 0	0 1 0 0
1 0 1	1 1 0 0
1 1 0	1 1 0 1
1 1 1	1 1 1 1

Design the encoder using diodes.

14.11 A ROM is required which will convert an integer N in the range 0 to 15 into a corresponding integer M such that

$$M = \text{Int} \left(255 \sin \frac{N\pi}{16} \right)$$

Construct the truth table of the ROM.

14.12 Design a (6×4)-bit parallel multiplier using ROMs and 2-bit adders.

14.13 Derive the transition table of the delay flip-flop.

14.14 A clocked SR flip-flop is used as in Fig. P14.14. Derive the transition table of the circuit.

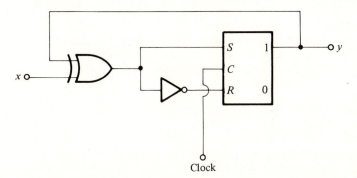

Clock   **Figure P14.14**

14.15 (*a*) Figure P14.15*a* shows a sequential circuit. Explain its operation.

(*b*) Repeat part (*a*) for the circuit of Fig. P14.15*b*.

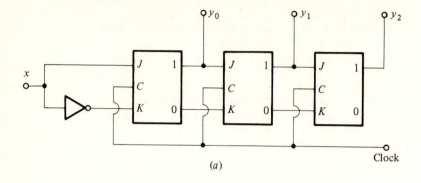

(a)

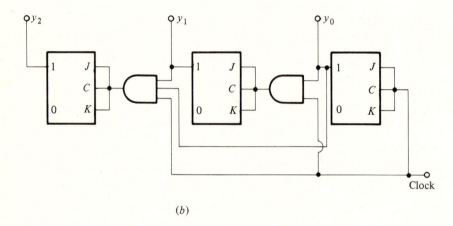

(b)

Figure P14.15

14.16 Form the following products by using the two's-complement multiplication algorithm described in Sec. 14.5:

(a) 0.703125(−0.34375)

(b) −0.546875(−0.21875)

14.17 A sixth-order filter is to be implemented by using three individual JKM sections connected in cascade. Signals and coefficients are to be in terms of 8- and 12-bit representations, respectively, and the sampling frequency is to be 24 kHz.

(a) Determine the length of the delay registers.

(b) Find the clock rate.

14.18 The filter of Prob. 14.17 is to be implemented using type 2 multiplexing.

(a) Find the length of the delay registers.

(b) Find the clock rate.

14.19 In the implementation of Prob. 14.17 it is found necessary to use additional delay flip-flops between multiplier sections, as depicted in Fig. P14.19, in order to compensate for the delay of the full adders.

(a) Find the length of the delay registers.

(b) Find the clock rate.

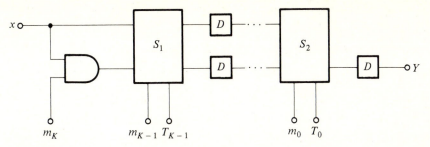

Figure P14.19

14.20 The compensation of Prob. 14.19 is found necessary in the implementation of Prob. 14.18.
  (a) Find the length of the delay registers.
  (b) Find the clock rate.

14.21 The maximum clock rate for standard T^2L is 20 MHz. Find the maximum number of second-order sections that can be effected by means of type 2 JKM multiplexing. Assume 8-bit signals, 12-bit coefficients, and a sampling frequency of 24 kHz. No delay compensation is to be employed in the multipliers.

14.22 Repeat Prob. 14.21, assuming that delay compensation is to be employed.

14.23 Find the widest signal bandwidth that can be processed by a twelfth-order type 2 JKM implementation. Assume standard T^2L with a maximum clock rate of 20 MHz. Signals and coefficients are to be in terms of 8- and 12-bit representations, respectively. Use delay compensation for the multipliers (see Fig. P14.19).

14.24 Design the ROM for the PL implementation of Fig. 14.31 if

$$H(z) = \frac{a_0 z^2 + a_1 z + a_2}{z^2 + b_1 z + b_2}$$

where
$$a_0 = a_2 = 0.3085386 \qquad a_1 = -0.6170772$$

$$b_1 = 0.2168171 \qquad b_2 = 0.4509715$$

Assume that $L = 7$ and quantize the coefficients by rounding.

14.25 The filter of Prob. 14.24 is to be constructed by means of T^2L integrated circuits. Find the widest signal bandwidth that can be processed. Assume that the access time of the ROM is 50 ns and that addition of two partial sums can be performed in 40 ns. The generation of F_i can be carried out concurrently with a corresponding partial addition.

ELLIPTIC FUNCTIONS

A.1 INTRODUCTION

The Jacobian elliptic functions are derived by employing the Legendre elliptic integral of the first kind. Their theory is quite extensive and is discussed in detail by Bowman [1] and Hancock [2, 3]. We provide here a brief, but for our purposes adequate, treatment of this theory [4].

A.2 ELLIPTIC INTEGRAL OF THE FIRST KIND

The elliptic integral of the first kind can be expressed as

$$u \equiv u(\phi, k) = \int_0^\phi \frac{d\theta}{\sqrt{1 - k^2 \sin^2 \theta}} \tag{A.1}$$

where $0 \leq k < 1$. The parameter k is called the *modulus*, and the upper limit of integration ϕ is called the *amplitude* of the integral. Evidently, for a real value of ϕ, $u(\phi, k)$ is real and represents the area bounded by the curve

$$I = \frac{1}{\sqrt{1 - k^2 \sin^2 \theta}}$$

and the vertical lines $\theta = 0$ and $\theta = \phi$. Plots of I and $u(\phi, k)$ for $k = 0.995$ are shown in Fig. A.1. The integrand I has minima equal to unity at $\theta = 0, \pi, 2\pi, \ldots$ and maxima equal to $1/\sqrt{1 - k^2}$ at $\theta = \pi/2, 3\pi/2, \ldots$. In effect, I is a periodic function of θ with a period π. The area bounded by lines $\theta = n\pi/2$ and

$\theta = (n + 1)\pi/2$ is constant for any n because of the symmetry of I and is equal to the area bounded by lines $\theta = 0$ and $\theta = \pi/2$. This area is referred to as the *complete elliptic integral of the first kind* and is given by

$$u\left(\frac{\pi}{2}, k\right) = K = \int_0^{\pi/2} \frac{d\theta}{\sqrt{1 - k^2 \sin^2 \theta}} \tag{A.2}$$

(see Fig. A.1).

As a consequence of the periodicity and symmetry of I we can write

$$u(n\pi + \phi_1, k) = 2nK + u(\phi_1, k) \qquad u\left(\frac{\pi}{2} + \phi_1, k\right) = 2K - u\left(\frac{\pi}{2} - \phi_1, k\right)$$

where $0 \le \phi_1 < \pi/2$. That is, the elliptic integral for a given k and any real ϕ can be determined from a table giving the values of the integral in the interval $0 \le \phi < \pi/2$.

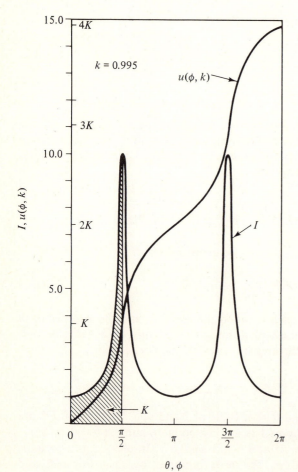

Figure A.1 Plots of I versus θ and $u(\phi, k)$ versus ϕ.

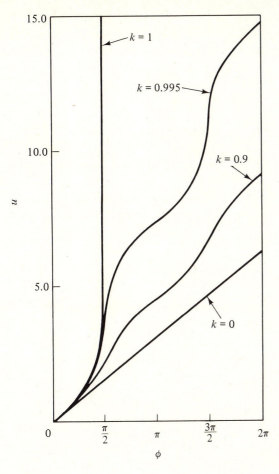

Figure A.2 Plots of u versus ϕ for various values of k.

If $k = 0$, Eq. (A.1) gives

$$u(\phi, 0) = \int_0^\phi d\theta = \phi$$

and if $k = 1$,

$$u(\phi, 1) = \int_0^\phi \frac{d\theta}{\cos \theta} = \ln \left[\tan \left(\frac{\pi}{4} + \frac{\phi}{2} \right) \right]$$

according to standard tables. Hence $u(\phi, 0)$ increases linearly with ϕ, whereas $u(\phi, 1)$ is discontinuous at $\phi = \pi/2$. For $0 \le \phi < \pi/2$

$$u(\phi, 0) \le u(\phi, k) \le u(\phi, 1)$$

as can be seen in Fig. A.2.

A.3 ELLIPTIC FUNCTIONS

Figure A.2 demonstrates a one-to-one correspondence between u and ϕ. Thus for a given pair of values (u, k) there corresponds a unique amplitude ϕ such that

$$\phi = f(u, k)$$

The Jacobian elliptic functions are defined as

$$\text{sn } (u, k) = \sin \phi \tag{A.3}$$

$$\text{cn } (u, k) = \cos \phi \tag{A.4}$$

$$\text{dn } (u, k) = \sqrt{1 - k^2 \sin^2 \phi} \tag{A.5}$$

Many of the properties of elliptic functions follow directly from the properties of trigonometric functions. For example, we can write

$$\text{sn}^2 (u, k) + \text{cn}^2 (u, k) = 1 \tag{A.6}$$

and $$k^2 \text{ sn}^2 (u, k) + \text{dn}^2 (u, k) = 1 \tag{A.7}$$

and so forth.

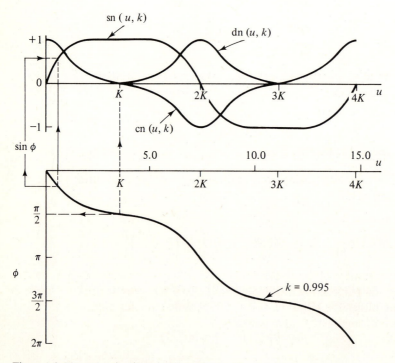

Figure A.3 Plots of sn (u, k), cn (u, k), and dn (u, k) versus u.

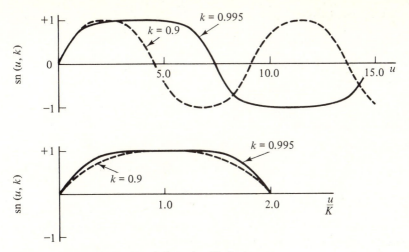

Figure A.4 Effect of variations in k on the elliptic sine.

Plots of the elliptic functions versus u can be constructed as in Fig. A.3. As can be seen, sn (u, k), cn (u, k) and dn (u, k) are periodic functions of u with periods of $4K$, $4K$, and $2K$, respectively, i.e.,

$$\text{sn } (u + 4mK, k) = \text{sn } (u, k) \tag{A.8}$$

$$\text{cn } (u + 4mK, k) = \text{cn } (u, k) \tag{A.9}$$

$$\text{dn } (u + 2mK, k) = \text{dn } (u, k) \tag{A.10}$$

Variations in k tend to change the shape and period of the elliptic functions, as illustrated in Fig. A.4.

If $k = 0$,

$$u(\phi, 0) = \phi$$

and so   sn $(u, 0) = \text{sn } (\phi, 0) = \sin \phi$ cn $(u, 0) = \text{cn } (\phi, 0) = \cos \phi$

i.e., the elliptic sine and cosine are generalizations of the conventional sine and cosine, respectively.

A.4 IMAGINARY ARGUMENT

Thus far the argument of the elliptic functions, namely u, has been assumed to be a real quantity. By performing the integration of Eq. (A.1) over an appropriate path in a complex plane, the elliptic integral can assume complex values. Let us consider the case of imaginary value whereby

$$jv = \int_0^{\psi} \frac{d\theta}{\sqrt{1 - k^2 \sin^2 \theta}} \tag{A.11}$$

As in Sec. A.3, we can define

$$\text{sn } (jv, k) = \sin \psi \qquad\qquad (A.12)$$

$$\text{cn } (jv, k) = \cos \psi \qquad\qquad (A.13)$$

$$\text{dn } (jv, k) = \sqrt{1 - k^2 \sin^2 \psi} \qquad\qquad (A.14)$$

These functions can be expressed in terms of elliptic functions which have real arguments, as we now show.

By applying the transformations

$$\sin \theta = j \tan \theta' \qquad \sin \psi = j \tan \psi' \qquad\qquad (A.15)$$

in Eq. (A.11), we have

$$jv = \int_0^{\psi'} \frac{j \, d\theta'}{\sqrt{1 - \sin^2 \theta' + k^2 \sin^2 \theta'}}$$

Alternatively

$$v = \int_0^{\psi'} \frac{d\theta'}{\sqrt{1 - (k')^2 \sin^2 \theta'}}$$

where k', given by

$$k' = \sqrt{1 - k^2}$$

is called the *complementary modulus*. Now from Sec. A.3

$$\text{sn } (v, k') = \sin \psi' \qquad\qquad (A.16)$$

$$\text{cn } (v, k') = \cos \psi' \qquad\qquad (A.17)$$

$$\text{dn } (v, k') = \sqrt{1 - (k')^2 \sin^2 \psi'} \qquad\qquad (A.18)$$

and, therefore, from Eqs. (A.12) to (A.18),

$$\text{sn } (jv, k) = j \tan \psi' = j \frac{\sin \psi'}{\cos \psi'} = \frac{j \, \text{sn } (v, k')}{\text{cn } (v, k')} \qquad\qquad (A.19)$$

$$\text{cn } (jv, k) = \frac{1}{\text{cn } (v, k')} \qquad\qquad (A.20)$$

$$\text{dn } (jv, k) = \frac{\text{dn } (v, k')}{\text{cn } (v, k')} \qquad\qquad (A.21)$$

By analogy with Eq. (A.2), the complementary complete integral of the first kind is given by

$$K' = \int_0^{\pi/2} \frac{d\theta}{\sqrt{1 - (k')^2 \sin^2 \theta}}$$

This has the same interpretation as K; that is, it is the quarter-period of sn (v, k') and cn (v, k') or the half-period of dn (v, k').

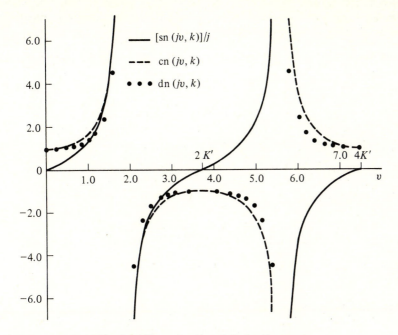

Figure A.5 Plots of $[\text{sn}\,(jv, k)]/j$, $\text{cn}\,(jv, k)$, and $\text{dn}\,(jv, k)$ versus v.

The functions $\text{sn}\,(jv, k)$, $\text{cn}\,(jv, k)$, and $\text{dn}\,(jv, k)$ are periodic functions of jv, as can be seen in Fig. A.5, with periods of $j2K'$, $j4K'$, and $j4K'$, respectively, i.e.,

$$\text{sn}\,(jv + j2nK', k) = \text{sn}\,(jv, k)$$

$$\text{cn}\,(jv + j4nK', k) = \text{cn}\,(jv, k)$$

$$\text{dn}\,(jv + j4nK', k) = \text{dn}\,(jv, k)$$

A.5 FORMULAS

Elliptic functions, like trigonometric functions, are interrelated by many useful formulas. The most basic one is the *addition formula*, which is of the form

$$\text{sn}\,(z_1 + z_2, k) = \frac{\text{sn}\,(z_1, k)\,\text{cn}\,(z_2, k)\,\text{dn}\,(z_2, k) + \text{cn}\,(z_1, k)\,\text{sn}\,(z_2, k)\,\text{dn}\,(z_1, k)}{D}$$

$$\text{(A.22)}$$

where

$$D = 1 - k^2\,\text{sn}^2\,(z_1, k)\,\text{sn}^2\,(z_2, k)$$

The variables z_1, z_2 can assume real or complex values. By using the above formula and Eqs. (A.6) and (A.7) we can show that

$$\text{cn}\,(z_1 + z_2,\, k) = \frac{\text{cn}\,(z_1,\, k)\,\text{cn}\,(z_2,\, k) - \text{sn}\,(z_1,\, k)\,\text{sn}\,(z_2,\, k)\,\text{dn}\,(z_1,\, k)\,\text{dn}\,(z_2,\, k)}{D}$$

(A.23)

$$\text{dn}\,(z_1 + z_2,\, k)$$
$$= \frac{\text{dn}\,(z_1,\, k)\,\text{dn}\,(z_2,\, k) - k^2\,\text{sn}\,(z_1,\, k)\,\text{sn}\,(z_2,\, k)\,\text{cn}\,(z_1,\, k)\,\text{cn}\,(z_2,\, k)}{D}$$

(A.24)

Another formula of interest is

$$\text{dn}^2\left(\frac{z}{2},\, k\right) = \frac{\text{dn}\,(z,\, k) + \text{cn}\,(z,\, k)}{1 + \text{cn}\,(z,\, k)}$$

(A.25)

A.6 PERIODICITY

In the preceding sections we have demonstrated that sn (z, k), where $z = u + jv$, has a real period of $4K$ if $v = 0$ and an imaginary period of $2K'$ if $u = 0$. In fact these are general properties for any value of v or u as can be easily shown. From the addition formula

$$\text{sn}\,(z + 4mK,\, k)$$
$$= \frac{\text{sn}\,(z,\, k)\,\text{cn}\,(4mK,\, k)\,\text{dn}\,(4mK,\, k) + \text{cn}\,(z,\, k)\,\text{sn}\,(4mK,\, k)\,\text{dn}\,(z,\, k)}{1 - k^2\,\text{sn}^2\,(z,\, k)\,\text{sn}^2\,(4mK,\, k)}$$

and since
$$\text{sn}\,(4mK,\, k) = \text{sn}\,(0,\, k) = 0$$
$$\text{cn}\,(4mK,\, k) = \text{cn}\,(0,\, k) = 1$$
$$\text{dn}\,(4mK,\, k) = \text{dn}\,(0,\, k) = 1$$

according to Eqs. (A.8) to (A.10), it follows that

$$\text{sn}\,(z + 4mK,\, k) = \text{sn}\,(z,\, k)$$

(A.26)

Similarly

$$\text{sn}\,(z + j2nK',\, k)$$
$$= \frac{\text{sn}\,(z,\, k)\,\text{cn}\,(j2nK',\, k)\,\text{dn}\,(j2nK',\, k) + \text{cn}\,(z,\, k)\,\text{sn}\,(j2nK',\, k)\,\text{dn}\,(z,\, k)}{1 - k^2\,\text{sn}^2\,(z,\, k)\,\text{sn}^2\,(j2nK',\, k)}$$

and from Eqs. (A.19) to (A.21)

$$\text{sn}\,(j2nK',\,k) = \frac{j\,\text{sn}\,(2nK',\,k')}{\text{cn}\,(2nK',\,k')} = 0$$

$$\text{cn}\,(j2nK',\,k) = \frac{1}{\text{cn}\,(2nK',\,k')} = (-1)^n$$

$$\text{dn}\,(j2nK',\,k) = \frac{\text{dn}\,(2nK',\,k')}{\text{cn}\,(2nK',\,k')} = (-1)^n$$

Hence we have

$$\text{sn}\,(z + j2nK',\,k) = \text{sn}\,(z,\,k) \tag{A.27}$$

Therefore, by combining Eqs. (A.26) and (A.27)

$$\text{sn}\,(z + 4mK + j2nK',\,k) = \text{sn}\,(z,\,k)$$

that is, sn $(z,\,k)$ is a doubly periodic function of z with a real period of $4K$ and an imaginary period of $2K'$.

The z plane can be subdivided into *period parallelograms* by means of lines

$$u = 4mK \qquad \text{and} \qquad jv = j2nK'$$

as illustrated in Fig. A.6. The specific parallelogram defined by vertices $(0, 0)$, $(4K, 0)$, $(4K, j2K')$, and $(0, j2K')$ is called the *fundamental period parallelogram*. If the value of sn $(z,\,k)$ is known for each and every value of z within this parallelogram and along any two adjacent sides, the function is known over the entire z plane.

The functions cn $(z,\,k)$ and dn $(z,\,k)$ can similarly be shown to be doubly periodic. The first has a real period of $4K$ and an imaginary period of $4K'$, whereas the second has a real period of $2K$ and an imaginary period of $4K'$.

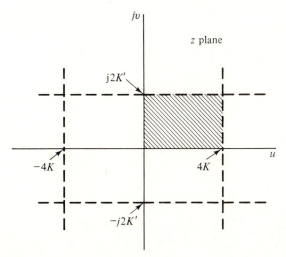

Figure A.6 Period parallelograms of sn (z, k).

A.7 TRANSFORMATION

The equation

$$\omega = \sqrt{k}\ \text{sn}\ (z, k) \tag{A.28}$$

is essentially a variable transformation which maps points in the z plane onto corresponding points in the ω plane. Let us examine the mapping properties of this transformation. These are required in the derivation of $F(\omega)$ in Sec. 5.5.

A point z_p as well as all points

$$z = z_p + 4mK + j2nK'$$

map onto a single point in the ω plane by virtue of the periodicity of sn (z, k). Hence only points in the fundamental period parallelogram need be considered. Three domains of $\sqrt{k}\ \text{sn}\ (z, k)$ are of interest as follows:

Domain 1: $z = u$ with $0 \leq u \leq K$

Domain 2: $z = K + jv$ with $0 \leq v \leq K'$

Domain 3: $z = u + jK'$ with $0 \leq u \leq K$

In domain 1

$$\omega = \sqrt{k}\ \sin\ (u, k)$$

If $u = 0$,

$$\omega = \sqrt{k}\ \sin\ (0, k) = 0$$

and if $u = K$,

$$\omega = \sqrt{k}\ \sin\ (K, k) = \sqrt{k}$$

i.e., Eq. (A.28) maps points on the real axis of the z plane between 0 and K onto points on the real axis of the ω plane between 0 and $\sqrt{k}$.

In domain 2 we have

$$\omega = \sqrt{k}\ \text{sn}\ (K + jv, k)$$

From the addition formula

$$\omega = \frac{\sqrt{k}\ \text{cn}\ (jv, k)\ \text{dn}\ (jv, k)}{1 - k^2\ \text{sn}^2\ (jv, k)} \tag{A.29}$$

since cn $(K, k) = 0$, and from Eqs. (A.19) to (A.21)

$$\omega = \frac{\sqrt{k}\ \text{dn}\ (v, k')}{\text{cn}^2\ (v, k') + k^2\ \text{sn}^2\ (v, k')}$$

Now from Eqs. (A.6) and (A.7)

$$\text{cn}^2\ (v, k') + k^2\ \text{sn}^2\ (v, k') = 1 - \text{sn}^2\ (v, k') + k^2\ \text{sn}^2\ (v, k')$$

$$= 1 - (k')^2\ \text{sn}^2\ (v, k') = \text{dn}^2\ (v, k')$$

Therefore, Eq. (A.29) simplifies to

$$\omega = \frac{\sqrt{k}}{\text{dn }(v, k')}$$

If $v = 0$,

$$\omega = \frac{\sqrt{k}}{\text{dn }(0, k')} = \sqrt{k}$$

and if $v = K'$,

$$\omega = \frac{\sqrt{k}}{\text{dn }(K', k')} = \frac{1}{\sqrt{k}}$$

For $v = K'/2$, the use of Eq. (A.25) yields

$$\omega = \frac{\sqrt{k}}{\text{dn }(K'/2, k')} = \sqrt{k}\left[\frac{1 + \text{cn }(K', k')}{\text{dn }(K', k') + \text{cn }(K', k')}\right]^{1/2} = 1$$

Thus Eq. (A.28) maps points on the line $z = K + jv$ for v between 0 and K' onto points on the real axis of the ω plane between $\sqrt{k}$ and $1/\sqrt{k}$; in particular, point $z = K + jK'/2$ maps onto point $\omega = 1$.

In domain 3 Eq. (A.28) becomes

$$\omega = \sqrt{k}\ \text{sn }(u + jK', k)$$

and as above, Eq. (A.22) yields

$$\omega = \frac{1}{\sqrt{k}\ \text{sn }(u, k)}$$

If $u = 0$,

$$\omega = \frac{1}{\sqrt{k}\ \text{sn }(0, k)} = \infty$$

and if $u = K$,

$$\omega = \frac{1}{\sqrt{k}\ \text{sn }(K, k)} = \frac{1}{\sqrt{k}}$$

i.e., points on line $z = u + jK'$ with u between 0 and K map onto the real axis of the ω plane between ∞ and $1/\sqrt{k}$.

By considering mirror-image points to those considered so far, the mapping depicted in Fig. A.7 can be completed, where points $A, B, \ldots$ map onto points A', $B', \ldots$.

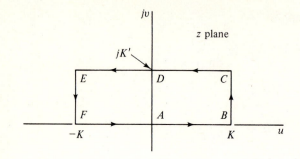

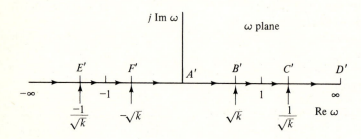

Figure A.7 Mapping properties of transformation $\omega = \sqrt{k}\,\mathrm{sn}\,(z, k)$.

A.8 SERIES REPRESENTATION

Elliptic functions, like other functions, can be represented in terms of series. From Ref. 3 or 4

$$\mathrm{sn}\,(z, k) = \frac{1}{\sqrt{k}}\frac{\theta_1(z/2K, q)}{\theta_0(z/2K, q)} \tag{A.30}$$

$$\mathrm{cn}\,(z, k) = \sqrt{\frac{k'}{k}}\frac{\theta_2(z/2K, q)}{\theta_0(z/2K, q)} \tag{A.31}$$

$$\mathrm{dn}\,(z, k) = \sqrt{k'}\frac{\theta_3(z/2K, q)}{\theta_0(z/2K, q)} \tag{A.32}$$

The parameter q is known as the *modular constant* and is given by

$$q = e^{-\pi K'/K}$$

The functions $\theta_0(z/2K, q)$ to $\theta_3(z/2K, q)$ are called *theta functions* and are given by

$$\theta_0\left(\frac{z}{2K}, q\right) = 1 + 2 \sum_{m=1}^{\infty} (-1)^m q^{m^2} \cos\left(2m\frac{\pi z}{2K}\right)$$

$$\theta_1\left(\frac{z}{2K}, q\right) = 2q^{1/4} \sum_{m=0}^{\infty} (-1)^m q^{m(m+1)} \sin\left[(2m+1)\frac{\pi z}{2K}\right]$$

$$\theta_2\left(\frac{z}{2K}, q\right) = 2q^{1/4} \sum_{m=0}^{\infty} q^{m(m+1)} \cos\left[(2m+1)\frac{\pi z}{2K}\right]$$

$$\theta_3\left(\frac{z}{2K}, q\right) = 1 + 2 \sum_{m=1}^{\infty} q^{m^2} \cos\left(2m\frac{\pi z}{2K}\right)$$

The above series converge rapidly and can be used to evaluate the elliptic functions to any desired degree of accuracy.

REFERENCES

1. F. Bowman, "Introduction to Elliptic Functions with Applications," Dover, New York, 1961.
2. H. Hancock, "Elliptic Integrals," Dover, New York.
3. H. Hancock, "Lectures on the Theory of Elliptic Functions," Dover, New York, 1958.
4. A. J. Grossman, Synthesis of Tchebyscheff Parameter Symmetrical Filters, *Proc. IRE*, vol. 45, pp. 454–473, April 1957.

B

COMPUTER PROGRAMS

B.1 INTRODUCTION

This appendix describes a number of computer programs which can be used to analyze and design recursive and nonrecursive filters.

B.2 SYSTEM CONFIGURATION

The programs have been written specifically for the 9825 Hewlett-Packard computing system. The minimum configuration is as follows:

Basic unit: 9825A
Internal storage: Standard (6844 bytes)
Output printer: HP 9871A, Opt. 025
ROM 1: Advance Programming, 98210A
ROM 2: Plotter, General I/0, 98212A
ROM 3: Matrix, 98211A
Language: HPL

B.3 PROGRAM DESCRIPTIONS

The programs are listed in Table B.1. Their descriptions follow. Listings of the programs and typical runs can be found in Sects. B.4 and B.5, respectively.

Table B.1 List of programs

File no.	Title
1–2	Digital-filter time-domain response
3–4	Frequency response
5	Plot
6	Recursive filters: prescribed specifications
7	Analog-filter approximations
8	Data file
9	Analog-filter transformations
10	Coefficients to zeros and poles
11	Bilinear transformation
12	Partial fractions
13	Impulse, unit-step, and sinusoid invariant methods
14	Matched-z transformation method
15–16	Design of nonrecursive filters
17	Fast Fourier transform

Files: 1 to 2 Digital-Filter Time-Domain Response

Computes and plots the impulse, unit-step, or sinusoidal response of cascade or parallel digital filters, using the transfer function $H(z)$.

Input Coefficients A_{JL}, B_{JL}, and H as in

$$H(z) = H \prod_{L=1}^{M} \frac{\sum_{J=0}^{N_L} A_{JL} z^J}{\sum_{J=0}^{D_L} B_{JL} z^J} \quad \text{or} \quad H(z) = \sum_{L=1}^{M} \frac{\sum_{J=0}^{N_L} A_{JL} z^J}{\sum_{J=0}^{D_L} B_{JL} z^J}$$

Output Time-domain response; plot.

Files 3 to 5 Frequency Response

Computes and plots amplitude, phase, or group-delay response of analog or digital, cascade or parallel filters using $H(s)$ or $H(z)$.

Input Coefficients A_{JL}, B_{JL}, and H (as above); ω_s, frequency range.

Output Gain, phase, or group delay; plot.

File 6 Recursive Digital Filters

Computes the necessary specifications for the normalized analog filter and the necessary parameters of the analog-filter transformations that will yield a digital filter satisfying prescribed specifications (see Chap. 8). A Butterworth, Tschebyscheff, or elliptic filter may be chosen.

Input ω_s, A_p, A_a; $\tilde{\Omega}_p$ and $\tilde{\Omega}_a$ (for LP and HP filters) or $\tilde{\Omega}_{p1}$, $\tilde{\Omega}_{p2}$, $\tilde{\Omega}_{a1}$, and $\tilde{\Omega}_{a2}$ (for BP and BS filters).

Output λ (for LP and HP filters) or ω_0 and B (for BP and BS filters); k (for elliptic filters), n, actual A_a (see Sec. 8.4).

File 7 Analog-Filter Approximations

Computes normalized LP Butterworth, Tschebyscheff, or elliptic transfer functions (see Chap. 5).

Input For Butterworth filters n; for Tschebyscheff filters n, A_p; for elliptic filters A_p, A_a, k.

Output n, Actual A_a (elliptic filters only). Option 1, zeros and poles: number of complex zero pairs, real zeros, complex pole pairs, real poles; zeros (R_I, I_I), poles (R'_I, I'_I), and H as in

$$H(s) = H \prod_{L=1}^{M} \frac{s - (R_I + jI_I)}{s - (R'_I + jI'_I)}$$

Option 2, coefficients: H, A_{IL}, and B_{IL} as in

$$H(s) = H \prod_{L=1}^{M} \frac{A_{0L} + A_{1L}s + A_{2L}s^2}{B_{0L} + B_{1L}s + B_{2L}s^2}$$

File 8 Data File

Used to store outputs of Programs 7, 9, and 11 to 14 (see Fig. B.1).

Files 9 and 10 Analog-Filter Transformations

Computes

$$H_X(\bar{s}) = H(s)\Big|_{s = f_X(\bar{s})}$$

(see Sec. 8.2).

Input Zeros and poles (see Option 1 of output in File 7, and note 5 below), λ (for LP and HP filters), or ω_0 and B (for BP and BS filters).

Output Option 1, zeros and poles: as for input. Option 2, coefficients: see File 7.

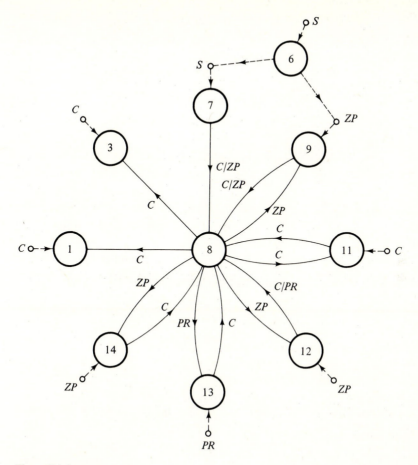

Figure B.1 Intercommunication links between files: C = coefficients, ZP = zeros and poles, PR = poles and residues, S = specifications, --- → manual entry.

File 11 Bilinear Transformation

Computes

$$H_D(z) = H(s)\Big|_{s=\frac{2}{T}\frac{z-1}{z+1}}$$

Input ω_s, coefficients.

Output Coefficients.

File 12 Partial Fractions

Computes

$$H(s) = H + \sum_{I=1}^{M} \frac{U_I + jV_I}{s - (P_I + jQ_I)}$$

or

$$H(s) = \sum_{L=1}^{M} \frac{A_{0L} + A_{1L}s + A_{2L}s^2}{B_{0L} + B_{1L}s + B_{2L}s^2}$$

for

$$H(s) = H \prod_{I=1}^{N} \frac{s - (R_I + jI_I)}{s - (R'_I + jI'_I)}$$

Input Zeros and poles.

Output Option 1, poles and residues: H, poles (R'_I, I'_I), residues (U_I, V_I). Option 2, coefficients.

File 13 Impulse-, Unit-Step-, and Sinusoid-Invariant Methods

Computes

$$H_D(z) = \sum_{L=1}^{M} \frac{A_{0L} + A_{1L}z + A_{2L}z^2}{B_{0L} + B_{1L}z + B_{2L}z^2}$$

for

$$H(s) = H + \sum_{I=1}^{M} \frac{U_I + jV_I}{s - (P_I + jQ_I)}$$

by means of the impulse-, unit-step-, or sinusoid-invariant method (see Sec. 7.3 and Probs. 7.5 and 7.6).

Input ω_s, poles and residues (see File 12).

Output Coefficients.

File 14 Matched-z Transformation Method

Computes

$$H_D(z) = \prod_{L=1}^{M} \frac{A_{0L} + A_{1L}z + A_{2L}z^2}{B_{0L} + B_{1L}z + B_{2L}z^2}$$

for

$$H(s) = H \prod_{I=1}^{N} \frac{s - (R_I + jI_I)}{s - (R'_I + jI'_I)}$$

by means of the matched-z transformation method (see Sec. 7.5).

Input ω_s, zeros and poles.

Output Coefficients

Files 15 and 16 Design of Nonrecursive Filters

Designs lowpass, highpass, bandpass, or bandstop filters using the Fourier-series method. The windows available are rectangular, Bartlett (see Prob. 9.6), Hann, Hamming, Blackman, and Kaiser (see Chap. 9).

Input Option 1 (all windows except Kaiser's): ω_s, cutoff frequencies. Option 2 (Kaiser's window): desired specifications, that is, ω_s, A_p, A_a; ω_p, and ω_a (for LP and HP filters) or ω_{p1}, ω_{p2}, ω_{a1}, ω_{a2} (for BP and BS filters).

Output Actual A_p, A_a, and N (option 2 only); $h(\pm n)$, $w(\pm n)h(\pm n)$; amplitude response; plot.

File 17 Fast Fourier Transform

Computes the DFT or IDFT of a given complex sequence using the decimation-in-time algorithm (see Sec. 13.11).

Input Number of elements; input sequence [Re $x(n)$, Im $x(n)$].

Output Output sequence [Re $X(k)$, Im $X(k)$, $|X(k)|$, arg $X(k)$].

Notes

1. Programs 1, 3, 7, 9, and 11 to 14 intercommunicate through File 8, as depicted in Fig. B.1.
2. Recursive filters satisfying prescribed specifications can be designed by running Programs 6, 7, 9, 11 in sequence.
3. Frequency is always in radians per second.
4. Gain or loss is always in decibels.
5. For programs 9, 12, and 14 enter zeros and poles with positive or zero imaginary parts, starting with zeros (for manual operation).

```
0: fmt 1,"FILE:1"
1: fmt 2,"DIGITAL-FILTER TIME-DOMAIN RESPONSE"
2: fmt 3,"==================================="
3: wtb 6,27,69
4: dim A$[1],D$[1],L$[11];rad
5: "-----------"→L$
6: wtb 6,14,27,84,27,76,int(912/64),int(912)
7: fmt ,3/,10x,z;wrt 6;wtb 6,27,77
8: wrt 6.1;wrt 6.2;wrt 6.3
9: ent "CASCADE or PARALLEL? c or p ",A$
10: cap(A$)→A$
11: if A$="C";sfg 1;fmt "CASCADE",z;wrt 6;gto +2
12: fmt "PARALLEL",z;wrt 6
13: fmt " FILTER";wrt 6
14: fmt "INPUT DATA:",/,cll;wrt 6,L$
15: if flg1;fmt "H(z)=H*H1(z)*H2(z)*...";wrt 6;gto
   +2
16: fmt "H(z)=H1(z)+H2(z)+...";wrt 6
17: fmt "H1(z)=(A01+A11z+A21z↑2)/(B01+B11z+B21z↑2)
";wrt 6
18: ent "DATA on FILE? y or n",D$;cap(D$)→D$
19: if D$="N";gto +4
20: dim N[10],D[10],H,S,A[0:2,10],B[0:2,10]
21: ina N:2,D:2
22: gto "LDF"
23: ent "Numb. of Sectns?",S
24: ent "Max numer. deg.?",U
25: ent "Max denom. deg:",B
26: if flg1;ent "Multiplier:",H
27: dim N[S],D[S],A[0:U,1:S],B[0:B,1:S]
28: 1→L;0→J
29: "SE":
30: if S=1;U→N[L];B→D[L];gto +5
31: fxd 0;spc ;prt "Section #:",L
32: ent "Numer. deg.:",N[L]
33: ent "Denom. deg.:",D[L]
34: 0→J
35: enp A[J,L];jmp (J+1→J)>N[L]
36: 0→J
37: enp B[J,L];jmp (J+1→J)>D[L]
38: L+1→L;if L<=S;gto "SE"
39: gto +3
40: "LDF":
41: ldf 8,H,S,A[*],B[*];2→B
42: ldf 2,0
```

```
0:  "FILE:2":
1:  "Cont. of File:1":
2:  for L=1 to S
3:  fmt 1,"Section #:",f1.0,/,cll
4:  fmt 2,"A",2f1.0,":",e15.6,5x,"B",2f1.0,":",e15.
6
5:  fmt 3,24x,"B",2f1.0,":",e15.6
6:  wrt 6.1,L,L$;for J=0 to N[L]
7:  wrt 6.2,J,L,A[J,L],J,L,B[J,L]
8:  D[L]→r1;A[J,L]/B[r1,L]→A[J,L];B[J,L]/B[r1,L]→B[
J,L];next J
9:  for J=N[L]+1 to D[L];wrt 6.3,J,L,B[J,L]
10: B[J,L]/B[D[L],L]→B[J,L];next J;next L
11: if flg1;fmt ,"H:",e14.6;wrt 6,H
12: dim X[0:B],Y[0:B,S],M$[100,4]
13: if flg1;for J=0 to N[1];HA[J,1]→A[J,1];next J
14: "RE":
15: fmt 2,cl,z
16: fmt 3,/,"SAMPLE",12x,"OUTPUT";wrt 6.3
17: fmt 4,f6.0,e18.6
18: fmt 6,"Input Freq., r/s:",e13.6,/,"Sampl. Freq
., r/s:",e13.6
19: for I=1 to 24;wrt 6.2,"=";next I;wtb 6,13,10
20: wtb 6,27,53
21: 0→N→W→X
22: 0→J
23: 0→X[J];jmp (J+1→J)>B
24: 1→L
25: 0→J
26: 0→Y[J,L];jmp (J+1→J)>B
27: L+1→L;if L<=S;gto -2
28: 0→J
29: ent "NUMB. OF SAMPLES",Z
30: dsp "CHOOSE INPUT";wait 750
31: ent "1:IMPULSE; 2:STEP; 3:SINUSOID",C
32: if C=1;fmt 5,"IMPULSE RESPONSE"
33: if C=2;fmt 5,"STEP RESPONSE"
34: if C=3;fmt 5,"SINUSOID RESPONSE"
35: wrt 6.5
36: if C#3;gto +3
37: ent "Signal Freq., r/s:",F,"Sampl. Freq., r/s:
",r1
38: wrt 6.6,F,r1;2π/r1→T
39: "INPUT":
40: 1→X[0]
41: if C=1;if N#0;0→X[0]
42: if C=3;sin(FNT)→X[0]
43: "OUTPUTS":
```

```
44: if flg1;gto "A"
45: gsb "PARALLEL"
46: 0→W
47: 1→L
48: W+Y[0,L]→W;jmp (L+1→L)>S
49: gto "B"
50: "A":
51: gsb "CASCADE"
52: Y[0,S]→W
53: "B":
54: wrt 6.4,N,W
55: fts (W)→M$[N+1]
56: 0→J
57: X[B-J-1]→X[B-J];jmp (J+1→J)>B-1
58: 1→L
59: 0→J
60: Y[B-J-1,L]→Y[B-J,L];jmp (J+1→J)>B-1
61: 0→Y[0,L];L+1→L;if L<=S;gto -2
62: N+1→N;if N<Z;gto "INPUT"
63: Z-1→P
64: wtb 6,12;ent "PLOT? y or n",D$;cap(D$)→D$
65: if D$="Y";gsb "PLOT"
66: ent "NEW RESPONSE? y or n",A$;cap(A$)→A$
67: if A$="Y";gto "RE"
68: end
69: "PARALLEL":
70: 1→L
71: 0→J
72: A[J,L]X[D[L]-J]+Y[0,L]→Y[0,L];jmp (J+1→J)>N[L]

73: 0→J
74: Y[0,L]-B[J,L]Y[D[L]-J,L]→Y[0,L];jmp (J+1→J)>D[
L]-1
75: L+1→L;if L<=S;gto -4
76: ret
77: "CASCADE":
78: 0→J
79: A[J,1]X[D[1]-J]+Y[0,1]→Y[0,1];jmp (J+1→J)>N[1]

80: 0→J
81: Y[0,1]-B[J,1]Y[D[1]-J,1]→Y[0,1];jmp (J+1→J)>D[
1]-1
82: if S=1;ret
83: 2→L
84: 0→J
85: Y[D[L]-J,L-1]A[J,L]+Y[0,L]→Y[0,L];jmp (J+1→J)>
N[L]
86: 0→J
```

```
87: Y[0,L]-B[J,L]Y[D[L]-J,L]→Y[0,L];jmp (J+1→J)>D[
L]-1
88: L+1→L;if L<=S;gto -4
89: ret
90: "PLOT":6→r0
91: 0→H
92: 9e99→M;-9e99→N
93: stf(M$[H+1])→r1;min(r1,M)→M;max(r1,N)→N;jmp (H
+1→H)'>P
94: if M>0;0→M
95: if N<0;0→N
96: "PS":2→B;4→W;5→r3;1→r4
97: wtb 6,27,79,int(r4*120/64),r4*120,int(r3*96/64
),r3*96
98: "SC":0→r1;P→r2;M→r3;N→r4
99: 120W/(r2-r1)→U
100: 96B/(r4-r3)→V
101: r1→X;r3→Y
102: "X":wtb 6,27,46,95,0,5,9
103: X→r3;X+120W/U→r4
104: r4-r3→r2
105: wtb 6,27,65,int((r3-X)U/64),int((r3-X)U),int(
(-Y)V/64),int((-Y)V)
106: r3→r5;wtb 6,43;wtb 6,8
107: wtb 6,27,114,int(r2U/64),int(r2U),0,0;wtb 6,4
3,8;jmp (r5+r2→r5)>=r4
108: "Y":(N-M)/10→r2
109: wtb 6,27,46,124,0,3,0
110: Y→r3;Y+96B/V→r4
111: wtb 6,27,65,int((-X)U/64),int((-X)U),int'(0),i
nt(0)
112: r3→r5;wtb 6,43;wtb 6,8
113: wtb 6,27,114,0,0,int(r2V/64),int(r2V);wtb 6,4
3,8;jmp (r5+r2→r5)>=r4
114: for I=1 to P+1;cll 'MV'(I-1,0);cll 'FPT'(I-1,
stf(M$[I]),124);next I
115: fmt 5,e13.6
116: cll 'MV'(0,M-2(N-M)/10);wtb 6,27,49;wrt 6.5,"
MIN-Y:",M
117: wtb 6,9;wrt 6.5,"MAX-Y:",N;wtb 6,9;wrt 6.5,"U
NITS/DIV:",(N-M)/10
118: wtb 6,12
119: ret
120: "MV":
121: wtb 6,27,65,int((p1-X)U/64),int((p1-X)U),int(
(p2-Y)V/64),int((p2-Y)V)
122: ret
123: "FPT":
```

```
124: wtb 6,27,97,int((p1-X)U/64),int((p1-X)U),int(
(p2-Y)V/64),int((p2-Y)V)
125: if p3=0;46→p3
126: if p3=46;wtb 6,27,82,0,0,0,6
127: wtb 6,p3;wtb 6,8
128: if p3=46;wtb 6,27,82,0,0,63,-6
129: ret


0: fmt 1,"FILE:3"
1: fmt 2,"FREQUENCY RESPONSE"
2: fmt 3,"==================="
3: wtb 6,27,69
4: dim A$[1],D$[1],F$[1],L$[11];rad
5: "------------"→L$
6: wtb 6,14,27,84,27,76,int(912/64),int(912)
7: fmt ,3/,10x,z;wrt 6;wtb 6,27,77
8: wrt 6.1;wrt 6.2;wrt 6.3
9: fmt 1,"Section #:",f1.0,/,cll
10: fmt 2,"A",2f1.0,":",e15.6,5x,"B",2f1.0,":",e15
.6
11: fmt 3,24x,"B",2f1.0,":",e15.6
12: ent "ANALOG or DIGITAL filter? a or d",F$
13: cap(F$)→F$;if F$="A";sfg 1;fmt "ANALOG ",z;wrt
 6;gto +2
14: fmt "DIGITAL ",z;wrt 6
15: ent "CASCADE or PARALLEL? c or p ",A$
16: cap(A$)→A$
17: if A$="P";sfg 3;fmt "PARALLEL",z;wrt 6;gto +2
18: fmt "CASCADE",z;wrt 6
19: fmt " FILTER";wrt 6
20: fmt "INPUT DATA:",/,cll;wrt 6,L$
21: if flg3;fmt "H(z)=H1(z)+H2(z)+...";wrt 6;gto +
2
22: fmt "H(z)=H*H1(z)*H2(z)...";wrt 6
23: fmt "H1(z)=(A01+A11z+A21z↑2)/(B01+B11z+B21z↑2)
";wrt 6
24: ent "INPUT DATA on FILE? y or n",D$;cap(D$)→D$

25: if D$="N";gto +4
26: dim N[10],D[10],C[0:2],H,S,A[0:2,10],B[0:2,10]

27: ina N:2,D:2
28: gto "LDF"
29: ent "Number of Sections?",S
30: ent "Max numer. degree?",U
31: ent "Max denomin. degree?",B
32: if not flg3;ent "Multiplier:",H
```

```
33: dim N[S],D[S],A[0:U,1:S],B[0:B,1:S]
34: dim C[0:max(U,B)]
35: 1→L;0→J
36: "SE":
37: if S=1;U→N[L];B→D[L];gto +4
38: fxd 0;spc ;prt "Section #",L
39: ent "Numerat. degree?",N[L]
40: ent "Denomin. degree?",D[L]
41: flt 6;0→J
42: enp A[J,L];jmp (J+1→J)>N[L]
43: 0→J
44: enp B[J,L];jmp (J+1→J)>D[L]
45: L+1→L;if L<=S;gto "SE"
46: gto +3
47: "LDF":
48: ldf 8,H,S,A[*],B[*]
49: for L=1 to S
50: wrt 6.1,L,L$;for J=0 to N[L]
51: wrt 6.2,J,L,A[J,L],J,L,B[J,L];next J
52: for J=N[L]+1 to D[L];wrt 6.3,J,L,B[J,L]
53: next J;next L
54: if not flg3;fmt ,"H:",e14.6;wrt 6,H
55: dim F[2],I[2],M[-2:106],P[2],R[2],W[3],Y[0:3],
R$[1]
56: ldf 4


0: "FILE:4":
1: "Cont. of FILE:3":
2: ldf 5,77,3
3: "RE":sfg 14;0→F→B;ina Y,M
4: fmt 1,"OUTPUT DATA:",/,cll
5: fmt 2,4x,"Freq., r/s",5x,z
6: fmt 3,"Gain,    dB"
7: fmt 4,"Phase, Deg."
8: fmt 5,"Delay, Sec."
9: fmt 6,2e15.6
10: enp "INIT.FREQ. r/s:",W[1],"FREQ.INCR.:",W[2],
"FINAL FREQ.:",W[3]
11: if (W[3]-W[1])/W[2]→r1;if frc(r1)=0;r1+1→r1
12: prt "# of points:",r1;if r1>50;beep;wait 500;b
eep
13: if r1>50;ent "Points>50;CONTINUE? y or n",R$
14: if cap(R$)="N";"Y"→R$;gto "RE"
15: if flg1;gto +3
16: enp "SAMPL. FREQ.:",r1;2π/r1→T
17: fmt "Sampl. Freq. r/s:",e13.6;wrt 6,r1
18: if W[1]=0;1e-9→W[1];W[3]+1e-9→W[3]
```

```
19: W[1]→W;ent "GAIN ,PHASE or DELAY? g, p or d",F
$;cap(F$)→F$
20: if F$="P" or F$="D";sfg 2
21: if F$="G";cfg 2
22: wrt 6.1,L$;wrt 6.2
23: if F$="G";wrt 6.3
24: if F$="P";wrt 6.4
25: if F$="D";wrt 6.5
26: W[3]→C;0→A;if F$="D";W-3W[2]→W;W[3]+6W[2]→C;-3
→A
27: "BN":A+1→A
28: if flg1;0→F[1];W→F[2];gto +2
29: cos(WT)→F[1];sin(WT)→F[2]
30: H→M;0→N;1→L;0→J
31: if flg3;0→U;0→V
32: "ST":
33: if flg3;1→M;0→N
34: A[J,L]→C[J];jmp (J+1→J)>N[L]
35: 0→J
36: if N[L]=0;MC[0]→M;gto +3
37: N[L]→Q;gsb "POLY"
38: MG→M;N+Z→N;0→J
39: B[J ,L]→C[J];jmp (J+1→J)>D[L]
40: 0→J
41: if D[L]=0;M/C[0]→M;gto +3
42: D[L]→Q;gsb "POLY"
43: M/G→M;N-Z→N
44: L+1→L;if flg3=0;gto +4
45: √((Ucos(V)+Mcos(N))↑2+(Usin(V)+Msin(N))↑2)→r2
46: Ucos(V)+Mcos(N)→r3;atn((Usin(V)+Msin(N))/r3)→V
;r2→U
47: if r3<0;V-π→V
48: if L<=S;gto "ST"
49: if flg3;U→M;V→N
50: if flg2;sin(N)→r4;cos(N)→r5;atn(r4/r5)→N;if r5
<0;if r4<0;N-π→N
51: if flg2;if r5<0;if r4>=0;π+N→N
52: if F$="D";cll 'PHC'(N);if A<4;gto +5
53: if F$="D" and A>=4;cll 'DELAY';gto +4
54: if F$="P";180N/π→M[A];gto +2
55: 20log(M)→M[A];if M[A]<-80;-80→M[A]
56: wrt 6.6,W,M[A]
57: W+W[2]→W;if W<=C;gto "BN"
58: A→P;if F$="D";A-6→P
59: wtb 6,12;ent "PLOT? y or n",D$;cap(D$)→D$
60: if D$="Y";gsb "PLT"
61: ent "NEW RESPONSE? y or n",A$;cap(A$)→A$
62: if A$="Y";gto "RE"
```

```
63:  end
64:  "POLY":
65:  C[Q]→P[1];0→P[2]
66:  1→I
67:  "LOOP":
68:  P[1]F[1]-P[2]F[2]→r1
69:  P[1]F[2]+P[2]F[1]→P[2]
70:  r1+C[Q-I]→P[1]
71:  I+1→I;if I<=Q;gto "LOOP"
72:  √(P[1]↑2+P[2]↑2)→G
73:  atn(P[2]/P[1])→Z
74:  if P[1]<0;if P[2]<0;Z-π→Z
75:  if P[1]<0;if P[2]>=0;π+Z→Z
76:  ret


0:  "FILE:5":
1:  "Cont. of FILE #4":
2:  "PHC":
3:  if p1>0 and F<0;B+1→B
4:  p1→F;-2Bπ+p1→M[A]
5:  ret
6:  "DELAY":
7:  (M[A]-9M[A-1]+45M[A-2]-45M[A-4]+9M[A-5]-M[A-6])
/60W[2]→Y[0]
8:  -Y[3]→M[A-6];Y[2]→Y[3];Y[1]→Y[2];Y[0]→Y[1]
9:  if A>=7;wrt 6.6,W-6W[2],M[A-6]
10: ret
11: "PT":
12: wtb 6,27,65,int((p1-X)U/64),int((p1-X)U),int((
p2-Y)V/64),int((p2-Y)V)
13: if p3=0;46→p3
14: if p3=46;wtb 6,27,82,0,0,0,6
15: wtb 6,p3;wtb 6,8
16: if p3=46;wtb 6,27,82,0,0,63,-6
17: ret
18: "PLT":
19: 1→A;9e99→M;-9e99→N
20: M[A]→r1;min(r1,M)→M;max(r1,N)→N;jmp (A+1→A)>P
21: if M=N;ret
22: "PS":
23: 4→B;5→W;3→r3;1→r4
24: wtb 6,27,79,int(r4*120/64),r4*120,int(r3*96/64
),r3*96
25: "SC":
26: 1→r1;P→r2;M→r3;N→r4
27: 120W/(r2-r1)→U
28: 96B/(r4-r3)→V
```

```
29: r1→X;r3→Y
30: "X":
31: M→r1;P/10→r2
32: wtb 6,27,46,95,0,5,9
33: X→r3;X+120W/U→r4
34: wtb 6,27,65,int(0),int(0),int((r1-Y)V/64),int(
(r1-Y)V)
35: r3→r5;wtb 6,43;wtb 6,8
36: wtb 6,27,114,int(r2U/64),int(r2U),0,0;wtb 6,43
,8;jmp (r5+r2→r5)>=r4
37: "Y":
38: 1→r1;(N-M)/10→r2
39: wtb 6,27,46,124,0,3,0
40: Y→r3;Y+96B/V→r4
41: wtb 6,27,65,int((r1-X)U/64),int((r1-X)U),int(0
),int(0)
42: r3→r5;wtb 6,43;wtb 6,8
43: wtb 6,27,114,0,0,int(r2V/64),int(r2V);wtb 6,43
,8;jmp (r5+r2→r5)>=r4
44: for I=1 to P;cll 'PT'(I,M[I],42);next I
45: fmt 7,e13.6,3x,e13.6
46: "mo":
47: 2→r1;M-2(N-M)/10→r2
48: wtb 6,27,65,int((r1-X)U/64),int((r1-X)U),int((
r2-Y)V/64),int((r2-Y)V)
49: wtb 6,27,49,13
50: wtb 6,9;wrt 6.7,"MIN-X:",W[1],"MIN-Y:",M
51: wtb 6,9;wrt 6.7,"MAX-X:",W[3],"MAX-Y:",N
52: wtb 6,9;wrt 6.7,"DIV-X:",(W[3]-W[1])/10,"DIV-:
Y",(N-M)/10
53: wtb 6,12;ret


0: fmt 1,"FILE:6"
1: fmt 2,"RECURSIVE FILTERS: PRESCRIBED SPECIFICAT
IONS"
2: fmt 3,"=========================================
===="
3: wtb 6,27,69
4: dim L$[11];"------------"→L$;rad
5: wtb 6,14,27,84,27,76,int(912/64),int(912)
6: fmt ,3/,10x,z;wrt 6;wtb 6,27,77
7: wrt 6.1;wrt 6.2;wrt 6.3
8: dim A[2],F[4],W[3],T[4],K[5],A$[2],F$[2]
9: ent "APPROXIMATION? BU,TS or EL",A$;cap(A$)→A$
10: if A$="BU";fmt "BUTTERWORTH ",z;wrt 6
11: if A$="TS";fmt "TSCHEBYSCHEFF ",z;wrt 6
12: if A$="EL";fmt "ELLIPTIC ",z;wrt 6
```

```
13: ent "TYPE OF FILTER? LP,HP,BP or BS",F$;cap(F$
)→F$
14: if F$="LP";fmt "LOWPASS ",z;wrt 6
15: if F$="HP";fmt "HIGHPASS ",z;wrt 6
16: if F$="BP";fmt "BANDPASS ",z;wrt 6
17: if F$="BS";fmt "BANDSTOP ",z;wrt 6
18: fmt "FILTER";wrt 6
19: fmt "INPUT DATA:",/,cll;wrt 6,L$
20: enp "Saml. Freq., r/s",T
21: fmt "Sampling Freq., r/s:",e13.6;wrt 6,T;2π/T→
T
22: enp "Ap,dB:",A[1],"Aa:",A[2]
23: fmt "Ap, dB:",f6.3,/,"Aa, dB:",f7.3;wrt 6,A[1]
,A[2]
24: 10↑(.1A[1])-1→W[1];(10↑(.1A[2])-1)/W[1]→D
25: if F$="BP" or F$="BS";gto "BPBS"
26: enp "Wp, r/s:",F[1],"Wa:",F[2]
27: fmt ,e13.6,/,e13.6;wrt 6,"Wp, r/s:",F[1],"Wa r
/s:",F[2]
28: tan(F[1]T/2)→L
29: L/tan(F[2]T/2)→K;if F$="HP";1/K→K
30: gto "AAA"
31: "BPBS":
32: enp "Wp1, r/s:",F[1],"Wp2:",F[2],"Wa1:",F[3],"
Wa2:",F[4]
33: fmt "Wp1, r/s:",e13.6,/,"Wp2, r/s:",e13.6;wrt
6,F[1],F[2]
34: fmt "Wa1, r/s:",e13.6,/,"Wa2, r/s:",e13.6;wrt
6,F[3],F[4]
35: for I=1 to 4;tan(F[I]T/2)→T[I];next I
36: T[2]-T[1]→K[3];T[1]T[2]→K[4];T[3]T[4]→K[5]
37: K[3]T[3]/(K[4]-T[3]↑2)→K[1]
38: K[3]T[4]/(T[4]↑2-K[4])→K[2]
39: if K[5]>=K[4];K[1]→K;if F$="BS";1/K[2]→K;gto +
2
40: K[2]→K;if F$="BS";1/K[1]→K
41: "AAA":
42: if A$="BU";gsb "NBU"
43: if A$="TS";gsb "NTS"
44: if A$="EL";gsb "NEL"
45: if F$="HP";2WL/T→L
46: if F$="LP";WT/2L→L
47: fmt "OUTPUT DATA:",/,cll;wrt 6,L$
48: if F$="LP" or F$="HP";fmt "Lamda:",13e.6;wrt 6
,L;gto +3
49: 2√K[4]/T→A;2K[3]/TW→B;if F$="BS";2K[3]W/T→B
50: fmt "Wo, r/s:",e13.6,/,"B,  r/s:",e13.6;wrt 6,
A,B
```

```
51: if A$="EL";fmt "k:",el3.6,"(selectivity)";wrt
6,K
52: fmt "n:",f2.0,12x,"(Order)",/,"Aa, dB:",f7.3,"
(Actual)";wrt 6,N,A[2]
53: wtb 6,12;end
54: "NBU":
55: log(D)/2log(1/K)+N;if frc(N)#0;int(N)+1+N
56: W[1]↑(1/2N)+W
57: 10log(1+(W/K)↑(2N))+A[2]
58: ret
59: "NTS":
60: 'acsh'(√D)+Z;Z/'acsh'(1/K)+N
61: if frc(N)#0;int(N)+1+N
62: N'acsh'(1/K)+X;.5(exp(X)+exp(-X))+X
63: 10log(1+W[1]X↑2)+A[2]
64: 1+W
65: ret
66: "acsh":
67: ln(p1+√(p1↑2-1))+Y
68: ret Y
69: "NEL":
70: √(1-K↑2)+H;(1-√H)/2(1+√H)+H;H+2H↑5+15H↑9+150H↑
13+Q
71: log(16D)/log(1/Q)+N;if frc(N)#0;int(N)+1+N
72: 10log(1+W[1]/16Q↑N)+A[2]
73: √K+W
74: ret


0: fmt 1,"FILE:7"
1: fmt 2,"ANALOG-FILTER APPROXIMATIONS"
2: fmt 3,"=============================="
3: wtb 6,27,69
4: dim D[0:3],T[0:1],R[20],I[20],C,D,E,F,H,S,A[0:2
,10],B[0:2,10]
5: dim P[2],Z[2],A$[2],L$[11],R$[1],O$[1];rad
6: "------------"+L$
7: for I=1 to 10;1+A[0,I];0+A[1,I]+A[2,I];next I
8: wtb 6,14,27,84,27,76,int(912/64),int(912)
9: fmt ,3/,10x,z;wrt 6;wtb 6,27,77
10: wrt 6.1;wrt 6.2;wrt 6.3
11: fmt 1,"OUTPUT DATA:",/,cll
12: fmt 2,"Section #:",f1.0,/,cll
13: fmt 3,"A",2f1.0,":",el3.6,5x,"B",2f1.0,":",el3
.6
14: fmt 6,"H(s)=H*H1(s)*H2(s)*..."
15: fmt 7,"H1(s)=(A0+A11s+A21s↑2)/(B0+B11s+B21s↑2)
"
```

```
16:  ent "TYPE OF APPROX.? BU, TS or EL",A$
17:  ent "OUT:Coefs or Zeros/Poles? c or z",O$
18:  fmt "INPUT DATA:",/,cll;wrt 6,L$
19:  cap(O$)→O$;cap(A$)→A$;if A$="BU";gto "BU"
20:  if A$="TS";gto "TS"
21:  fmt "ELLIPTIC APPROXIMATION";wrt 6
22:  ent "Ap, dB:",D[1],"Aa, dB:",D[2],"k:",K
23:  fmt "Ap, dB:",f6.3,/,"Aa, dB:",f7.3,/,"k:",e13
.6
24:  wrt 6,D[1],D[2],K
25:  √(1-K↑2)→E
26:  (1-√E)/2(1+√E)→E
27:  E+2E↑5+15E↑9+150E↑13→Q
28:  10↑(.1D[1])-1→R;(10↑(.1D[2])-1)/R→D
29:  log(16D)/log(1/Q)→N;if frc(N)#0;int(N)+1→N
30:  10log(1+R/16Q↑N)→D[3]
31:  wrt 6.1,L$
32:  fmt "Order:",f2.0,/,"Actual Aa:",f8.3;wrt 6,N,
D[3]
33:  if frc(N/2)=0;sfg 1;N/2→S;0→G;gto +2
34:  (N+1)/2→S;1→G
35:  10↑(.05D[1])→r1;ln((r1+1)/(r1-1))/2N→L
36:  1→T[0]
37:  1→M
38:  "B":
39:  2ML→X
40:  T[0]→r2
41:  2*(-1)↑MQ↑(M↑2)'csh'(X)+T[0]→T[0]
42:  M+1→M
43:  if abs(T[0]-r2)>1e-6;gto "B"
44:  L→X
45:  'snh'(X)→T[1]
46:  1→M
47:  "C":
48:  T[1]→r2
49:  (2M+1)L→X
50:  (-1)↑MQ↑(M(M+1))'snh'(X)+T[1]→T[1]
51:  M+1→M
52:  if abs(T[1]-r2)>1e-6;gto "C"
53:  2T[1]Q↑.25→T[1]
54:  T[1]/T[0]→D[0]
55:  if flg1=0;D[0]→H→B[0,1];1→B[1,1];0→B[2,1]
56:  if flg1=0;1→A[0,1];0→A[1,1]→A[2,1];gto +2
57:  1→H
58:  1→I
59:  "F":
60:  if flg1;I-.5→J;gto +2
61:  I→J
```

```
62:  1→M
63:  1→T[0]
64:  "D":
65:  T[0]→r2
66:  2MπJ/N→X
67:  2*(-1)↑MQ↑(M↑2)cos(X)+T[0]→T[0]
68:  M+1→M
69:  if abs(T[0]-r2)>1e-6;gto "D"
70:  1→M
71:  sin(πJ/N)→T[1]
72:  "E":
73:  T[1]→r2
74:  (2M+1)πJ/N→X
75:  (-1)↑MQ↑(M(M+1))sin(X)+T[1]→T[1]
76:  M+1→M
77:  if abs(T[1]-r2)>1e-6;gto "E"
78:  2T[1]Q↑.25→T[1]
79:  T[1]/T[0]→Z
80:  √((1+KD[0]↑2)(1+D[0]↑2/K))→W
81:  √((1-KZ↑2)(1-Z↑2/K))→V
82:  1+D[0]↑2Z↑2→U
83:  D[0]V/U→A
84:  ZW/U→B
85:  Z↑2(A↑2+B↑2)H→H
86:  (1/Z)↑2→A[0,I+G];1→A[2,I+G]
87:  A↑2+B↑2→B[0,I+G];2A→B[1,I+G];1→B[2,I+G]
88:  I+1→I
89:  if I<=N/2;gto "F"
90:  if flg1;H10↑(-D[1]/20)→H
91:  gto "OUT"
92:  "snh":
93:  (exp(p1)-exp(-p1))/2→X
94:  ret X
95:  "csh":
96:  (exp(p1)+exp(-p1))/2→X
97:  ret X
98:  "asnh":
99:  ln(p1+√(p1↑2+1))→Y
100: ret Y
101: ret
102: "BU":
103: fmt ,"BUTTERWORTH APPROXIMATION";wrt 6
104: ent "Order:",N
105: fmt "Order:",f1.0;wrt 6,N;wrt 6.1,L$
106: if frc(N/2)=0;N/2→S; π/2N→T;gto +2
107: 0→T;(N+1)/2→S
108: for I=1 to S
109: -cos(T)→R[I]→P[1];sin(T)→I[I]→P[2]
```

```
110:  if I[I]=0;-R[I]→B[0,I];1→B[1,I];0→B[2,I];gto
      +2
111:  R[I]↑2+I[I]↑2→B[0,I];-2R[I]→B[1,I];1→B[2,I]
112:  T+π/N→T;next I
113:  1→H
114:  gto "OUT"
115:  "TS":
116:  fmt "TSCHEBYSCHEFF APPROXIMATION";wrt 6
117:  ent "Order:",N,"Ap, dB:",A
118:  fmt "Order:",f1.0,/,"Ap, dB:",f6.3;wrt 6,N,A
119:  √(10↑(.1A)-1)→E;1/E→E
120:  (N+1)/2→S;if frc(N/2)=0;N/2→S
121:  0→I;1→H;wrt 6.1,L$
122:  for K=S to 1 by -1;I+1→I
123:  (2K-1)π/2N→T;-sin(T)'snh'('asnh'(E)/N)→R[I]
124:  cos(T)'csh'('asnh'(E)/N)→I[I]
125:  if I[I]=0;-R[I]→B[0,I];1→B[1,I];0→B[2,I];HB[0
      ,I]→H;gto +2
126:  R[I]↑2+I[I]↑2→B[0,I];-2R[I]→B[1,I];1→B[2,I];H
      B[0,I]→H
127:  next K
128:  if frc(N/2)=0;10↑(-A/20)H→H
129:  "OUT":
130:  ent "RECORD OUTPUT? y or n",R$;cap(R$)→R$
131:  if O$="Z";ldf 10
132:  wrt 6.6;wrt 6.7
133:  for I=1 to S;wrt 6.2,I,L$
134:  for J=0 to 2;wrt 6.3,J,I,A[J,I],J,I,B[J,I]
135:  next J;next I
136:  fmt "H:",e13.6;wrt 6,H
137:  if R$="Y";rcf 8,H,S,A[*],B[*]
138:  wtb 6,12;end


0:  fmt 1,"FILE:9"
1:  fmt 2,"ANALOG-FILTER TRANSFORMATIONS"
2:  fmt 3,"==============================="
3:  wtb 6,27,69
4:  dim R[20],I[20],C,D,E,F,H,S,A[0:2,10],B[0:2,10]

5:  dim A$[1],F$[2],L$[11],O$[1],R$[1],M[20]
6:  "------------"→L$
7:  wtb 6,14,27,84,27,76,int(912/64),int(912)
8:  fmt ,3/,10x,z;wrt 6;wtb 6,27,77
9:  wrt 6.1;wrt 6.2;wrt 6.3
10: fmt 1,"Section #:",f1.0,/,cll
11: fmt 2,"A",2f1.0,":",e13.6,5x,"B",2f1.0,":",e13
    .6
```

```
12: ent "FILTER TYPE:LP,HP,BP,BS?",F$
13: cap(F$)→F$;rad
14: if F$="LP";fmt "LOWPASS",z;wrt 6
15: if F$="HP";fmt "HIGHPASS",z;wrt 6
16: if F$="BP";fmt "BANDPASS",z;wrt 6
17: if F$="BS";fmt "BANDSTOP",z;wrt 6
18: fmt " FILTER";wrt 6
19: fmt "INPUT DATA:",/,cll;wrt 6,L$
20: ent "INPUT DATA ON FILE?y or n",A$
21: ent "OUT:Coefs or Zeros/Poles? c or z",O$
22: ent "RECORD OUTPUT? y or n",R$
23: cap(R$)→R$;cap(O$)→O$
24: cap(A$)→A$;if A$="N";gto +2
25: ldf 8,R[*],I[*],C,D,E,F,H;gto +3
26: enp "# Clx zero pairs:",C,"# Re zeros:",D
27: enp "# Clx pole pairs :",E,"# Re poles:",F
28: C+D→K;E+F→L;K+L→G;2C+D→N;2E+F→M;if A$="Y";gto
+5
29: "ENTER ZEROS & POLES WITH POSITIVE OR":
30: "ZERO IMAGINARY PART; START WITH ZEROS":
31: for J=1 to G;enp R[J],I[J];next J
32: enp "Multiplier:",H
33: fmt "# Complex zero pairs:",f2.0,/,"# Real zer
os:",f2.0;wrt 6,C,D
34: fmt "# Complex pole pairs:",f2.0,/,"# Real pol
es:",f2.0;wrt 6,E,F
35: fmt "H:",el3.6;wrt 6,H
36: fmt "ZEROS:";wrt 6;for J=1 to G
37: if J=K+1;fmt "POLES:";wrt 6
38: fmt el3.6," +j",el3.6;wrt 6,R[J],I[J];next J
39: for J=1 to G;R[J]↑2+I[J]↑2→M[J];next J
40: if F$#"LP" and F$#"HP";gto +3
41: if F$="LP" or F$="HP";enp "Lamda:",W
42: fmt "Lamda:",el3.6;wrt 6,W;gto +3
43: enp "Centre Freq. r/s:",W,"Bandwidth:",B;spc
44: fmt "Centre Freq. r/s:",el3.6,/,"Bandwidth r/s
:",el3.6;wrt 6,W,B
45: if F$="LP";1/W→W
46: 1→S
47: if F$#"LP";gto +3
48: if K<L;for J=1 to L-K;1→A[0,S];0→A[1,S];0→A[2,
S];S+1→S;next J
49: 1→J;gto "C"
50: if F$#"BP";gto +4
51: 1→J
52: if M>N;0→A[0,S];1→A[1,S];0→A[2,S];S+1→S;jmp (J
+1→J)>M-N
53: 1→J;gto "A"
```

```
54: 1→J
55: "D":
56: if I[J]=0 and J<=K;-R[J]H→H
57: if I[J]#0 and J<=K;M[J]H→H
58: if I[J]=0 and J>K;-H/R[J]→H
59: if I[J]#0 and J>K;H/M[J]→H
60: R[J]/M[J]→R[J];I[J]/M[J]→I[J];1/M[J]→M[J]
61: J+1→J;if J<=G;gto "D"
62: if F$="BS";gto +6
63: 1→J
64: if F>D;0→A[0,S];1→A[1,S];0→A[2,S];S+1→S;jmp (J
+1→J)>F-D
65: 1→J
66: if E>C;0→A[0,S];0→A[1,S];1→A[2,S];S+1→S;jmp (J
+1→J)>E-C
67: 1→J;gto "C"
68: 1→J
69: if M>N;W↑2→A[0,S];0→A[1,S];1→A[2,S];S+1→S;jmp
(J+1→J)>M-N
70: 1→J
71: "A":
72: if J=K+1;1→S
73: if I[J]#0;gto +6
74: W↑2→r4;-R[J]B→r5;1→r6
75: if J<=K;r4→A[0,S];r5→A[1,S];r6→A[2,S]
76: if J>K;r4→B[0,S];r5→B[1,S];r6→B[2,S]
77: J+1→J;S+1→S;if J<=G;gto "A"
78: gto "CO"
79: BR[J]/2→U;BI[J]/2→V
80: R[J]↑2-I[J]↑2→r1;2I[J]R[J]→r2
81: 4W↑2→r3;√((B↑2r1-r3)↑2+(r2B↑2)↑2)→P
82: atn(r2B↑2/(B↑2r1-r3))→Q;if r2<=0;Q+π→Q
83: √Pcos(Q/2)/2→X;√Psin(Q/2)/2→Z
84: (U+X)↑2+(V+Z)↑2→r4;-2(U+X)→r5;1→r6
85: (U-X)↑2+(V-Z)↑2→r7;-2(U-X)→r8;1→r9
86: if J<=K;r4→A[0,S];r5→A[1,S];r6→A[2,S];S+1→I;r7
→A[0,I];r8→A[1,I];r9→A[2,I]
87: if J>K;r4→B[0,S];r5→B[1,S];r6→B[2,S];S+1→I;r7→
B[0,I];r8→B[1,I];r9→B[2,I]
88: J+1→J;S+2→S;if J<=G;gto "A"
89: "CO":
90: if F$="BP";HB↑(M-N)→H
91: gto "RC"
92: "C":
93: if J=K+1;1→S
94: if I[J]=0;-WR[J]→r4;1→r5;0→r6;gto +2
95: M[J]W↑2→r4;-2WR[J]→r5;1→r6
96: if J<=K;r4→A[0,S];r5→A[1,S];r6→A[2,S]
```

```
97:  if J>K;r4+B[0,S];r5+B[1,S];r6+B[2,S]
98:  J+1+J;S+1+S;if J<=G;gto "C"
99:  if F$="LP";HW+(M-N)+H
100: "RC":
101: S-1+S
102: fmt "OUTPUT DATA:",/,cll;wrt 6,L$
103: if O$="Z";ldf 10
104: fmt "H(s)=H*H1(s)*H2(s)...";wrt 6
105: fmt "H1(s)=(A01+A11s+A21s+2)/(B01+B11s+B21s+2
)";wrt 6
106: for I=1 to S;wrt 6.1,I,L$
107: for J=0 to 2;wrt 6.2,J,I,A[J,I],J,I,B[J,I]
108: next J;next I
109: fmt ,"H:",e13.6;wrt 6,H
110: if R$="Y";rcf 8,H,S,A[*],B[*]
111: wtb 6,12;end


0:  "FILE:10":
1:  "Subroutine for FILES:7,9":
2:  "COEFFICIENTS TO ZEROS/POLES":
3:  fmt 1,e13.6," +j",e13.6
4:  0+C+D+E+F;1+J
5:  for I=1 to S
6:  if A[0,I]=0 and A[2,I]=0;1+D+D;0+R[J]+I[J];1+J+
J
7:  if A[0,I]=0 and A[1,I]=0;2+D+D;0+R[J]+I[J]+R[J+
1]+I[J+1];2+J+J
8:  4A[0,I]A[2,I]-A[1,I]+2+r1;if r1>0;1+C+C;-A[1,I]
/2A[2,I]+R[J]
9:  if r1>0;√r1/2A[2,I]+I[J];J+1+J
10: next I
11: for I=1 to S
12: if B[2,I]=0;F+1+F;-B[0,I]/B[1,I]+R[J];0+I[J];J
+1+J
13: 4B[0,I]B[2,I]-B[1,I]+2+r2;if r2>0;E+1+E;-B[1,I
]/2B[2,I]+R[J]
14: if r2>0;√r2/2B[2,I]+I[J];J+1+J
15: next I
16: fmt "# Complex zero pairs:",f1.0;wrt 6,C
17: fmt "# Real zeros:",f1.0;wrt 6,D
18: fmt "# Complex pole pairs:",f1.0;wrt 6,E
19: fmt "# Real poles:",f1.0;wrt 6,F
20: fmt "H:",13e.6;wrt 6,H;C+D+K;K+E+F+G
21: fmt "ZEROS:";wrt 6;for I=1 to K;wrt 6.1,R[I],I
[I];next I
22: fmt "POLES:";wrt 6;for I=K+1 to G;wrt 6.1,R[I]
,I[I];next I
```

```
23: if R$="Y";rcf 8,R[*],I[*],C,D,E,F,H
24: wtb 6,12;end


0: fmt 1,"FILE:11"
1: fmt 2,"BILINEAR TRANSFORMATION"
2: fmt 3,"========================="
3: wtb 6,27,69
4: dim N[10],D[10],H,S,A[0:2,10],B[0:2,10]
5: dim A$[1],D$[1],L$[11];rad
6: "-----------"→L$
7: wtb 6,14,27,84,27,76,int(912/64),int(912)
8: fmt ,3/,10x,z;wrt 6;wtb 6,27,77
9: wrt 6.1;wrt 6.2;wrt 6.3
10: fmt 1,"Section #:",f1.0,//,cll
11: fmt 2,"A",2f1.0,":",e15.6,5x,"B",2f1.0,":",e15
.6
12: ent "CASCADE or PARALLEL? c or p ",A$
13: cap(A$)→A$
14: if A$="C";sfg 1;fmt "CASCADE",z;wrt 6;gto +2
15: fmt "PARALLEL",z;wrt 6
16: fmt " FILTER";wrt 6
17: fmt "INPUT DATA:",/,cll;wrt 6,L$
18: if flg1;fmt "H(s)=H*H1(s)*H2(s)*...";wrt 6;gto
 +2
19: fmt "H(s)=H1(s)+H2(s)+...";wrt 6
20: fmt "H1(s)=(A01+A11s+A21s↑2)/(B01+B11s+B21s↑2)
";wrt 6
21: enp "Sampl. Freq., r/s:",W;2π/W→T
22: fmt "Sampl. Freq. r/s:",e13.6;wrt 6,W
23: ent "DATA on FILE? y or n",D$;cap(D$)→D$
24: if D$="N";gto +2
25: gto "LDF"
26: ent "Numb. of Sectns?",S
27: if flg1;ent "Multiplier:",H
28: 1→L;0→J
29: "SE":
30: fxd 0;spc ;prt "Section #:",L
31: 0→J
32: enp A[J,L];jmp (J+1→J)>2
33: 0→J
34: enp B[J,L];jmp (J+1→J)>2
35: L+1→L;if L<=S;gto "SE"
36: gto +3
37: "LDF":
38: ldf 8,H,S,A[*],B[*]
39: gsb "WRT"
40: for L=1 to S
```

```
41: if B[2,L]#0;gto +6
42: 2A[1,L]/T→r7;A[0,L]-r7→r1;A[0,L]+r7→r2
43: 2B[1,L]/T→r8;B[0,L]-r8→r3;B[0,L]+r8→r4
44: r1/r4→A[0,L];r2/r4→A[1,L];0→A[2,L]
45: r3/r4→B[0,L];1→B[1,L];0→B[2,L]
46: gto +9
47: A[0,L]-2A[1,L]/T+4A[2,L]/T↑2→r1
48: 2(A[0,L]-4A[2,L]/T↑2)→r2
49: A[0,L]+2A[1,L]/T+4A[2,L]/T↑2→r3
50: B[0,L]-2B[1,L]/T+4B[2,L]/T↑2→r4
51: 2(B[0,L]-4B[2,L]/T↑2)→r5
52: B[0,L]+2B[1,L]/T+4B[2,L]/T↑2→r6
53: r1/r6→A[0,L];r2/r6→A[1,L];r3/r6→A[2,L]
54: r4/r6→B[0,L];r5/r6→B[1,L];1→B[2,L]
55: next L
56: ent "RECORD DATA? y or n",D$;cap(D$)→D$
57: if D$="N";gto +2
58: rcf 8,H,S,A[*],B[*]
59: fmt ,"OUTPUT DATA:",/,cll;wrt 6,L$
60: if flg1;fmt "H(z)=H*H1(z)*H2(z)*...";wrt 6;gto
 +2
61: fmt "H(z)=H1(z)+H2(z)+...";wrt 6
62: fmt "H1(z)=(A01+A11z+A21z↑2)/(B01+B11z+B21z↑2)
";wrt 6
63: gsb "WRT"
64: wtb 6,12
65: end
66: "WRT":
67: for L=1 to S
68: wrt 6.1,L,L$;for J=0 to 2
69: wrt 6.2,J,L,A[J,L],J,L,B[J,L]
70: next J;next L
71: if flg1;fmt ,"H:",e14.6;wrt 6,H
72: ret


0: fmt 1,"FILE:12"
1: fmt 2,"PARTIAL FRACTIONS"
2: fmt 3,"================="
3: wtb 6,27,69
4: dim R[20],I[20],C,D,E,F,H,S,A[0:2,10],B[0:2,10]
,M[10]
5: dim Y,Z,P[10],Q[10],U[10],V[10],T[10]
6: dim A$[1],L$[11],O$[1],R$[1]
7: "-----------"→L$
8: wtb 6,14,27,84,27,76,int(912/64),int(912)
9: fmt ,3/,10x,z;wrt 6;wtb 6,27,77
10: wrt 6.1;wrt 6.2;wrt 6.3
```

```
11: fmt "INPUT DATA:",/,cll;wrt 6,L$
12: ent "INPUT DATA ON FILE? y or n",A$;cap(A$)→A$

13: if A$="N";gto +2
14: ldf 8,R[*],I[*],C,D,E,F,H;gto +3
15: enp "# Complex zero pairs:",C,"# Real zeros:",
D
16: enp "# Complex pole pairs :",E,"# Real poles:"
,F
17: C+D→K;E+F→L;K+L→G;2C+D→N;2E+F→M;if cap(A$)="Y"
;gto +5
18: "ENTER ZEROS & POLES WITH POSITIVE OR":
19: "ZERO IMAGINARY PART; START WITH ZEROS":
20: for J=1 to G;enp R[J],I[J];next J
21: enp "Multiplier:",H
22: fmt "# Complex zero pairs:",f2.0,/,"# Real zer
os:",f2.0;wrt 6,C,D
23: fmt ,"# Complex pole pairs:",f2.0,/,"# Real po
les:",f2.0;wrt 6,E,F
24: fmt "Multiplier:",e13.6;wrt 6,H
25: fmt "ZEROS:";wrt 6;for J=1 to G
26: if J=K+1;fmt "POLES:";wrt 6
27: fmt e13.6," +j",e13.6;wrt 6,R[J],I[J];next J
28: for J=1 to K;R[J]↑2+I[J]↑2→M[J];next J
29: fmt "OUTPUT DATA:",/,cll;wrt 6,L$
30: ent "OUT:Coefs or Poles/Resds? c or p",O$
31: ent "RECORD OUTPUT? y or n",R$
32: cap(O$)→O$;cap(R$)→R$
33: for I=1 to L;R[K+I]→P[I];I[K+I]→Q[I];P[I]↑2+Q[
I]↑2→T[I]
34: 0→R[K+I]→I[K+I];next I
35: for I=1 to L
36: H→U[I];0→V[I]
37: if Q[I]#0;gto "CXP"
38: for J=1 to K
39: if I[J]=0;U[I](P[I]-R[J])→U[I];gto +2
40: U[I](P[I]↑2-2R[J]P[I]+M[J])→U[I]
41: next J
42: for J=1 to L
43: if P[J]=P[I] and Q[J]=Q[I];gto +3
44: if Q[J]=0;U[I]/(P[I]-P[J])→U[I];gto +2
45: U[I]/(P[I]↑2-2P[J]P[I]+T[J])→U[I]
46: next J
47: gto "A"
48: "CXP":
49: for J=1 to K
50: if I[J]=0;cll 'CX*'(P[I]-R[J],Q[I]);gto +2
51: cll 'CX*'(P[I]↑2-Q[I]↑2-2R[J]P[I]+M[J],2(P[I]Q
```

```
   [I]-R[J]Q[I]))
52: next J
53: for J=1 to L
54: if P[J]=P[I] and Q[J]=Q[I];U[I]→r];V[I]/2Q[I]→
U[I];-r]/2Q[I]→V[I];gto +3
55: if Q[J]=0;cll 'CX/'(P[I]-P[J],Q[I]);gto +2
56: cll 'CX/'(P[I]↑2-Q[I]↑2-2P[J]P[I]+T[J],2(P[I]Q
[I]-P[J]Q[I]))
57: next J
58: "A":
59: if Q[I]#0;gto +3
60: U[I]→A[0,I];0→A[1,I];0→A[2,I]
61: -P[I]→B[0,I];1→B[1,I];0→B[2,I];gto +3
62: (-2)(U[I]P[I]+V[I]Q[I])→A[0,I];2U[I]→A[1,I];0→
A[2,I]
63: T[I]→B[0,I];-2P[I]→B[1,I];1→B[2,I]
64: next I
65: if M#N;gto +3
66: H→A[0,I];0→A[1,I];0→A[2,I]
67: 1→B[0,I];0→B[1,I];0→B[2,I]
68: L→Z→S;if M=N;L+1→S
69: if O$="P";gsb "OUT1"
70: if O$="C";gsb "OUT2"
71: wtb 6,12;end
72: "CX*":
73: U[I]p]-V[I]p2→p3;V[I]p]+U[I]p2→p4
74: p3→U[I];p4→V[I];ret
75: "CX/":
76: p]↑2+p2↑2→p3;(U[I]p]+V[I]p2)/p3→p4
77: (V[I]p]-U[I]p2)/p3→p5
78: p4→U[I];p5→V[I];ret
79: "OUT1":
80: fmt ,12x,"POLES",24x,"RESIDUES";wrt 6
81: fmt 1,e13.6," j",e13.6,3x,e13.6," j",e13.6
82: for I=1 to Z;wrt 6.1,P[I],Q[I],U[I],V[I]
83: next I;if R$="Y";rcf 8,Y,Z,P[*],Q[*],U[*],V[*]

84: 0→Y;if M=N;H→Y;fmt "Constant:",e13.6;wrt 6,H
85: ret
86: "OUT2":
87: fmt "H(z)=H1(z)+H2(z)+...";wrt 6
88: fmt "H](z)=(A01+A11z+A21z↑2)/(B01+B11z+B21z↑2)
";wrt 6
89: fmt 2,"Section #:",f1.0,/,cll
90: fmt 3,"A",2f1.0,":",e13.6,5x,"B",2f1.0,":",e13
.6
91: for I=1 to S;wrt 6.2,I,L$
92: for J=0 to 2;wrt 6.3,J,I,A[J,I],J,I,B[J,I]
```

```
93: next J;next I
94: if R$="Y";rcf 8,H,S,A[*],B[*]
95: ret


0: fmt 1,"FILE:13"
1: fmt 2,"IMPULSE ,UNIT-STEP & SINUSOID INVARIANT
METHODS"
2: wtb 6,27,69
3: fmt 3,"=========================================
======="
4: dim H,S,A[0:2,10],B[0:2,10]
5: dim Y,Z,P[10],Q[10],U[10],V[10]
6: dim C[2],D[2],E[2]
7: dim A$[1],B$[1],L$[11]
8: "-----------"→L$
9: wtb 6,14,27,84,27,76,int(912/64),int(912)
10: fmt ,3/,10x,z;wrt 6;wtb 6,27,77
11: wrt 6.1;wrt 6.2;wrt 6.3
12: fmt "INPUT DATA:",/,cll;wrt 6,L$
13: rad;enp "SAMP. FREQ., r/s:",T
14: fmt "Sampling Freq., r/s:",e13.6;wrt 6,T;2π/T→
T
15: ent "IMPLS, UNSTP or SINSD? i, u or s",A$
16: ent "INPUT DATA ON FILE? y or n",B$
17: cap(A$)→A$;cap(B$)→B$
18: if' A$="I";fmt "IMPULSE ",z;wrt 6;gto +4
19: if A$="U";fmt "UNIT-STEP ",z;wrt 6;gto +3
20: fmt "SINUSOID ",z;wrt 6
21: enp "DESIGN FREQ., r/s:",W;cos(WT)→C;sin(WT)→S

22: fmt "INVARIANT METHOD";wrt 6
23: if A$="S";fmt "Design Freq., r/s:",e13.6;wrt 6
,W
24: if B$="Y";gto +6
25: ent "# SECTIONS:",Z;for I=1 to Z
26: enp P[I],Q[I],U[I],V[I];next I
27: ent "CONSTANT(if num. deg.=den. deg):",Y
28: if flg13;0→Y;cfg 13
29: gto +2
30: ldf 8,Y,Z,P[*],Q[*],U[*],V[*]
31: if Y#0 and A$="I";prt "Impls. Inv. meth. does
not work in this case";end
32: fmt ,12x,"Poles",24x,"Residues";wrt 6
33: fmt 1,e13.6," j",e13.6,3x,e13.6," j",e13.6
34: for I=1 to Z;wrt 6.1,P[I],Q[I],U[I],V[I];next
I
35: if Y#0;fmt "Constant:",e13.6;wrt 6,Y
```

```
36: for I=1 to Z
37: exp(TP[I])→r1;r1cos(TQ[I])→E[1];r1sin(TQ[I])→E
[2]
38: if Q[I]=0;-E[1]→B[0,I];1→B[1,I];0→B[2,I];gto +
2
39: E[1]↑2+E[2]↑2→B[0,I];-2E[1]→B[1,I];1→B[2,I]
40: if A$="U";gto "USP"
41: if A$="S";gto "SIN"
42: if Q[I]=0;0→A[0,I];TU[I]→A[1,I];0→A[2,I];gto "
Z"
43: 0→A[0,I];-2T(U[I]E[1]+V[I]E[2])→A[1,I];2TU[I]→
A[2,I];gto "Z"
44: "USP":
45: 1-E[1]→C[1];-E[2]→C[2];cll 'mlt*'(U[I],V[I],C[
1],C[2],C[1],C[2])
46: cll 'div*'(C[1],C[2],-P[I],-Q[I],C[1],C[2])
47: if Q[I]=0;C[1]→A[0,I];0→A[1,I]→A[2,I];gto "Z"
48: cll 'mlt*'(C[1],C[2],E[1],-E[2],r1,r2)
49: -2r1→A[0,I];2C[1]→A[1,I];0→A[2,I];gto "Z"
50: "SIN":
51: -SP[I]+W(E[1]-C)→C[1];-SQ[I]+WE[2]→C[2]
52: cll 'mlt*'(U[I],V[I],C[1],C[2],C[1],C[2])
53: SP[I]-WC→D[1];SQ[I]→D[2]
54: cll 'mlt*'(E[1],E[2],D[1],D[2],D[1],D[2])
55: W+D[1]→D[1];cll 'mlt*'(U[I],V[I],D[1],D[2],D[1
],D[2])
56: cll 'mlt*'(P[I],Q[I],P[I],Q[I],r1,r2)
57: r1+W↑2→r1;Sr1→r1;Sr2→r2
58: cll 'div*'(C[1],C[2],r1,r2,C[1],C[2])
59: cll 'div*'(D[1],D[2],r1,r2,D[1],D[2])
60: if Q[I]=0;D[1]→A[0,I];C[1]→A[1,I];0→A[2,I];gto
 "Z"
61: 2C[1]→A[2,I];cll 'mlt*'(C[1],C[2],E[1],-E[2],r
1,r2)
62: 2(D[1]-r1)→A[1,I];cll 'mlt*'(D[1],D[2],E[1],-E
[2],r1,r2)
63: -2r1→A[0,I]
64: "Z":
65: next I
66: Z→S;if Y#0;1+S→S;Y→A[0,S];1→B[0,S];0→A[1,S]→A[
2,S]→B[1,S]→B[2,S]
67: "OUT":
68: fmt "OUTPUT DATA:",/,cll;wrt 6,L$
69: ent "RECORD OUTPUT DATA? y or n",B$
70: if cap(B$)="Y";rcf 8,H,S,A[*],B[*]
71: fmt "H[z]=H1[z]+H2[z]+...";wrt 6
72: fmt "H1[z]=(A01+A11z+A21z↑2)/(B01+B11z+B21z↑2)
";wrt 6
```

```
73: fmt 1,"Section #:",f1.0,/,cll
74: fmt 2,"A",2f1.0,":",el3.6,5x,"B",2f1.0,":",el3
 6
75: for I=1 to S;wrt 6.1,I,L$
76: for J=0 to 2;wrt 6.2,J,I,A[J,I],J,I,B[J,I]
77: next J;next I
78: wtb 6,12;end
79: "mlt*":
80: p1p3-p2p4→p7
81: p1p4+p2p3→p6
82: p7→p5;ret
83: "div*":
84: p1p3+p2p4→p7
85: p2p3-p1p4→p8
86: p3↑2+p4↑2→p9
87: p7/p9→p5;p8/p9→p6;ret


0: fmt 1,"FILE:14"
1: fmt 2,"MATCHED-Z TRANSFORMATION METHOD"
2: fmt 3,"================================="
3: wtb 6,27,69
4: dim R[20],I[20],C,D,E,F,H,S,A[0:2,10],B[0:2,10]

5: dim P[10],Q[10]
6: dim A$[2],F$[2],L$[11],R$[1]
7: "------------"→L$;rad
8: wtb 6,14,27,84,27,76,int(912/64),int(912)
9: fmt ,3/,10x,z;wrt 6;wtb 6,27,77
10: wrt 6.1;wrt 6.2;wrt 6.3
11: fmt 1,"Section #:",f1.0,/,cll
12: fmt 2,"A",2f1.0,":",el3.6,5x,"B",2f1.0,":",el3
 6
13: ent "APPROXIMATION? BU, TS or EL",A$;cap(A$)→A
$
14: if A$="BU";fmt "BUTTERWORTH ",z;wrt 6
15: if A$="TS";fmt "TSCHEBYSCHEFF ",z;wrt 6
16: if A$="EL";fmt "ELLIPTIC ",z;wrt 6
17: ent "TYPE OF FILTER? LP,HP,BP or BS",F$
18: cap(F$)→F$;if F$="LP";fmt "LOWPASS ",z;wrt 6
19: if F$="HP";fmt "HIGHPASS ",z;wrt 6
20: if F$="BP";fmt "BANDPASS ",z;wrt 6
21: if F$="BS";fmt "BANDSTOP ",z;wrt 6
22: fmt "FILTER";wrt 6
23: fmt "INPUT DATA:",/,cll;wrt 6,L$
24: enp "SAMPLING FREQ., r/s:",T
25: fmt "Sampling Freq., r/s:",el3.6;wrt 6,T;2π/T→
T
```

```
26: ent "INPUT DATA ON FILE?y or n",R$;cap(R$)→R$
27: if R$="N";gto +2
28: ldf 8,R[*],I[*],C,D,E,F,H;gto +3
29: enp "# Complex zero pairs:",C,"# Real zeros:",
D
30: enp "# Complex pole pairs :",E,"# Real poles:"
,F
31: C+D→K;E+F→S;K+S→G;2C+D→N;2E+F→M;if R$="Y";gto
+5
32: "ENTER ZEROS & POLES WITH POSITIVE OR":
33: "ZERO IMAGINARY PART; START WITH ZEROS":
34: for J=1 to G;enp R[J],I[J];next J
35: enp "Multiplier:",H
36: fmt "# Complex zero pairs:",f2.0,/,"# Real zer
os:",f2.0;wrt 6,C,D
37: fmt "# Complex pole pairs:",f2.0,/,"# Real pol
es:",f2.0;wrt 6,E,F
38: fmt "H:",e13.6;wrt 6,H
39: fmt "ZEROS:";wrt 6;for J=1 to G
40: if J=K+1;fmt "POLES:";wrt 6
41: fmt e13.6," +j",e13.6;wrt 6,R[J],I[J];next J
42: for I=1 to S;R[K+I]→P[I];I[K+I]→Q[I];next I
43: 1→L;0→Q;if F$="BS";gto +3
44: if A$#"EL";cll 'BUTS';gto +5
45: gsb "EL"
46: for I=L to S;if I[I-Q]=0;gto +2
47: exp(TR[I-Q])→B;B↑2→A[0,I];-2Bcos(TI[I-Q])→A[1,
I];1→A[2,I]
48: next I
49: for I=1 to S;if Q[I]#0;gto +2
50: -exp(TP[I])→B[0,I];1→B[1,I];0→B[2,I];gto +2
51: exp(TP[I])→B;B↑2→B[0,I];-2Bcos(TQ[I])→B[1,I];1
→B[2,I]
52: next I
53: 1→H;1→A;if F$="HP";-1→A
54: if F$="BP" or F$="BS";gto +3
55: for I=1 to S
56: (B[0,I]+AB[1,I]+B[2,I])H/(A[0,I]+AA[1,I]+A[2,I
])→H;next I
57: "OUT":
58: fmt "OUTPUT DATA:",/,cll;wrt 6,L$
59: ent "RECORD OUTPUT DATA? y or n",R$
60: if cap(R$)="Y";rcf 8,H,S,A[*],B[*]
61: fmt "H[z]=H*H1[z]*H2[z]*...";wrt 6
62: fmt "H1[z]=(A01+A11z+A21z↑2)/(B01+B11z+B21z↑2)
";wrt 6
63: fmt 1,"Section #:",f1.0,/,cll
64: fmt 2,"A",2f1.0,":",e13.6,5x,"B",2f1.0,":",e13
```

```
    6
65: for I=1 to S;wrt 6.1,I,L$
66: for J=0 to 2;wrt 6.2,J,I,A[J,I],J,I,B[J,I]
67: next J;next I
68: fmt "H:",e13.6;wrt 6,H
69: if F$="BP" or F$="BS";fmt "(Readjust H for zer
o min loss)";wrt 6
70: wtb 6,12;end
71: "BUTS":
72: if F$="BP";gto "BP"
73: if frc(M/2)=0;M/2→P;gto +2
74: 1→A[0,1]→A[1,1];0→A[2,1];2→L;(M+1)/2→P;if F$="
HP";-1→A[0,1]
75: for I=L to P;1→A[0,I]→A[2,I];2→A[1,I];if F$="H
P";-2→A[1,I]
76: next I;ret
77: "BP":
78: if frc(M/4)=0;M/4→P;gto +2
79: -1→A[0,1];0→A[1,1];1→A[2,1];2→L;(M+2)/4→P
80: for I=L to P;1→A[0,I]→A[2,I];-2→A[1,I];next I
81: for I=P+1 to S;1→A[0,I]→A[2,I];2→A[1,I];next I
;ret
82: "EL":
83: if F$="BP";gto +3
84: if frc(M/2)#0;2→L;1→Q;1→A[0,1]→A[1,1];0→A[2,1]
;if F$="HP";-1→A[0,1];0→Q
85: ret
86: if frc(M/4)#0;2→L;-1→A[0,1];0→A[1,1];1→A[2,1]
87: ret


0: fmt 1,"FILE:15"
1: fmt 2,"DESIGN OF NONRECURSIVE FILTERS"
2: fmt 3,"=============================="
3: wtb 6,27,69
4: dim A$[1],D$[1],F$[2],L$[11],R$[1],W$[2];"-----
-------"→L$;rad
5: wtb 6,14,27,84,27,76,int(912/64),int(912)
6: fmt ,3/,10x,z;wrt 6;wtb 6,27,77
7: wrt 6.1;wrt 6.2;wrt 6.3
8: dim A[2],F[4],I[2],W[3],M[100]
9: "CHOOSE THE KAISER WINDOW FOR PRESCRIBED SPECS"
:
10: ent "WINDOW:R, BA, HN, HM, BL or KA",W$;cap(W$
)→W$
11: if W$="R";fmt "RECTANGULAR ",z;wrt 6
12: if W$="BA";fmt "BARTLETT ",z;wrt 6
13: if W$="HN";fmt "HAN ",z;wrt 6
```

```
14: if W$="HM";fmt "HAMMING ",z;wrt 6
15: if W$="BL";fmt "BLACKMAN ",z;wrt 6
16: if W$="KA";fmt "KAISER ",z;wrt 6
17: fmt "WINDOW";wrt 6
18: ent "TYPE OF FILTER? LP,HP,BP or BS",F$;cap(F$
)→F$
19: if F$="LP";fmt "LOWPASS ",z;wrt 6
20: if F$="HP";fmt "HIGHPASS ",z;wrt 6
21: if F$="BP";fmt "BANDPASS ",z;wrt 6
22: if F$="BS";fmt "BANDSTOP ",z;wrt 6
23: fmt "FILTER";wrt 6
24: fmt "INPUT DATA:",/,cll;wrt 6,L$
25: ent "Saml. Freq. r/s",W
26: fmt "Sampling Freq., r/s:",e13.6;wrt 6,W;2π/W→
T
27: if W$#"KA";gto "A"
28: ent "Ap,dB:",A[1],"Aa:",A[2]
29: fmt "Ap, dB:",f6.3,/,"Aa, dB:",f7.3;wrt 6,A[1]
,A[2]
30: if F$="BP" or F$="BS";gto +3
31: ent "Wp, r/s:",F[1],"Wa:",F[2];abs(F[2]-F[1])→
B
32: fmt ,e13.6,/,e13.6;wrt 6,"Wp, r/s:",F[1],"Wa r
/s:",F[2];gto +5
33: ent "Wp1,r/s:",F[1],"Wp2:",F[2],"Wa1:",F[3],"W
a2:",F[4]
34: fmt "Wp1,r/s:",e13.6,/,"Wp2,r/s:",e13.6;wrt 6,
F[1],F[2]
35: fmt "Wa1,r/s:",e13.6,/,"Wa2,r/s:",e13.6;wrt 6,
F[3],F[4]
36: min(abs(F[1]-F[3]),abs(F[4]-F[2]))→B
37: 10↑(-.05A[2])→r1;10↑(.05A[1])→r2;min(r1,(r2-1)
/(r2+1))→r3;-20log(r3)→A
38: if F$="LP";F[1]+B/2→F[2];0→F[1]
39: if F$="HP";F[1]-B/2→F[1];W/2→F[2]
40: if F$="BP";F[1]-B/2→F[1];F[2]+B/2→F[2]
41: if F$="BS";F[1]+B/2→F[1];F[2]-B/2→F[2]
42: fmt "Actual Ap, dB:",e13.6,/,"Actual Aa, dB:",
e13.6
43: wrt 6,20log((1+r3)/(1-r3)),A
44: if A>50;.1102(A-8.7)→E
45: if A>=21 and A<=50;.5842(A-21)↑.4+.07886(A-21)
→E
46: if A<21;0→E
47: .9222→D;if A>21;(A-7.95)/14.36→D
48: cll 'I(x)'(E,I[2])
49: int(DW/B+1)→N;if frc(N/2)=0;N+1→N
50: gto "OUT"
```

```
51: "A":ent N;if frc(N/2)=0;N+1→N
52: if F$="BP" or F$="BS";gto +4
53: ent "CUTOFF FREQ. r/s:",F[1];fmt "Cutoff Freq.
 r/s:",e13.6
54: wrt 6,F[1];if F$="LP";F[1]→F[2];0→F[1];gto +5
55: W/2→F[2];gto +4
56: ent "LOW CUTOFF FREQ., r/s",F[1],"HIGH CUTOFF
FREQ.:",F[2]
57: fmt "Low Cutoff Freq. r/s:",e13.6,/,"High Cuto
ff Freq. r/s:",e13.6
58: wrt 6,F[1],F[2]
59: "OUT":fmt "N:",f4.0;wrt 6,N
60: fmt "OUTPUT DATA:",/,cll;wrt 6,L$
61: fmt "+n or -n",5x,"h(n)",9x,"w(n)h(n)";wrt 6
62: fmt 1,f4.0,e17.6,z
63: fmt 2,e15.6
64: (N-1)/2→C
65: dim H[0:C]
66: (F[2]-F[1])T/π→H[0];if F$="BS";1-H[0]→H[0]
67: wrt 6.1,0,H[0];wrt 6.2,H[0]
68: for I=1 to C
69: (sin(F[2]IT)-sin(F[1]IT))(1/πI)→H[I];if F$="BS
";-H[I]→H[I]
70: wrt 6.1,I,H[I]
71: if W$="BA";H[I](1-abs(I/C))→H[I]
72: if W$="HN";H[I](.5+.5cos(πI/C))→H[I]
73: if W$="HM";H[I](.54+.46cos(πI/C))→H[I]
74: if W$="BL";H[I](.42+.5cos(πI/C)+.08cos(2πI/C))
→H[I]
75: if W$#"KA";gto +3
76: E√(1-(I/C)↑2)→X;cll 'I(x)'(X,I[1])
77: H[I]I[1]/I[2]→H[I]
78: wrt 6.2,H[I];2H[I]→H[I]
79: next I
80: ldf 16
81: end
82: "I(x)":
83: if p1=0;1→p2;ret
84: 1→S;1→U;0→V
85: 2+V→V;Up1↑2/V↑2→U;S+U→S
86: if U>=2e-8S;gto -1
87: S→p2;ret


0: "FILE:16":
1: "Cont. of FILE:15":
2: ldf 5,24,3
3: fmt "AMPLITUDE RESPONSE:",/,2cll;wrt 6,L$,L$
```

```
4: fmt ,4x,"Freq., r/s",5x,"Gain    dB";wrt 6
5: fmt 1,2e15.6
6: "RE":sfg 14;0→A
7: enp "INIT.FREQ. r/s:",W[1],"FREQ.INCR.:",W[2],"
FINAL FREQ.:",W[3]
8: if (W[3]-W[1])/W[2]→r1;if frc(r1)=0;r1+1→r1
9: prt "# of points:",r1;if r1>50;beep;wait 500;be
ep
10: if r1>50;ent "Points>50;CONTINUE? y or n",R$
11: if cap(R$)="N";"Y"→R$;gto "RE"
12: if W[1]=0;1e-9→W[1];W[3]+1e-9→W[3]
13: W[1]→W
14: "BN":A+1→A;0→J→S
15: S+H[J]cos(WJT)→S;jmp (J+1→J)>C
16: 20log(abs(S))→M[A];if M[A]<-80;-80→M[A]
17: wrt 6.1,W,M[A]
18: W+W[2]→W;if W<=W[3];gto "BN"
19: A→P;wtb 6,12;ent "PLOT? y or n",D$;cap(D$)→D$
20: if D$="Y";gsb "PLT"
21: ent "NEW RESPONSE? y or n",A$;cap(A$)→A$
22: if A$="Y";gto "RE"
23: end


0: fmt 1,"FILE:17"
1: fmt 2,"FAST FOURIER TRANSFORM"
2: fmt 3,"======================";rad
3: wtb 6,27,69
4: dim B[0:10],A$[1],L$[13]
5: "--------------"→L$
6: wtb 6,14,27,84,27,76,int(912/64),int(912)
7: fmt ,3/,5x,z;wrt 6;wtb 6,27,77
8: wrt 6.1;wrt 6.2;wrt 6.3
9: fmt 1,f4.0,4e15.6
10: fmt 2,f4.0,2e15.6
11: fmt 3,"No. of elements:",f3.0
12: fmt 4,3x,"n",6x,"REAL X",7x,"IMAGINARY X"
13: fmt 5,3x,"k",6x,"REAL X",7x,"IMAGINARY X",5x,"
MAGNITUDE",5x,"ANGLE,deg"
14: enp "N:",N
15: int(ln(N)/ln(2)+1e-6)→R;if 2↑R#N;prt "N NOT a
power of 2";gto -1
16: dim X[0:N-1,2],Y[0:N-1,2],W[0:N/2-1,2]
17: wrt 6.3,N
18: fmt "INPUT ARRAY:",/,cl3;wrt 6,L$;wrt 6.4
19: for K=0 to N-1;spc ;enp X[K,1],X[K,2]
20: wrt 6.2,K,X[K,1],X[K,2]
21: X[K,1]→Y[K,1];X[K,2]→Y[K,2];next K
```

```
22: ent "INVERT? y or n",A$
23: cap(A$)→A$;if A$="N";gto +4
24: "INVRT":
25: for K=0 to N-1
26: X[K,1]/N→Y[K,1];-X[K,2]/N→Y[K,2];next K
27: "RE-ORDER":
28: for K=0 to N-1;-1→I;K→L
29: I+1→I;L/2→M
30: if 2int(M)=L;M→L;0→B[I];gto +2
31: int(M)→L;1→B[I]
32: if L>0;gto -3
33: for J=I+1 to R-1;0→B[J];next J;R-1→I
34: 0→C;for J=0 to I;C+B[J]2↑(I-J)→C;next J
35: Y[K,1]→X[C,1];Y[K,2]→X[C,2];next K
36: for S=1 to R
37: "W ARRAY":
38: 2↑(S-1)→L;2↑(R-S)→P
39: for K=0 to L-1
40: -PK2π/N→r1;cos(r1)→W[K,1];sin(r1)→W[K,2];next
K
41: "BUTTERFLIES":
42: for I=0 to N-2L by 2L;for K=0 to L-1
43: I+K→J;J+L→M
44: X[M,1]W[K,1]-X[M,2]W[K,2]→r1
45: X[M,1]W[K,2]+X[M,2]W[K,1]→r2
46: X[J,1]-r1→X[M,1];X[J,2]-r2→X[M,2]
47: X[J,1]+r1→X[J,1];X[J,2]+r2→X[J,2]
48: next K;next I;next S
49: fmt "OUTPUT ARRAY:",/,c13;wrt 6,L$;wrt 6.5
50: for K=0 to N-1
51: X[K,1]→Y[K,1];X[K,2]→Y[K,2]
52: if A$="Y";-X[K,2]→X[K,2]
53: sfg 14;√(X[K,1]↑2+X[K,2]↑2)→B;deg
54: atn(X[K,2]/X[K,1])→A
55: if X[K,1]<0 and X[K,2]>=0;A+180→A
56: if X[K,1]<0 and X[K,2]<0;A-180→A
57: wrt 6.1,K,X[K,1],X[K,2],B,A;next K
58: if A$="Y";gto +5
59: ent "CHECK? y or n",A$
60: cap(A$)→A$;if A$="N";gto +3
61: fmt "CHECK";wrt 6
62: rad;gto "INVRT"
63: wtb 6,12;end
```

B.5 TYPICAL RUNS

```
FILE:1
DIGITAL-FILTER TIME-DOMAIN RESPONSE
===================================
CASCADE FILTER
INPUT DATA:
-----------
H(z)=H*H1(z)*H2(z)*...
H1(z)=(A01+A11z+A21z↑2)/(B01+B11z+B21z↑2)
Section #:1
-----------
A01:    1.381727E 01        B01:    2.699994E-01
A11:    2.545597E 01        B11:   -9.085607E-01
A21:    1.381727E 01        B21:    1.000000E 00
Section #:2
-----------
A02:    2.468324E 00        B02:    5.313024E-01
A12:    2.744249E 00        B12:   -6.610971E-01
A22:    2.468324E 00        B22:    1.000000E 00
Section #:3
-----------
A03:    1.745879E 00        B03:    8.408148E-01
A13:    1.171356E 00        B13:   -4.795876E-01
A23:    1.745879E 00        B23:    1.000000E 00
H:   2.225780E-04

SAMPLE                OUTPUT
=========================
IMPULSE RESPONSE
        0        1.325320E-02
        1        7.520248E-02
        2        2.075813E-01
        3        3.495776E-01
        4        3.655365E-01
        5        1.857260E-01
        6       -6.848712E-02
        7       -1.792566E-01
        8       -7.526075E-02
        9        8.276229E-02
       10        1.036325E-01
         .            .

         .            .

         .            .
```

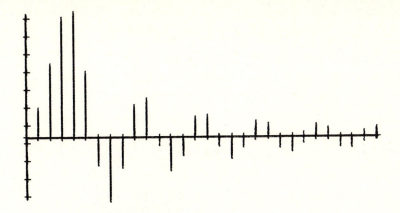

MIN-Y:-1.792570E-01
MAX-Y: 3.655360E-01
UNITS/DIV: 5.447930E-02

```
FILE:3
FREQUENCY RESPONSE
==================
ANALOG CASCADE FILTER
INPUT DATA:
-----------
H(z)=H*H1(z)*H2(z)...
H1(z)=(A01+A11z+A21z↑2)/(B01+B11z+B21z↑2)
Section #:1
-----------
A01:   1.434825E 01      B01:   1.320934E-01
A11:   0.000000E 00      B11:   4.646803E-01
A21:   1.000000E 00      B21:   1.000000E 00
Section #:2
-----------
A02:   2.231643E 00      B02:   4.970516E-01
A12:   0.000000E 00      B12:   2.671900E-01
A22:   1.000000E 00      B22:   1.000000E 00
Section #:3
-----------
A03:   1.320447E 00      B03:   7.829546E-01
A13:   0.000000E 00      B13:   1.076335E-01
A23:   1.000000E 00      B23:   1.000000E 00
Section #:4
-----------
A04:   1.128832E 00      B04:   8.987104E-01
A14:   0.000000E 00      B14:   2.681721E-02
A24:   1.000000E 00      B24:   1.000000E 00
H:  8.627105E-04
OUTPUT DATA:
-----------
     Freq., r/s      Gain,    dB
     1.000000E-09   -9.999983E-01
     2.500000E-02   -9.812484E-01
     5.000000E-02   -9.260114E-01
     7.500000E-02   -8.373783E-01
     1.000000E-01   -7.206563E-01
     1.250000E-01   -5.835340E-01
     1.500000E-01   -4.361604E-01
     1.750000E-01   -2.909635E-01
     2.000000E-01   -1.619944E-01
           .              .
           .              .
           .              .
```

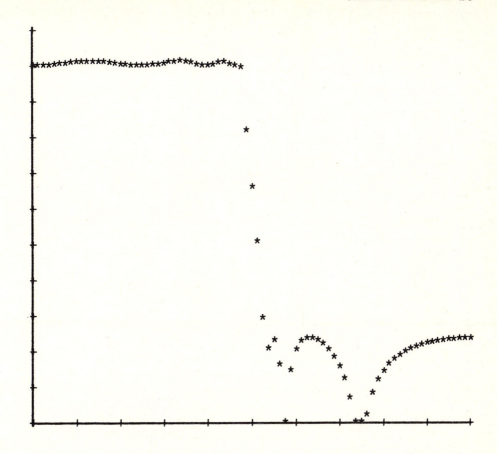

MIN-X: 1.000000E-09   MIN-Y:-8.000000E 01
MAX-X: 2.000000E 00   MAX-Y:-4.011885E-03
DIV-X: 2.000000E-01   DIV-:Y 7.999599E 00

OUTPUT DATA:

Freq., r/s	Delay, Sec.
1.000000E-09	4.222677E 00
1.250000E-02	4.228936E 00
2.500000E-02	4.247687E 00
3.750000E-02	4.278845E 00
5.000000E-02	4.322259E 00
6.250000E-02	4.377695E 00
7.500000E-02	4.444814E 00
8.750000E-02	4.523143E 00
1.000000E-01	4.612041E 00
1.125000E-01	4.710664E 00
1.250000E-01	4.817924E 00

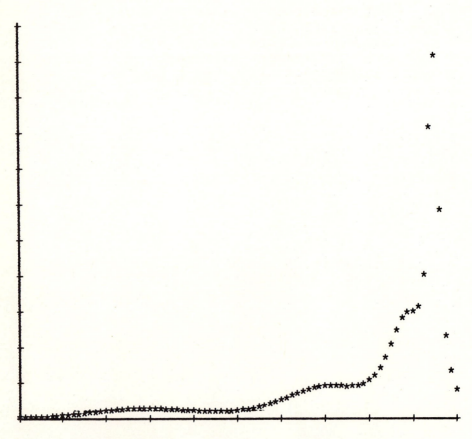

MIN-X: 1.000000E-09   MIN-Y: 4.222677E 00
MAX-X: 1.000000E 00   MAX-Y: 7.628205E 01
DIV-X: 1.000000E-01   DIV-:Y 7.205937E 00

```
FILE:6
RECURSIVE FILTERS: PRESCRIBED SPECIFICATIONS
============================================
ELLIPTIC BANDPASS FILTER
INPUT DATA:
------------
Sampling Freq., r/s: 6.000000E 03
Ap, dB: 1.000
Aa, dB: '45.000
Wp1, r/s: 9.000000E 02
Wp2, r/s: 1.100000E 03
Wa1, r/s: 8.000000E 02
Wa2, r/s: 1.200000E 03
OUTPUT DATA:
------------
Wo, r/s: 1.098609E 03
B,  r/s: 3.719263E 02
k: 5.159572E-01(selectivity)
n: 4             (Order)
Aa, dB: 50.636 (Actual)
```

```
FILE:7
ANALOG-FILTER APPROXIMATIONS
============================
INPUT DATA:
------------
ELLIPTIC APPROXIMATION
Ap, dB: 1.000
Aa, dB: 45.000
k: 8.000000E-01
OUTPUT DATA:
------------
Order: 6
Actual Aa:  53.920
H(s)=H*H1(s)*H2(s)*...
H1(s)=(A0+A11s+A21s↑2)/(B0+B11s+B21s↑2)
Section #:1
------------
A01: 1.199341E 01       B01: 1.592827E-01
A11: 0.000000E 00       B11: 5.039614E-01
A21: 1.000000E 00       B21: 1.000000E 00
Section #:2
------------
A02: 2.000130E 00       B02: 5.555270E-01
A12: 0.000000E 00       B12: 2.533954E-01
A22: 1.000000E 00       B22: 1.000000E 00
Section #:3
------------
A03: 1.302358E 00       B03: 7.976494E-01
A13: 0.000000E 00       B13: 6.522405E-02
A23: 1.000000E 00       B23: 1.000000E 00
H: 2.013515E-03
```

```
FILE:7
ANALOG-FILTER APPROXIMATIONS
=============================
INPUT DATA:
-----------
ELLIPTIC APPROXIMATION
Ap, dB: 0.100
Aa, dB: 30.000
k: 8.000000E-01
OUTPUT DATA:
-----------
Order: 5
Actual Aa:  31.492
# Complex zero pairs:2
# Real zeros:0
# Complex pole pairs:2
# Real poles:1
H:1.172666E-01
ZEROS:
 0.000000E 00 +j 1.638633E 00
 0.000000E 00 +j 1.152012E 00
POLES:
-6.655549E-01 +j 0.000000E 00
-3.499603E-01 +j 7.599004E-01
-7.581614E-02 +j 9.440900E-01
```

```
FILE:9
ANALOG-FILTER TRANSFORMATIONS
================================
BANDPASS FILTER
INPUT DATA:
------------
# Complex zero pairs: 2
# Real zeros: 0
# Complex pole pairs: 2
# Real poles: 1
H: 1.172666E-01
ZEROS:
  0.000000E 00 +j 1.638633E 00
  0.000000E 00 +j 1.152012E 00
POLES:
-6.655549E-01 +j 0.000000E 00
-3.499603E-01 +j 7.599004E-01
-7.581614E-02 +j 9.440900E-01
Centre Freq. r/s: 1.000000E 03
Bandwidth r/s: 2.000000E 02
OUTPUT DATA:
------------
H(s)=H*H1(s)*H2(s)...
H1(s)=(A01+A11s+A21s^2)/(B01+B11s+B21s^2)
Section #:1
------------
A01: 0.000000E 00        B01: 1.000000E 06
A11: 1.000000E 00        B11: 1.331110E 02
A21: 0.000000E 00        B21: 1.000000E 00
Section #:2
------------
A02: 1.385800E 06        B02: 1.164075E 06
A12: 0.000000E 00        B12: 7.529868E 01
A22: 1.000000E 00        B22: 1.000000E 00
Section #:3
------------
A03: 7.216050E 05        B03: 8.590514E 05
A13: 0.000000E 00        B13: 6.468544E 01
A23: 1.000000E 00        B23: 1.000000E 00
Section #:4
------------
A04: 1.258469E 06        B04: 1.207490E 06
A14: 0.000000E 00        B14: 1.658848E 01
A24: 1.000000E 00        B24: 1.000000E 00
Section #:5
------------
A05: 7.946164E 05        B05: 8.281641E 05
A15: 0.000000E 00        B15: 1.373798E 01
A25: 1.000000E 00        B25: 1.000000E 00
H: 2.345332E 01
```

```
FILE:9
ANALOG-FILTER TRANSFORMATIONS
==============================
BANDPASS FILTER
INPUT DATA:
-----------
# Complex zero pairs: 2
# Real zeros: 0
# Complex pole pairs: 2
# Real poles: 1
H: 1.172666E-01
ZEROS:
 0.000000E 00 +j 1.638633E 00
 0.000000E 00 +j 1.152012E 00
POLES:
-6.655549E-01 +j 0.000000E 00
-3.499603E-01 +j 7.599004E-01
-7.581614E-02 +j 9.440900E-01
Centre Freq. r/s: 1.000000E 03
Bandwidth r/s: 2.000000E 02
OUTPUT DATA:
-----------
# Complex zero pairs:4
# Real zeros:1
# Complex pole pairs:5
# Real poles:0
H:2.345332E 01
ZEROS:
 0.000000E 00 +j 0.000000E 00
 0.000000E 00 +j 1.177200E 03
 0.000000E 00 +j 8.494734E 02
 0.000000E 00 +j 1.121815E 03
 0.000000E 00 +j 8.914126E 02
POLES:
-6.655549E 01 +j 9.977827E 02
-3.764934E 01 +j 1.078266E 03
-3.234272E 01 +j 9.262858E 02
-8.294238E 00 +j 1.098827E 03
-6.868990E 00 +j 9.100093E 02
```

```
FILE:11
BILINEAR TRANSFORMATION
=======================
CASCADE FILTER
INPUT DATA:
-----------
H(s)=H*H1(s)*H2(s)*...
H1(s)=(A01+A11s+A21s↑2)/(B01+B11s+B21s↑2)
Sampl. Freq. r/s: 1.000000E 01
Section #:1
-----------
A01:    1.000000E 00        B01:    1.494328E 00
A11:    0.000000E 00        B11:    1.558128E 00
A21:    0.000000E 00        B21:    1.000000E 00
Section #:2
-----------
A02:    1.000000E 00        B02:    5.276205E 00
A12:    0.000000E 00        B12:    6.453978E-01
A22:    0.000000E 00        B22:    1.000000E 00
H:   7.026960E 00
OUTPUT DATA:
-----------
H(z)=H*H1(z)*H2(z)*...
H1(z)=(A01+A11z+A21z↑2)/(B01+B11z+B21z↑2)
Section #:1
-----------
A01:    6.029137E-02        B01:    4.019487E-01
A11:    1.205827E-01        B11:   -1.041568E 00
A21:    6.029137E-02        B21:    1.000000E 00
Section #:2
-----------
A02:    5.726495E-02        B02:    7.647138E-01
A12:    1.145299E-01        B12:   -5.561473E-01
A22:    5.726495E-02        B22:    1.000000E 00
H:   7.026960E 00
```

```
FILE:12
PARTIAL FRACTIONS
=================
INPUT DATA:
------------
# Complex zero pairs: 3
# Real zeros: 0
# Complex pole pairs: 3
# Real poles: 0
Multiplier: 2.013515E-03
ZEROS:
  0.000000E 00 +j 3.463150E 00
  0.000000E 00 +j 1.414260E 00
  0.000000E 00 +j 1.141209E 00
POLES:
-2.519807E-01 +j 3.094971E-01
-1.266977E-01 +j 7.344894E-01
-3.261203E-02 +j 8.925166E-01
OUTPUT DATA:
------------
```

$H(z) = H1(z) + H2(z) + \ldots$

$H1(z) = (A01 + A11 z + A21 z^2) / (B01 + B11 z + B21 z^2)$

```
Section #:1
------------
A01: 1.884987E-01        B01: 1.592827E-01
A11: 7.311049E-02        B11: 5.039614E-01
A21: 0.000000E 00        B21: 1.000000E 00
Section #:2
------------
A02:-1.887357E-01        B02: 5.555270E-01
A12:-1.308653E-01        B12: 2.533954E-01
A22: 0.000000E 00        B22: 1.000000E 00
Section #:3
------------
A03: 3.633917E-02        B03: 7.976494E-01
A13: 5.609852E-02        B13: 6.522406E-02
A23: 0.000000E 00        B23: 1.000000E 00
Section #:4
------------
A04: 2.013515E-03        B04: 1.000000E 00
A14: 0.000000E 00        B14: 0.000000E 00
A24: 0.000000E 00        B24: 0.000000E 00
```

```
FILE:12
PARTIAL FRACTIONS
================
INPUT DATA:
-----------
# Complex zero pairs: 3
# Real zeros: 0
# Complex pole pairs: 3
# Real poles: 0
Multiplier: 2.013515E-03
ZEROS:
0.000000E 00  +j  3.463150E 00
0.000000E 00  +j  1.414260E 00
0.000000E 00  +j  1.141209E 00
POLES:
-2.519807E-01  +j  3.094971E-01
-1.266977E-01  +j  7.344894E-01
-3.261203E-02  +j  8.925166E-01
OUTPUT DATA:
------------
          POLES                           RESIDUES
-2.519807E-01  j  3.094971E-01    3.655524E-02  j-2.747624E-01
-1.266977E-01  j  7.344894E-01   -6.543264E-02  j 1.171939E-01
-3.261203E-02  j  8.925166E-01    2.804926E-02  j-1.933279E-02
Constant: 2.013515E-03
```

FILE:13
IMPULSE ,UNIT-STEP & SINUSOID INVARIANT METHODS
===

INPUT DATA:

Sampling Freq., r/s: 1.000000E 02
IMPULSE INVARIANT METHOD
            Poles   Residues
-9.238795E-01 j 3.826834E-01   4.619398E-01 j-1.115221E 00
-3.826834E-01 j 9.238795E-01   -4.619398E-01 j 1.913417E-01
OUTPUT DATA:

H[z]=H1[z]+H2[z]+...
H1[z]=(A01+A11z+A21z↑2)/(B01+B11z+B21z↑2)
Section #:1

A01: 0.000000E 00   B01: 8.903879E-01
A11:-5.158013E-02   B11:-1.886662E 00
A21: 5.804907E-02   B21: 1.000000E 00
Section #:2

A02: 0.000000E 00   B02: 9.530486E-01
A12: 5.521264E-02   B12:-1.949195E 00
A22:-5.804907E-02   B22: 1.000000E 00

```
FILE:14
MATCHED-Z TRANSFORMATION METHOD
================================
ELLIPTIC LOWPASS FILTER
INPUT DATA:
------------
Sampling Freq., r/s: 7.500000E 00
# Complex zero pairs: 3
# Real zeros: 0
# Complex pole pairs: 3
# Real poles: 0
H: 6.713267E-03
ZEROS:
  0.000000E 00 +j 3.463150E 00
  0.000000E 00 +j 1.414260E 00
  0.000000E 00 +j 1.141209E 00
POLES:
-4.754168E-01 +j 3.635543E-01
-2.211582E-01 +j 7.982251E-01
-5.443745E-02 +j 9.275597E-01
OUTPUT DATA:
------------
H[z]=H*H1[z]*H2[z]*...
H1[z]=(A01+A11z+A21z↑2)/(B01+B11z+B21z↑2)
Section #:1
------------
A01: 1.000000E 00        B01: 4.508735E-01
A11: 1.942528E 00        B11:-1.281134E 00
A21: 1.000000E. 00       B21: 1.000000E 00
Section #:2
------------
A02: 1.000000E 00        B02: 6.903517E-01
A12:-7.529504E-01        B12:-1.303834E 00
A22: 1.000000E 00        B22: 1.000000E 00
Section #:3
------------
A03: 1.000000E 00        B03: 9.128252E-01
A13:-1.153491E 00        B13:-1.362371E 00
A23: 1.000000E 00        B23: 1.000000E 00
H: 8.677238E-03
```

```
FILE:15
DESIGN OF NONRECURSIVE FILTERS
==============================
KAISER WINDOW
BANDSTOP FILTER
INPUT DATA:
-----------
Sampling Freq., r/s: 2.000000E 03
Ap, dB: 0.200
Aa, dB: 45.000
Wp1,r/s: 2.000000E 02
Wp2,r/s: 7.000000E 02
Wa1,r/s: 4.000000E 02
Wa2,r/s: 6.000000E 02
Actual Ap, dB: 9.768972E-02
Actual Aa, dB: 4.500000E 01
N:   53
OUTPUT DATA:
-----------
+n or -n       h(n)           w(n)h(n)
    0       6.000000E-01    6.000000E-01
    1      -5.853711E-02   -5.838878E-02
    2       2.879140E-01    2.850043E-01
    3       9.162457E-02    8.955115E-02
    4      -7.568267E-02   -7.265842E-02
    5       0.000000E 00    0.000000E 00
    6      -3.665779E-02   -3.342482E-02
    7      -7.706715E-02   -6.793100E-02
    8       2.338723E-02    1.981787E-02
    9       4.106542E-02    3.326176E-02
   10       0.000000E 00    0.000000E 00
   11       3.359898E-02    2.442022E-02
   12       1.559149E-02    1.063287E-02
   13      -4.149770E-02   -2.637443E-02
   14      -1.571048E-02   -9.239111E-03
   15       0.000000E 00    0.000000E 00
   16      -1.892067E-02   -9.305437E-03
   17       1.616904E-02    7.178245E-03
   18       3.199044E-02    1.269576E-02
   19      -3.080900E-03   -1.081275E-03
   20       0.000000E 00    0.000000E 00
   21       2.787481E-03    7.365323E-04
   22      -2.617400E-02   -5.864448E-03
   23      -1.195103E-02   -2.227488E-03
   24       1.261378E-02    1.910196E-03
   25       0.000000E 00    0.000000E 00
   26       8.459489E-03    7.645398E-04
```

AMPLITUDE RESPONSE:
--- --- --- --- --- --- --- --- ---

Freq., r/s	Gain dB
1.000000E-09	-9.069174E-03
1.000000E 01	-6.709181E-03
2.000000E 01	-7.028869E-04
3.000000E 01	6.088970E-03
4.000000E 01	1.009410E-02
5.000000E 01	8.665152E-03
6.000000E 01	1.731035E-03
7.000000E 01	-7.475457E-03
8.000000E 01	-1.349282E-02
9.000000E 01	-1.156604E-02
1.000000E 02	-1.130596E-03
.	.
.	.
.	.

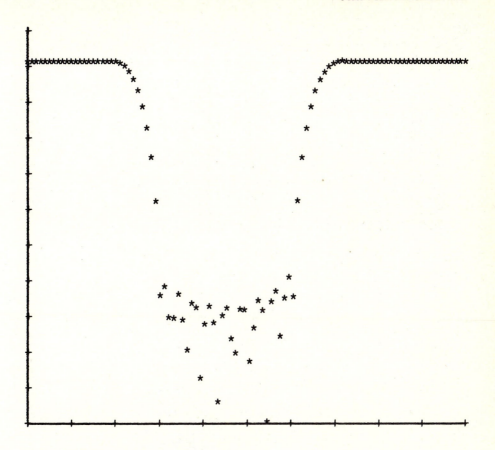

MIN-X: 1.000000E-09   MIN-Y:-7.303967E 01
MAX-X: 9.900000E 02   MAX-Y: 5.297553E-02
DIV-X: 9.900000E 01   DIV-:Y 7.309265E 00

```
FILE:17
FAST FOURIER TRANSFORM
=======================
No. of elements: 16
INPUT ARRAY:
------------

 n      REAL X          IMAGINARY X
 0   0.0000000E 00      0.0000000E 00
 1   3.826834E-01       0.0000000E 00
 2   7.071068E-01       0.0000000E 00
 3   9.238795E-01       0.0000000E 00
 4   1.000000E 00       0.0000000E 00
 5   9.238795E-01       0.0000000E 00
 6   7.071068E-01       0.0000000E 00
 7   3.826834E-01       0.0000000E 00
 8   0.0000000E 00      0.0000000E 00
 9   0.0000000E 00      0.0000000E 00
10   0.0000000E 00      0.0000000E 00
11   0.0000000E 00      0.0000000E 00
12   0.0000000E 00      0.0000000E 00
13   0.0000000E 00      0.0000000E 00
14   0.0000000E 00      0.0000000E 00
15   0.0000000E 00      0.0000000E 00
```

OUTPUT ARRAY:

k	REAL X	IMAGINARY X	MAGNITUDE	ANGLE, deg
0	5.027339E 00	0.000000E 00	5.027339E 00	0.000000E 00
1	8.000000E−12	−4.000000E 00	4.000000E 00	−9.000000E 01
2	−1.765367E 00	0.000000E 00	1.765367E 00	1.800000E 02
3	−2.561144E−12	8.310159E−09	8.310159E−09	9.001766E 01
4	−4.142136E−01	0.000000E 00	4.142136E−01	1.800000E 02
5	2.561144E−12	6.153016E−08	6.153016E−08	8.999762E 01
6	−2.346331E−01	0.000000E 00	2.346331E−01	1.800000E 02
7	−8.000000E−12	1.114000E−07	1.114000E−07	9.000411E 01
8	−1.989122E−01	0.000000E 00	1.989122E−01	1.800000E 02
9	−8.000000E−12	−1.114000E−07	1.114000E−07	−9.000411E 01
10	−2.346331E−01	0.000000E 00	2.346331E−01	1.800000E 02
11	2.561144E−12	−6.153016E−08	6.153016E−08	−8.999762E 01
12	−4.142136E−01	0.000000E 00	4.142136E−01	1.800000E 02
13	−2.561144E−12	−8.310159E−09	8.310159E−09	−9.001766E 01
14	−1.765367E 00	0.000000E 00	1.765367E 00	1.800000E 02
15	8.000000E−12	4.000000E 00	4.000000E 00	9.000000E 01

INDEX